Yeast

a practical approach

Edited by

I Campbell

Department of Brewing and Biological Sciences, Heriot-Watt University, Edinburgh EH1 1HX, UK

J H Duffus

Department of Brewing and Biological Sciences, Heriot-Watt University, Edinburgh EH1 1HX, UK

IRL PRESS
OXFORD · WASHINGTON DC

IRL Press Limited
PO Box 1,
Eynsham,
Oxford OX8 1JJ,
England

British Library Cataloguing in Publication Data

Yeast.
1. Yeasts
I. Campbell, I. (Iain) *1937-*,   II. Duffus, J.H.   III. Series.
589.2'33

ISBN 0-947946-80-2 (hardbound)
ISBN 0-947946-79-9 (softbound)

Typeset by Infotype and printed by Information Printing Ltd, Oxford, England.

Yeast

CHESTER COLLEGE LIBRARY

This book is to be returned on or before the
last date stamped below.

TITLES PUBLISHED IN
THE
PRACTICAL APPROACH
SERIES

Series editors:
Dr. D. Rickwood
Department of Biology, University of Essex
Wivenhoe Park, Colchester, Essex C04 3SQ, UK
Dr. B. D. Hames
Department of Biochemistry, University of Leeds
Leeds LS2 9JT, UK

Affinity chromatography

Animal cell culture

Biochemical toxicology

Biological membranes

Carbohydrate analysis

Centrifugation (2nd Edition)

DNA cloning

Drosophila

Electron microscopy
in molecular biology

Gel electrophoresis of nucleic acids

Gel electrophoresis of proteins

H.p.l.c. of small molecules

Human cytogenetics

Human genetic diseases

Immobilised cells and enzymes

Iodinated density-gradient media

Lymphocytes

Lymphokines and interferons

Mammalian development

Microcomputers in biology

Mitochondria

Mutagenicity testing

Neurochemistry

Nucleic acid and
protein sequence analysis

Nucleic acid hybridisation

Oligonucleotide synthesis

Photosynthesis:
energy transduction

Plant cell culture

Prostaglandins
and related substances

Spectrophotometry
and spectrofluorimetry

Steroid hormones

Teratocarcinomas
and embryonic stem cells

Transcription and translation

Virology

Preface

Yeasts, and especially strains of *Saccharomyces cerevisiae,* are important experimental organisms for modern biological research, with good reason. Most yeasts are safe to handle and easy to grow, or can even be obtained in bulk as baker's or brewer's yeast if necessary. As eukaryotic cells they provide convenient models for most of the activities of cells of higher forms of life. This has been particularly valuable in molecular biology, and genetics in particular, with the added advantage that yeasts can be grown as haploid or diploid cells, or indeed as higher ploidies as required.

Finding good descriptions of fundamental methodology for yeasts may be difficult: papers on yeasts are published in a wide range of literature. Also, the brevity required in the standard literature may require the omission of useful hints which are important for beginners. To remedy the lack of a single source-book of methods for the study of yeasts, David Rickwood suggested to us that one volume of the 'Practical Approach' series should be devoted to the subject. Our interest in the project was reinforced by the enthusiasm of the contributors, most of whom commented that a collection of such methods was long overdue. Our choice of the range of chapters was obviously a personal one, but is intended to cover all of the principal areas of interest to those using, or about to use, yeasts as experimental organisms.

Inevitably there are minor variations between laboratories in their experimental methods, which set us an editorial problem when techniques were independently described by authors of different chapters. When (rarely!) different contributors gave the same description, only one was necessary. Normally, the descriptions differed, and in the belief that following the author's routine precisely was essential for successful repetition of the work, many methods have been repeated in different chapters, with authors' individual variations. A description of the techniques for preparation of spheroplasts and protoplasts in four different chapters is an extreme case!

Our grateful thanks are due to the authors for their willingness to distil their laboratory experience into contributions for this book. We also acknowledge with thanks the assistance of the staff of IRL Press at all stages of preparation and production.

I.Campbell
J.H.Duffus

Contributors

I.Campbell
Department of Brewing and Biological Sciences, Heriot-Watt University, Edinburgh EH1 1HX, UK

B.J.Catley
Department of Brewing and Biological Sciences, Heriot-Watt University, Edinburgh EH1 1HX, UK

V.M.Darley-Usmar
Department of Biochemistry, Wellcome Research Laboratories, Langley Court, Beckenham, Kent BR3 3BS, UK

B.Dujon
Centre de Genetique Moleculaire, Gif-sur-Yvette, France 91190

B.Feinberg
Biological Chemistry Department, University of California, Irvine, CA 92717, USA

J.R.Johnston
Department of Bioscience and Biotechnology, University of Strathclyde, Glasgow G1 1XW, UK

D.Lohr
Department of Chemistry, Arizona State University, Tempe, AZ 85287, USA

C.S.McLaughlin
Biological Chemistry Department, University of California, Irvine, CA 92717, USA

J.M.Mitchison
Department of Zoology, University of Edinburgh, West Mains Road, Edinburgh EH9 3JT, UK

D.Rickwood
Department of Biology, University of Essex, Wivenhoe Park, Colchester CO4 3SQ, UK

A.H.Rose
Department of Biological Sciences, University of Bath, Claverton Down, Bath BA2 7AY, UK

D.M.Spencer
Department of Life Sciences, Goldsmiths' College, University of London, Rachel McMillan Building, Creek Road, London SE8 3BU, UK

J.F.T.Spencer
Department of Life Sciences, Goldsmiths' College, University of London, Rachel McMillan Building, Creek Road, London SE8 3BU, UK

E.Streiblová
*Institute of Microbiology, Czechoslovak Academy of Sciences, Videnska 1083,
142 20 Praha, Czechoslovakia*

F.J.Veazey
*Department of Biological Sciences, University of Bath, Claverton Down, Bath
BA2 7AY, UK*

Contents

5. YEAST GENETICS, MOLECULAR ASPECTS 107
J.R.Johnston

Abbreviations

ADH	alcohol dehydrogenase
BSA	bovine serum albumin
5-BU	5-bromouracil
5-BUdR	5-deoxyuridine
CAPS	cyclohexylaminopropane sulphonic acid
Con A	concanavalin A
DAPI	4′,6′-diamidino-2-phenylindole
DBM	diazobenzyloxymethyl
DCCD	N,N'-dicyclohexyl-carbodiimide
DEPC	diethylpyrocarbonate
DMSO	dimethylsulphoxide
dTMP	2′-deoxythymidine 5′-monophosphate
DTT	dithiothreitol
EES	ethylethane sulphonate
EMS	ethylmethane sulphonate
FID	flame ionization detector
FIGE	field-inversion gel electrophoresis
FITC	fluorescein isothiocyanate
MMS	methylmethane sulphonate
MNNG	N-methyl-N'-nitro-N-nitrosoguanidine
mtDNA	mitochondrial DNA
NA	nitrous acid
NTP	nucleoside triphosphate
OFAGE	orthogonal-field-alteration gel electrophoresis
PBS	phosphate-buffered saline
PEG	polyethylene glycol
PEPCK	phosphoenolpyruvate carboxykinase
PMSF	phenylmethylsulphonyl fluoride
RC	respiratory-competent
RCR	respiratory control ratio
RD	respiratory-deficient
TCA	trichloroacetic acid
TEMED	N,N,N',N-tetramethylethylenediamine
VLP	virus-like particles
YCp	yeast centromeric plasmid
YEp	yeast episomal plasmid
YEP	yeast extract−peptone
YIp	yeast integrating plasmid
YNB	Yeast−Nitrogen Base
YRp	yeast replicating plasmid

CHAPTER 1

Culture, storage, isolation and identification of yeasts

I.CAMPBELL

1. INTRODUCTION

Many of the techniques described in this book recommend specific strains from recognized culture collections. It is advisable to use these strains rather than other isolates of the same species, to ensure the expected results. For other methods there is more opportunity to use different yeast strains, species or even genera. Therefore the first part of this chapter explains the maintenance of cultures obtained from collections; procedures for isolating and identifying yeasts follow later.

2. STOCK CULTURES

Culture collections normally supply micro-organisms in a freeze-dried state, with instructions for reconstitution of the cultures. Yeasts are normally best reactivated in malt extract broth. After opening the ampoule (across the cotton wool plug, which acts as a filter during loss of vacuum on opening) the rounded end of the tube can be treated as a small cotton-plugged test tube. Addition of $1-2$ ml of sterile malt extract broth reactivates the dried yeast which, after incubation for 24 h at an appropriate temperature, usually 25°C, can be plated on an appropriate solid medium. Commercially available malt extract agar is usually the most suitable. Most yeasts form satisfactory colonies in $2-3$ days at 25°C. Some yeasts are psychrophilic, and require lower incubation temperatures. Such yeasts may occur in refrigerated foodstuffs, or in fresh or sea water. There are few yeasts for which 25°C is too low; although many yeasts, not necessarily pathogenic, can grow at 37°C, this higher temperature is often above the optimum for efficient growth.

2.1 Preservation of stock cultures

Storage on agar slopes requires no laboratory equipment other than a refrigerator and is, moreover, convenient if inocula are frequently required. Suitable storage media are malt extract agar, Sabouraud glucose agar and MYGP medium (malt extract, yeast extract, glucose, peptone), but the last is not commercially available. Note that bacteriological peptones are not normally suitable for yeast growth, and media manufacturers supply mycological peptone which gives better results. Apart from nutritional differences, the more acid pH (usually $\sim 5.0-5.5$ when reconstituted) is preferable for fungi, including yeasts.

Screw-capped McCartney bottles are a useful size of container (~ 25 ml; $6-8$ ml gives a good surface when sloped). We have suffered from poor survival rates in bottles fitted with plastic caps; aluminium caps with rubber liners are essential for a 6-month storage period. Our storage system is to prepare a fresh culture at 6-monthly intervals, inoculating two slopes which are incubated at 25°C for 24 h. Longer incubation is unnecessary, and increases the risk of mutation, or loss of the culture over prolonged storage. The screw caps are tightened at the end of the incubation and bottles are stored at 0−4°C. Only one bottle is used as a source of routine inocula over the 6-month storage period; the other is preserved unopened to provide the inoculum for the next pair of slopes for the following 6 months. Some yeasts, and in particular, *Brettanomyces* species, require subculture at more frequent intervals. The best incubation temperature is usually 25°C, but individual requirements can be decided with a knowledge of the properties of the species concerned.

Our yeast culture collection is stored freeze-dried (Appendix II) as a safeguard against loss of the 'working' stock cultures on agar slopes, and cultures which are not required regularly are stored only in the freeze-dried state to avoid labour-intensive subculturing.

3. CULTURE METHODS

Commonly used media for cultivation of yeasts include:

(i)   malt extract broth or agar,
(ii)   Sabouraud's broth or agar and
(iii)   Wickerham's medium (Appendix I).

Media (i) and (ii) are available from various suppliers; (iii) is less widely available but is produced commercially by Difco. Wickerham's medium was originally intended for identification of yeasts, by testing their ability to grow on defined media with specified compounds as sole source of carbon or nitrogen. In addition to this use in identification (*Table 1*), it is useful as a standard synthetic medium for yeast growth under defined conditions. Wickerham's medium is commonly used as Yeast Nitrogen Base, to which individual carbon compounds are added to provide a complete medium, and Yeast Carbon Base, to which the user adds the specific nitrogen source. The basal medium contains the necessary inorganic salts and pure vitamins for yeast growth, and either glucose or ammonium sulphate (1,2).

Malt extract broth is, in effect, a dilute version of brewers' malt wort, supplemented with mycological peptone. Even so, the sugar concentration is sufficiently high that yeasts can ferment the medium. *Saccharomyces* species are unable to metabolize sugars, particularly glucose, oxidatively above approximately 1%: even under aerobic conditions the sugar is metabolized by the fermentative pathway. Therefore in preparation of media for aerobic growth it is important not to exceed 1% of fermentable sugar.

The media named above are normally adjusted to pH 5.5, the optimum for the majority of yeasts. Acid production during growth will lower the pH, which under anaerobic conditions may fall to pH 4 or less.

Most methods described in later chapters require aerobically-grown cells. The most common system for aerobic growth of yeasts uses liquid medium in conical flasks on a shaking platform, for example orbital incubator. With 100 ml volumes of medium in 300 ml conical flasks, shaking at 150−200 r.p.m. provides very efficient aeration.

Under these conditions, the generation time is about 2 h at 25°C and a satisfactory cell yield is achieved in 36−48 h. For larger amounts, 700−1000 ml in 2 litre flasks gives equivalent aeration. Many yeasts settle on overnight cold storage to leave a clear supernatant medium, but usually centrifugation is advisable.

4. ISOLATION OF YEASTS

The use of specified strains from culture collections is recommended wherever possible to make full use of the published literature on these strains. However, it may be necessary on occasions to isolate yeasts; if so, the following procedures may prove useful.

Although particularly associated with plant life, yeasts are present in most environments in association with other micro-organisms, particularly bacteria and filamentous fungi. A few types of rapidly-spreading fungi (e.g. *Rhizopus* species) may prove troublesome by over-growing yeast colonies, but generally the faster growth of yeasts allows isolation from mixtures of fungi without any requirement for selective inhibitors. If an anti-mould inhibitor is necessary, use 250 μg/ml sodium propionate (1). Malt extract agar is a useful medium for isolation of most yeasts, and its relatively low pH (5.5) is inhibitory to many types of bacteria. Isolation from specialized environments may require specially formulated media, for example yeasts from confectionery, jams, sugar-processing and other high-sugar environments should be isolated on media with the sugar content increased to 45−60% (3).

Reduction of the pH of malt extract agar to 3.5 by addition of the calculated quantity of lactic, citric, phosphoric or hydrochloric acid, has been recommended for the production of a selective medium of enhanced anti-bacterial effect. The instructions for such adjustment are normally included with the manufacturer's data on malt extract agar. Unfortunately, this pH is too low for reliable isolation of all yeasts, and the acidulants themselves may have different effects on yeasts. Therefore addition of antibiotics may be preferred for the formulation of anti-bacterial selective media: addition of 60 μg/ml penicillin + 100 μg/ml streptomycin together to malt extract or other agar medium gives a good selective effect. Alternatively, 100 μg/ml chlortetracycline or oxytetracycline added to agar media gives good results. These various selective media were discussed by Davenport (1). Note also that yeasts should be isolated or counted on solid media by spread-plate methods; the pour-plate method is unsatisfactory for the following reasons:

(i)   possible lethal effect of hot molten agar,
(ii)   the difficulty of distinguishing colonies growing in the depths of the agar, and
(iii)   the possibility that some aerobic yeasts may not grow in the depths of the agar.

5. IDENTIFICATION OF YEASTS

Newly-isolated yeasts have to be identified, and even yeasts originally obtained from a culture collection should be examined from time to time to ensure that they are indeed the correct organisms and not contaminants or unwanted mutants.

Classification and identification of yeasts is based on the methods for both filamentous fungi and bacteria. Yeasts are fungi, and their classification into families and genera is based on the morphology of vegetative cells and spores, if formed (*Figure 1*). Species

Table 1. Identication key to common ascosporogenous and non-sporing yeast genera (modified and simplified from Kreger-van Rij, 1984).

Methods and media required for identification

1. Grow in malt extract broth, 3 days at 25°C; note method of multiplication (*Figure 1a*).
2. Grow on 0.5% sodium acetate agar, examine for spores. Incubate at 25°C for up to 2 weeks, examining wet film at intervals (*Figure 1b,c*).
3. Grow on corn meal agar; 3 days at 25°C; examine for mycelium or pseudomycelium, or arthrospores (*Figure 1d*).
4. Grow in glucose/yeast extract broth (3% glucose, 0.5% yeast extract, + bromocresol purple to colour, pH 5.8–6.0) 3 days at 25°C; examine for fermentation (+ = strong, w = weak, − = no fermentation).
5. Grow on glucose/nitrate agar (Wickerham's Yeast Carbon Base + 0.5% KNO_3 or $NaNO_3$); 3 days at 25°C; examine for growth. N.B. use a small speck of culture as inoculum, and inoculate a known culture of *Saccharomyces cerevisiae* as a negative control (+ = growth, − = no growth).

Identification of genera

1. Spindle-or needle-shaped ascospores (*Figure 1b*), multilateral budding (*Figure 1a*).
    a.   Pseudomycelium, fermentation w/−, NO_3 −   *Metschnikowia*
    b.   True mycelium, fermentation w, NO_3 −   *Nematospora*
2. Spherical, oval, reniform, hat or Saturn ascospores (*Figure 1b,c*)
    a.   Vegetative growth by binary fission, may or may not form true/pseudomycelium, fermentation +   *Schizosaccharomyces*
    b.   Vegetative growth by polar budding
        (i)   Hat spores, or warty, rough spores formed in mother cell   *Hanseniaspora*
        (ii)   Rough, warty spores formed in bud   *Nadsonia*
        (iii)   Spherical spores, conjugating in pairs in ascus   *Saccharomycodes*
    c.   Vegetative growth by multilateral budding
        (i)   No pseudomycelium, spherical or oval spores, conjugated asci, fermentation w/−, NO_3 −   *Debaryomyces*
        (ii)   Pseudomycelium, or possibly true mycelium, spherical, hat or Saturn spores, liberated. Fermentation w/−, NO_3 −   *Pichia*
        (iii)   As *Pichia*, but Saturn spores, liberated, conjugated asci   *Schwanniomyces*
        (iv)   Pseudomycelium, or possibly true mycelium, hat or Saturn spores, liberated, fermentation w/−, NO_3 + NO_3 +   *Hansenula*
        (v)   No pseudomycelium, spherical, oval or reniform spores, liberated, fermentation + , NO_3 −   *Kluyveromyces*
        (vi)   Pseudomycelium may be formed, ascospores spherical or oval, not liberated, fermentation + , NO_3 −   *Saccharomyces*
        (vii)   As *Saccharomyces*, but conjugated asci   *Zygosaccharomyces*
        (viii)   As *Saccharomyces*, but ascospores formed after conjugation between mother cell and bud   *Torulaspora*
3. Ascospores formed in separate ascus (*Figure 1c*), multilateral budding
                Lipomyces
4. No sexual spores
    a.   Ballistospores formed (*Figure 1e*)
        (i)   Red or pink pigment, pseudo- or true mycelium may be formed, fermentation −, NO_3 +/−   *Sporobolomyces*
        (ii)   No pigment, otherwise as *Sporobolomyces*   *Bullera*
    b.   Polar budding
        (i)   Pseudomycelium +/−, fermentation +, NO_3 −   *Kloeckera*
        (ii)   As *Kloeckera*, but no fermentation   *Schizoblastosporion*
    c.   Triangular or tetrahedral cells, budding at points, fermentation −, NO_3 −   *Trigonopsis*
    d.   Ogive cells or multilateral budding, acetic acid formed (recognizable by aroma), pseudo- or true mycelium, fermentation +/w, NO_3 +/−   *Brettanomyces*

e. Multilateral budding
  (i) Red, pink, orange or yellow pigment, normally grows as mucoid colonies (extracellular slime), fermentation −, NO_3 +/−, no growth on inositol (Wickerham's Yeast Nitrogen Base + 1% inositol)   ***Rhodotorula***
  (ii) Normally not pigmented, grows as mucoid colonies (capsule or slime), fermentation −, NO_3 +/−, inositol −   ***Cryptococcus***
  (iii) No pigment, true mycelium with arthrospores (*Figure 1d*), asexual endospores may be formed, fermentation w/−, NO_3 −   ***Trichosporon***
  (iv) No pigment, or brown pigment (rare), may or may not form true or pseudomycelium, fermentation +/w/−, NO_3 −   ***Candida***

Identification of species of strongly-fermenting perfect yeasts

Glucose	Galactose	Sucrose	Maltose	Lactose	Lysine	Pseudo-mycelium	
F	F	F/A	A/−	F/A	A	+/−	***Kluyveromyces marxianus***
F	F/A/−	F	F	−	−	+/−	***Saccharomyces cerevisiae***
F	F	F	−	−	−	−	***S. exiguus***
F	F	−	−	−	−	−	***Torulaspora delbrueckii***
F	−	F/A/−	−	−	A	+/−	***Zygosaccharomyces bailii***
F	F	F	F	−	A	+/−	***Z. fermentati***
F	−	f/A/−	F/A	−	A	+/−	***Z. rouxii***

F = strong fermentation; f = weak fermentation; A = aerobic growth; − = no reaction.

are classified and identified by physiological tests, similar to those used in bacterial identification. The definitive classification is currently that of Kreger-van Rij (4), but some experience of yeast identification is required for application of the system. Sporulation of yeasts is unreliable, especially in laboratory cultures, and beginners may have difficulty in recognizing the different types of spore. An alternative scheme by Barnett *et al.* (5), although using the same classification as Kreger-van Rij (4), requires physiological tests only and is easier to operate, particularly by those with experience of bacterial rather than fungal identification.

A simplified version of the Barnett system is conveniently available commercially from API Laboratory Products: a test strip of 20 media, including a negative control. The growth and no-growth reactions are converted to a numerical code and the species corresponding to that number is identified from the key provided by API. The API identification system is primarily designed for identification of yeasts of medical importance, but is also suited to identification of yeasts of non-human sources. Identification by API test strips is expensive and may be unsuitable for establishments with a requirement to identify large numbers of yeast isolates.

Finally, for no effort other than postage and payment of the required fee, cultures will be identified by national culture collections. In the UK, yeast cultures should be sent, after appropriate preliminary arrangement with the Curator, to the National Collection of Yeast Cultures.

The following identification scheme, although not intended to substitute fully for the methods of Barnett *et al.* (5) or Kreger-van Rij (4), is nevertheless useful for the common yeasts of the food and fermentation industries, and has sufficed for class use in this Department over many years. The version in *Table 1* includes the most recent changes in nomenclature introduced by Kreger-van Rij (4). Physiological tests are limited to aerobic growth on KNO_3 as sole nitrogen source and growth on glucose, galactose, sucrose, maltose and lactose as carbon sources. The sugars are prepared in a liquid medium of 3% sugar, 0.5% yeast extract, adjusted to pH 5.8−6.0; sufficient

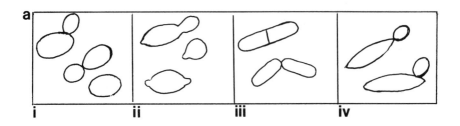

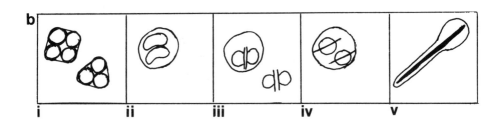

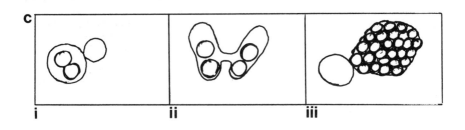

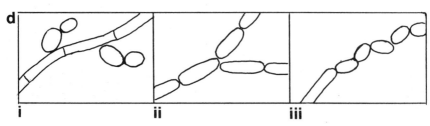

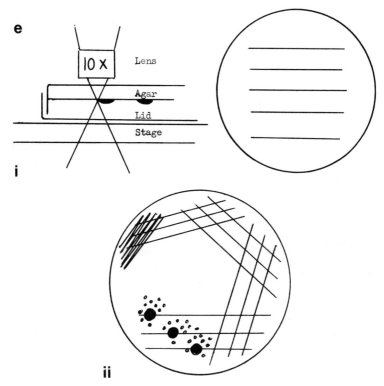

Figure 1. Appearance of vegetative yeast cells and spores. **a**, vegetative growth of yeasts: (i) multilateral budding; (ii) polar budding; (iii) binary fission; (iv) 'Ogive cells' (pointed ends). **b**, appearance of ascospores of various yeasts; (i) spherical (*Saccharomyces*); (ii) reniform (*Kluyveromyces*); (iii) hat; (iv) saturn (*Hansenula, Pichia*); (v) needle (*Metschnikowia*). **c**, alternative forms of ascus: (i) conjugated ascus, from nuclear fusion of parent cell and bud (*Debaryomyces, Torulaspora*); (ii) conjugated ascus, from fusion of two independent cells (*Zygosaccharomyces*); (iii) ascus of *Lipomyces*: separate structure from vegetative cells. **d**, filamentous forms: (i) septate mycelium, with occasional budding cells; (ii) Pseudomycelium; (iii) tip of septate mycelium fragmented into arthrospores. **e**, examination of plate cultures. (i) mycelium/pseudomycelium/arthrospores. Grow 4−5 cultures as streaks on corn meal agar and examine the edge of the line of growth by the ×10 objective (i.e. ×100 magnification). (ii) Evidence of ballistospore production. These spores are difficult to recognize microscopically. Ballistospores are ejected from mature colonies and fall on the medium nearby. After 3 days the original colonies have produced ballistospores, and by 4−5 days incubation these germinated ballistospores have produced 'satellite' colonies as shown.

bromocresol purple solution is added to give a faint purple colour (6). The medium is dispensed in 5 ml amounts in test tubes with inverted Durham tubes to trap CO_2 if produced by fermentation. The tubes show the following reactions: fermentation (medium changed to yellow, gas in Durham tube), aerobic growth without fermentation (yellow colour but no gas in tube) or no growth. Aerobic growth may be limited to the surface layer of the medium only, with a yellow surface layer of medium overlying a zone of intermediate colour of the pH change purple-yellow. Faint turbidity without any change of colour of the indicator should be ignored, being due to limited yeast growth on intracellular storage compounds. A negative control without any sugar, that is containing 0.5% yeast extract alone, is advisable.

6. REFERENCES

1. Davenport,R.R. (1980) In *Biology and Activities of Yeast*. Skinner,F.A., Passmore,S.M. and Davenport,R.R. (eds), Academic Press, London, p. 261.
2. Wickerham,L.J. (1951) *Technical Bulletin no. 1029*, US Department of Agriculture, Washington DC. (The essential information of this reference is available in the literature of the manufacturers of dehydrated Wickerham's Yeast Carbon Base and Yeast Nitrogen Base).
3. Tilbury,R.H. (1980) In *Biology and Activities of Yeasts*. Skinner,F.A., Passmore,S.M. and Davenport,R.R. (eds), Academic Press, London, p. 153.
4. Kreger-van Rij,N.J.W. (1980) *The Yeasts, a Taxonomic Study*. 3rd edition, Elsevier-North Holland, Amsterdam (see also her chapter in *Biology and Activities of Yeasts*, p. 29).
5. Barnett,J.A., Payne,R.W. and Yarrow,D. (1983) *Yeasts: Characteristics and Identification*. Cambridge University Press, Cambridge.
6. Campbell,I. and Brudzynski,A. (1966) *J. Inst. Brew.*, **72**, 556.

7. SUPPLIERS' ADDRESSES

Culture media: see Appendix I

Yeast culture collections: see Appendix III

API identification kits: API System SA, La Balme les Grottes, 38390 Montalieu-Vercieu, France and API Laboratory Products Ltd, Grafton Way, Basingstoke, Hampshire RG22 6HY, UK

CHAPTER 2

Cytological methods

EVA STREIBLOVÁ

1. INTRODUCTION

Modern cell biology approaches the problem of the cell at all levels of its organization to understand how it manages its biosynthetic activities for constructing copies of itself. This development may well lead to a renaissance of yeast cytology. Novel themes, such as cytoskeletal structures and their function and expression of spatial information are emerging from current work with eukaryotic cells. Those who recognize the opportunities offered by the modern approaches concentrate on yeast, the best understood cellular model system (1) among eukaryotes. This organism opens up possibilities for convergence of cytology with other areas of biology, especially with genetics and biochemistry.

This chapter gives a condensed description of cytological methods that were selected with respect to simple manipulation, reliability and perspectives in research at the yeast population level. Classical cytological methods are being replaced by new techniques and methods partially adopted from related areas of biology, such as immunology. Optical and fluorescence methods are emphasized here as they are relatively non-invasive and do not affect cell physiology. These methods will continue to map the yeast cell by resolving unknown features and discovering new relationships which will then be analysed in detail at the submicroscopic level. Electron microscopic methods, although indispensable in their realm, are limited in that their use necessarily results in an irreversible cessation of cell activities and one should realize this when interpreting results obtained under these limitations.

2. DIMENSIONS AND LIMITS IN YEAST CYTOLOGY

Cytological techniques make it possible to study yeast cells at different dimensional levels. On the whole, the boundaries between the levels are given by the resolving power of the magnifying instruments. Yeast colonies can be examined without such instruments. The obvious purpose of microscopical techniques used in yeast cytology is to magnify and resolve the details of yeast cell architecture. Beyond the organization discernible with light microscopic methods, there is an organized arrangement of subcellular, macromolecular and molecular complexes ranging from 1 to 200 nm which compose the cell as a unit — this level is studied by electron microscopy and X-ray diffraction (*Table 1*).

3. METHODS FOR MORPHOLOGICAL AND DEVELOPMENTAL STUDIES

In yeast, microscopic preparations are extensively used to examine the morphological

Table 1. Optical methods for yeast cytology.

Organizational level	Dimensions	Object to study	Magnifying system
Macroscopic	$>10\ \mu m$	colonies, slants	eye, magnifying glass
Microscopic	$10-0.2\ \mu m$	cell topography, some organelles	various types of light microscopes including fluorescence microscope
Submicroscopic	$0.2\ \mu m - 1\ nm$	cellular components,	polarization microscope,
Molecular		macromolecular configurations	electron microscope
Amicroscopic	$<1\ nm$	molecular configurations, arrangement of atoms	X-ray diffraction

features of samples taken under various experimental conditions and when estimating cell number and cell size. Although the preparation of wet mounts is essentially a simple technique it may be useful to add a few notes for routine investigations.

In special cases, direct examination of individual cells is required for observing cells in their natural state. For this purpose, cultivation techniques were developed to permit direct microscopic examination with minimum disturbance of single-cell morphology, arrangement of conidia and mycelium development. Work under sterile conditions to avoid contamination is a prerequisite for successful microculturing of yeast.

3.1 Wet mounts

A small amount of yeast cells taken from a centrifuged sample grown in liquid medium or on agar is transferred with a sterile loop to a microscopic slide where it is mixed with a drop of mounting fluid (generally tap water or 0.9% NaCl). Usually, a coverslip is placed immediately over the preparation which is then ready for examination by bright-field or phase-contrast microscopy.

Some experience is required for preparing yeast suspensions at suitable density. Practice is also needed to remove the superfluous liquid from a wet mount. A piece of filter paper applied to the edge of the coverslip or gentle pressure on a filter paper placed over the coverslip usually brings the cells into one optical plane. Generally, not all areas of the preparation are satisfactory and it is necessary to search for places of suitable density and thinness. The easiest way to make the cells stationary is to allow the mounted specimen to dry out for a few minutes at room temperature before examination. When the cells are still difficult to immobilize, the slide or coverslip can be thinly smeared with egg white, egg albumin or gelatin (Section 5.3) before mounting the suspension to be examined.

When no examination of living organisms is required, the yeast suspension is air-dried on a coverslip or slide before mounting in a drop of fluid. Smears mounted on the coverslip bring the preparation closer to the objective lens.

Specimens of suitable thinness tend to dry out quickly and it is advisable to seal the specimen unless it is examined immediately. Conveniently, using a flamed scalpel, molten paraffin wax is placed along the edges of the coverslip to seal the preparation. Nail varnish applied around the coverslip with the brush fitted to the cap is usually

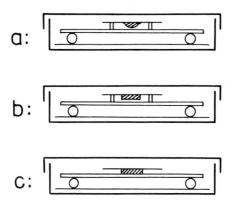

Figure 1. Slide and coverslip method for examining yeast morphology: **(a)** hanging drop; **(b)** agar block; **(c)** slide culture.

equally effective. Sealing has the additional advantage of fixing the mount for repeated examination.

It is frequently essential to observe the state of the culture at the instant of sampling. In this case, add 4.5 ml of formaldehyde – saline solution (37 ml of 40% formaldehyde solution and 6.3 ml of 0.9% NaCl) to 0.5 ml of the yeast sample to arrest the development of the culture. This treatment stops growth of yeast cells rapidly, as verified by Pringle and Moor (2).

3.2 Hanging drop methods

Hanging drop methods are useful to check the morphogenesis of individual cells and their mutual arrangement. Hanging drops (*Figure 1a*) are mostly examined using bright-field illumination; phase-contrast is less effective. On the other hand, hanging agar blocks (*Figure 1b*) are suitable for observations with phase optics using oil-immersion.

3.2.1 *Hanging drop cultures*

(i)   Mix thoroughly a small yeast inoculum with 1 ml of sterile 20% gelatin molten at 30°C on a sterile slide.
(ii)   Place one drop of the mixture on a sterile coverslip, invert it over the circular depression of a cavity slide, or over a suitable glass or plastic ring.
(iii)   Seal the preparation to the slide with vaseline or nail varnish.
(iv)   Transfer the assembly to a Petri dish and incubate.

3.2.2 *Hanging agar block cultures*

(i)   Spread a clear agar layer about 2 mm deep in a small sterile Petri dish.
(ii)   Inoculate sparsely when set.
(iii)   Excise a small square of the inoculated agar with a sterile scalpel and place it in the centre of a sterile (flamed) coverslip.
(iv)   Place the coverslip agar over a glass or plastic ring of thickness about the depth of the agar layer.

(v)   Seal the specimen to the slide with a small drop of water.

(vi)   Incubate the hanging culture, supported on a V-shaped glass rod placed in a sterile
Petri dish containing a filter paper soaked in 20−50% glycerol.

3.3 Slide culture methods

These procedures permit direct microscopic observation of developmental stages of
the living yeast culture (*Figure 1c*).

One of the simplest procedures consists of placing a sterile coverslip at the periphery
of a slant or developing yeast colony growing on a thin film of nutrient agar in a Petri
dish. This method gives good results with yeast species forming mycelium and pseudomycelium.

A refined procedure of slide cultures which can be used for a wide range of species
is the method improved by Robinow (3). Nuclear division and sporulation resolved
in delicate detail are examples of excellent results obtained using this technique.

(i)   Coat sterile glass slides with molten nutrient agar.

(ii)   Inoculate along a few cross channels made with a fine glassfibre covered with
yeast suspension.

(iii)   Cut the inoculated agar to a narrow agar block of about 2 × 4 mm with a sterile
scalpel and remove the excess agar.

(iv)   Clean the underside of the slide.

(v)   Place a large coverslip extending beyond the agar cube and seal the chamber
with paraffin wax; the closed cultivation space assures supply of nutrients and
optimal gas exchange.

(vi)   Observe the cells that are generally well spaced in the lines across the agar square.

We prefer the method described below since the yeast culture makes the maximum
use of nutrient agar during growth.

(i)   Pour a thin layer of nutrient agar into warmed Petri dishes containing microscopic
slides (usually 10 ml per dish).

(ii)   Inoculate over the surface of the slide and mount the coverslip in place.

(iii)   Make a few 'channels' under the coverslip with a sterile needle to facilitate
aeration.

(iv)   Invert the dish and put a sterile moist filter paper at the bottom to avoid desiccation.

(v)   Turn the dish again after culturing, excise the whole microscopic slide with the
agar film using a sterile scalpel, clean the underside of the slide and install the
specimen on the microscopic stage for observation.

A remarkable microculture chamber for examining development of fungi, including
yeast, was constructed by Cole and Kendrick (4). The chamber is based on a slide 75 mm
× 50 mm × 1 mm. A hole 18−20 mm in diameter is drilled through the slide and
a slot approximately 1 mm wide is cut from the hole to the periphery of the slide. A
60 × 24 mm no. 1 coverslip covers the hole and the slot is sealed to the upper side
of the slide with nail varnish. A paper card former is placed across the middle of the
shallow circular chamber and the system is sterilized. Sterile molten nutrient agar is
injected with a micropipette into the semicircular chamber on the side away from the

slot and the medium is allowed to set. After removal of the paper the vertical surface of the agar is inoculated with yeast to be investigated. Subsequently, the chamber is covered by sealing the second 60 × 24 mm coverslip over the top of the hole and slot. Excellent illustrations and detailed discussions of the applications of this technique can be found in the book by Cole and Samson (5).

3.3.1 *Staining slide cultures*

Staining with lactophenol cotton blue gives satisfactory durable slides. The stain is commercially available (Gurr) but can be prepared as follows.

Lactophenol solution: 10 g of phenol (pure crystals), 10 g of lactic acid, 20 g of glycerol, 10 ml of water. Cotton blue is dissolved in this solution. Heating is required to dissolve the phenol, then 0.05 g of cotton blue is added per 100 ml of lactophenol.

(i)   Dry the slide culture in an incubator at 37°C for 24 h.
(ii)   Coat the slides with collodion (25% w/v in ethanol:ether 1:1).
(iii)   Stain with lactophenol−cotton blue solution for 10−15 min.
(iv)   Drain without washing.
(v)   Wash in a graded ethyl alcohol series (25%, 50%, 75%, absolute alcohol) until the slide is only faintly stained.
(vi)   Mount in Canada balsam.

4. LIGHT MICROSCOPIC TECHNIQUES

Within the visible spectrum of light with wavelengths between 400 and 700 nm, the human eye is able to detect variations in intensity and colour (wavelength). Since yeast cells are essentially transparent to the visible spectrum of light and even structures such as vacuoles, nuclei and mitochondria show little contrast, basically two approaches are possible to obtain a clear definition of cellular detail.

One way is to increase the contrast of the living cells by optical means. Both phase-contrast and interference microscopy are based on the amplification of phase differences caused by differences in refractive index within various structures of the observed specimen. The yeast cell as a phase specimen is shown in *Figure 2*.

The second possibility for increasing contrast is to use dyes that selectively stain the required structures or compartments of the cells. Normally, observations are made after fixation to preserve the organization of the cell in as natural a state as possible. In coloured specimens high contrast is introduced by light absorption (*Figure 3*).

Invisible wavelengths flanking the visible spectrum are also used in light microscopic techniques with yeast. This is made possible either by adding to the specimen fluoro-chromes that emit light in the visible region of the spectrum when irradiated with UV. light (see Section 6) or by using appropriate photographic emulsions sensitive to wave-lengths in the visible range (UV and IR microscopy).

4.1 **Bright-field microscopy**

Microscopic examination with visible light yields only limited information when applied to unstained and relatively structureless yeast cells. Using stained cells, however, obser-vations can visualize important structural detail. At any rate, the resolution of the light

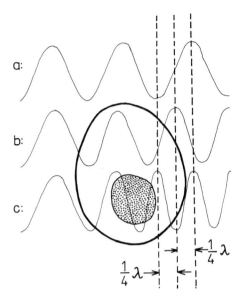

Figure 2. The yeast cell as phase specimen: relative to the mountant **(a)**, the phase change (retardation) of a light ray traversing the cytoplasm **(b)** is less pronounced than the retardation caused by the structure that is less transparent than the cytoplasm **(c)**.

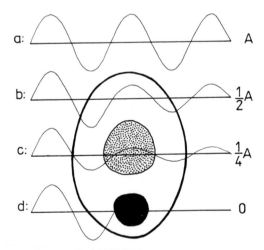

Figure 3. The yeast cell as amplitude specimen: relative to the unabsorbed ray traversing the transparent mountant **(a)**, the cytoplasm **(b)** and a faintly stained structure **(c)** partially absorb the light ray; there is no exit ray on a structure containing light-absorbing pigment or other coloured body **(d)**.

microscope depends on correctly adjusted illumination which substantially influences the performance of the instrument.

The principles of bright-field imaging are beyond the scope of this chapter, but conditions for Köhler illumination (*Figure 4*) are generally the same for all research microscopes and the adjustments described below should be carried out routinely. For

14

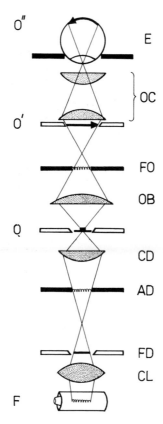

Figure 4. Köhler illumination principle: the filament of the light source (F) is imaged by the collector lens (CL) in the lower focal plane of the condenser with aperture (substage) diaphragm (AD), then again in the back focal plane of the objective (FO), and finally in the pupil beyond the ocular. Similarly, the field diaphragm (FD) is imaged by the condenser (CD) in the specimen plane (O), with the objective (OB) in the plane of the intermediate image (O'), and with the ocular (OC) in the final image plane (O").

further details see *Light Microscopy—a Practical Approach* (1988) A.J.Lacey, (ed.).

(i)   Accurately centre the lamp with associated field and collector lenses.

(ii)   Place the specimen on the stage and bring it into focus using a dry objective.

(iii)   Close the field diaphragm and move the condenser up and down until the field diaphragm is imaged in the specimen field.

(iv)   Centre the image of the field diaphragm.

(v)   Focus the image of the lamp filament on the lower surface of the substage diaphragm by moving the source of light until a clear image of the filament is obtained.

(vi)   Check again that the image of the field diaphragm is in the centre of the field and sharply defined.

(vii)   Open the field diaphragm until its image just disappears from view.

(viii) Remove the eyepiece and adjust the substage aperture so that 2/3 of the objective aperture is illuminated.

(ix)   Replace the eyepiece and bring the specimen into focus.

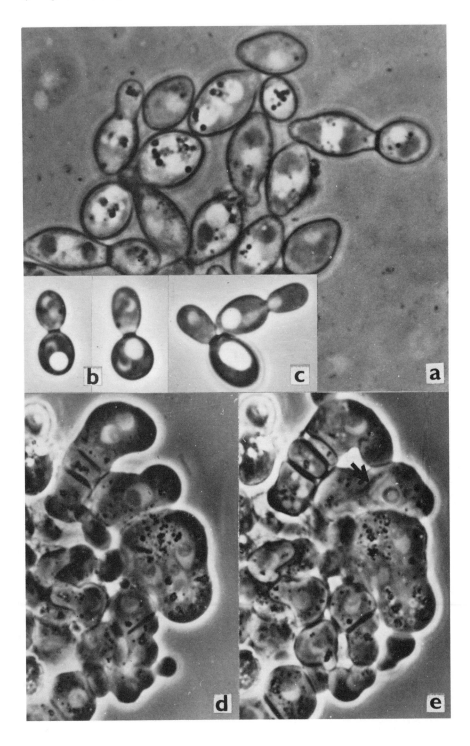

4.2 Phase-contrast microscopy

Phase optics make it possible to examine unstained living yeast cells with good contrast over a long period of time without harming them. When simply observing specimens mounted in a gelatin film (*Figure 5a,b,c*) or when microculturing different yeast species in gelatin (Section 3.3; *Figure 5d,e*) subtle cytologic detail can be seen in the course of nuclear and cell division.

The phase contrast device separates the most lateral bright rays passing through the objective with respect to the central light path passing through the object. A phase difference is introduced between the two sets of light rays and their paths are recombined by superposition or subtraction, substantially enhancing the contrast of the rather transparent yeast cell in the final image (*Figure 6*).

Usually, the phase microscope is an ordinary microscope set up for bright-field, with a phase ring fixed in the lower focal plane of the objective and a substage annulus fixed in the lower focal plane of the condenser.

(i)   Set up the microscope for bright-field Köhler illumination but with the phase objective and phase condenser in the bright-field position.

(ii)   Focus on the specimen.

(iii)   Replace the eyepiece with the auxiliary telescope which is supplied with the phase-contrast equipment.

(iv)   Focus the telescope on the phase ring in the back focal plane of the objective — a clearly defined dark ring is seen against the bright background.

(v)   Bring the appropriate substage annulus for the objective in use into register.

(vi)   Look down the telescope; the field is now dark except for the brightly illuminated ring of the substage aperture.

(vii)   Manipulate the substage annulus until the phase ring and the substage annulus are exactly concentric. Important: the brightly illuminated substage annulus must fall completely within the boundaries of the phase ring so that no illumination is visible in the plane of view (*Figure 7*).

(viii) Remove the telescope and replace the eyepiece.

(ix)   Insert a green filter and observe the specimen.

In bright or negative contrast the two sets of rays are added and the cells appear brighter than the background (*Figure 8a*). In dark or positive contrast the two sets of rays are subtracted, making the image of the cells darker than that of the surrounding medium (*Figure 8b*).

4.2.1 *Mounting yeast for phase-contrast observation*

Phase optics work best when the yeast suspension is mounted in a medium of higher

Figure 5. Phase-contrast images of yeast species in gelatin medium show well resolved cell organelles and dividing nuclei: **(a)** *Saccharomycodes ludwigii* ($\times 2000$) (E.Streiblová); **(b)** *Hansenula saturnus* n_D (refractive index of the mountant) $= 1.375$ ($\times 2250$) (R.Mueller); **(c)** *Saccharomyces fragilis* $n_D = 1.378$ ($\times 2250$) (R. Mueller). Time-lapse phase-contrast micrographs of *Schizosaccharomyces versatilis* microcultured in gelatin: **(d)** 1 h after inoculation one sees nuclei with large nucleoli; **(e)** the same field 30 min later — note the progression in septation and nuclear division, intranuclear spindle is some distance from the nucleolus (arrow) is clearly resolved; **(d)** and **(e)** $\times 1700$ (A.Svoboda).

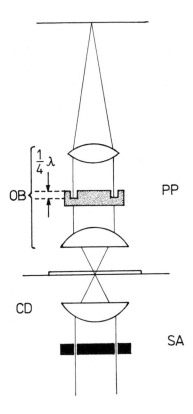

Figure 6. The light path in a phase-contrast microscope: two components of the phase microscope — the phase ring (or phase plate, PP) and the substage annulus (SA) — introduce interference phenomena bringing about an enhanced contrast of the specimen.

Figure 7. The correct adjustment of the substage annulus to the phase plate for setting up phase-contrast illumination.

refractive index. Usually gelatin is used for this purpose, as originally proposed by Mueller (6). For routine observations use coverslips or slides coated with a thin film of 20−25% gelatin; with a pipette place a small drop of suspension on the gelatin film, cover with a coverslip and observe immediately.

Polyvinylpyrrolidone is an obvious candidate for microculturing temperature-sensitive yeast mutants growing at 37°C.

4.3 Dark-field microscopy

The method is based on light-scattering phenomena observed at the boundaries between phases of different diffractive index. High contrast is essential at the periphery of the

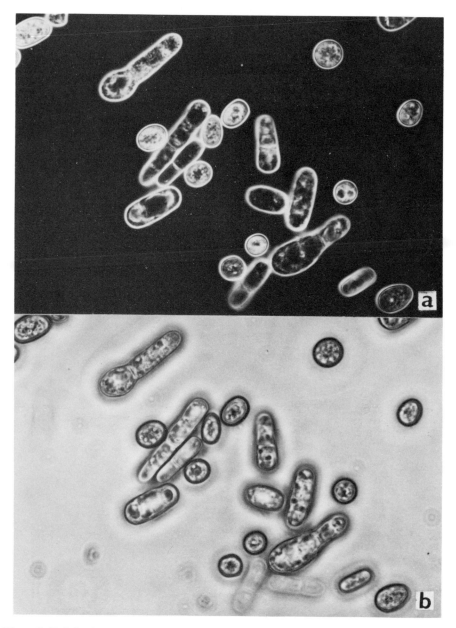

Figure 8. Variation in image characteristics obtained with negative and positive phase-contrast: **(a)** with subtractive superposition of light waves (negative contrast) the image contrast is reversed, causing those parts of the cells of *Endomyces magnusii* formerly dark to appear bright; **(b)** with additive superposition of light waves (positive contrast) cells in the same field as **(a)** appear darker than the background and details of the cell interior are well resolved, ×600 (J.Fiala).

object and loss of definition inside is unavoidable. In yeast cells, structures such as cell wall, lipid droplets and vacuoles appear bright whereas the cytoplasm is dark (*Figure 9a*). Subcellular fractions containing isolated yeast components that are below the

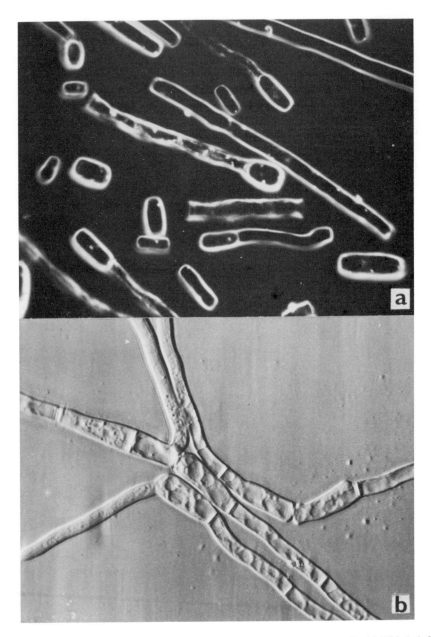

Figure 9. Dark-field and interference-microscopy micrographs of *Endomyces vernalis*. **(a)** With dark-field illumination walls and granules inside the cells are sharply outlined but the cytoplasm is poorly resolved. **(b)** Interference illumination reveals continuous changes in refractive index and accentuates differences in thickness of the various structures in the cells, ×600 (J.Fiala).

resolving power of the light microscope can be detected but not resolved.

For dark-field illumination the condenser and the objective are arranged relative to each other in such a way that none of the direct (zero) rays are allowed to enter the

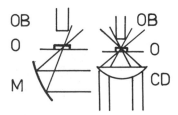

Figure 10. Principle of dark-field illumination: mirror (M) or dark-field condenser (CD) illuminate the specimen (O) obliquely and no direct light enters the objective (OB).

objective and the image is formed entirely by diffracted rays produced by the specimen (*Figure 10*).

Normal objectives are used for low magnification dark-field illumination, but oil immersion examination requires special objectives with an iris diaphragm. Dark-field condensers are of the reflecting type, with numerical aperture ranging from 1.20 to 1.45.

(i)   Remove the objective and condenser and ensure that the light illuminates the microscope tube intensely and uniformly.

(ii)   Install the appropriate objective and the dark-field condenser.

(iii)   Place a large drop of immersion oil carefully on the top surface of the condenser lens.

(iv)   Place the preparation in position and move the condenser upwards until the oil drop just contacts the underside of the slide.

(v)   Observe the specimen with a low-power objective ($\times 10$).

(vi)   Adjust the condenser vertically so as to produce a circle of light in the specimen plane; this may usually be observed by examining the preparation from the side.

(vii)   Open the field diaphragm so that the specimen field is just illuminated.

(viii) Focus and observe the specimen.

4.4 Interference microscopy

This technique is based on principles similar to those of phase-contrast microscopy. Small and continuous changes in refractive index within the object can be detected, whereas phase-contrast reveals only sharp phase changes of transmitted radiation.

Interference microscopy has given important quantitative data concerning dry mass, nucleic acid, water and lipid content. The potential of this method for yeast cytology (*Figure 9b*) is not yet fully explored.

5. STAINING METHODS

In light microscopy, contrast can be enhanced by the use of dyes which are bound to different parts of the cell, producing areas distinguished by their different colours. In yeast cytology these methods are less important at present than in the past since phase-contrast and fluorescence techniques have rendered many classical staining procedures redundant.

The following selection of staining methods should be sufficient to cover the special needs of yeast cytology. Whatever method is to be used, practice is essential and the results will improve with experience.

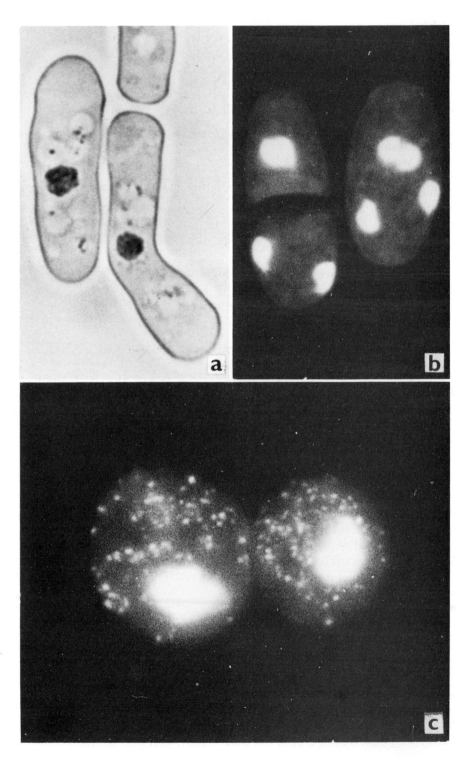

5.1 **Vital staining**

Staining with dyes such as methylene blue, benzidine blue, Nile blue, rose Bengal, neutral red and Janus green, which have no toxic effect at concentrations ranging from 0.1 to 100 μg/ml may, in special cases, go beyond a simple description of what is coloured in the yeast cell. Some of these dyes (e.g. Janus green) can be added to the nutrient medium and their fate followed during development (7).

5.1.1 *Vital staining of mitochondria*

In living yeast cells mitochondria may be observed under certain circumstances by phase contrast. Vital and supravital examination of mitochondria is possible with Janus green B, which stains these organelles greenish blue. This is due to the action of the mitochondrial cytochrome oxidase system which maintains the dye in the coloured form. In the surrounding cytoplasm the dye is in the colourless reduced state.

(i)   Dissolve 5 mg of Janus green B in 100 ml of 1% glucose and aerate by passing air through the staining solution for 30 min.

(ii)   Mount a drop of yeast suspension and mix with an equal volume of a fresh Janus green solution without putting on the coverslip.

(iii)   Observe immediately with a *water-immersion* objective ($\times$40 or $\times$70).

5.1.2 *Vital staining of vacuoles*

The yeast vacuolar system is involved in import, export and intracellular re-cycling of various compounds. This is particularly evident by vital staining with neutral red. The dye, actively taken up by the vacuolar system, stains the vacuoles reddish-purple. Within 15−20 min the stain in the vacuoles tends to fade, the dye is released from the vacuole and stains the cytoplasm brownish-pink. Interesting differences in coloration with neutral red are noted when staining young and old yeast cells.

(i)   Prepare a 1% aqueous solution of neutral red.

(ii)   Mix the yeast suspension with an equal volume of the staining solution, apply a coverslip.

(iii)   Observe immediately.

5.1.3 *Vital staining of nuclei*

Phase-contrast optics can give spectacular images of living nuclei. Lomofungin, an antibiotic belonging to the phenazine dyes, contrasts red chromosome-like configurations in living yeast nuclei. The antibiotic is dissolved in dimethyl sulphoxide and added either to the nutrient medium before culturing in final concentrations ranging from 20 to 100 μg/ml, or used as aqueous staining solution for preparing specimens (*Figure 11a* and ref. 8).

Figure 11. Micrographs of yeast cells using DNA-specific probes. (a) Living cells of *Schizosaccharomyces versatilis* are stained with the drug lomofungin. Coloration within the nuclei probably represents distribution of chromatin, $\times$2200 (M.Kopecká). (b) Cells of *Endomyces magnusii* stained with the drug mithramycin. Interphase nuclei display a strong fluorescence but details are not resolved, $\times$1800 (J.Hašek). (c) Both nuclear and mtDNA are visualized in *Saccharomyces uvarum* using the DNA-specific probe DAPI. Nuclei show small constrictions at the periphery following treatment with MBC. DNA−DAPI complexes in the mitochondria show up as bright dots visible at different focal levels, $\times$4000 (J.Hašek and J.Svobodová).

5.2 **Staining of unfixed cells**

5.2.1 *Staining glycogen*

Glycogen deposits composed of polymerized glucose molecules, and thus an important reserve of energy, are invisible in living yeast cells. After treatment with iodine solution, the polymers precipitate in the form of small granules. Proteins stain yellow; glycogen gives a reddish-brown colour. Stained glycogen deposits temporarily disappear after a brief heating of the mounts to about 70°C.

(i)   Dissolve 1 g of iodine and 2 g of potassium iodide in about 25 ml of distilled water and make up to 100 ml with water when fully dissolved.
(ii)   Prepare smears by spreading a thick yeast suspension into a thin film with the edge of a coverslip.
(iii)   Fix by prolonged air-drying.
(iv)   Stain with the iodine solution.
(v)   Put a coverslip over the preparation and observe immediately.

5.2.2 *Staining volutin*

Volutin is another polymeric compound used as reserve by the yeast cell. Phosphoric acid groups make the substance readily stainable with basophilic dyes such as methylene blue or toluidine blue. Vacuoles are stained dark blue or violet; the cytoplasm is lighter blue.

(i)   Make up the stain using 10 g of methylene blue, 1 ml of 1% aqueous KOH, 30 ml of absolute alcohol and 100 ml of distilled water.
(ii)   Prepare and air-dry the smears.
(iii)   Stain for 30 min with methylene blue solution.
(iv)   Bleach the specimen in 1% H_2SO_4 for 1 min.
(v)   Rinse with water, apply a coverslip and examine.

5.2.3 *Staining lipid granules*

Lipid granules are easily seen by phase-contrast microscopy as refractive bodies but they cannot easily be differentiated from other types of granules. Staining with Sudan black is based on simple diffusion and accumulation of the dyes in the interior of lipid droplets. Staining with Sudan black B gives the best contrast and is usually satisfactory. Lipid granules are bluish-grey in contrast to the pale pink cytoplasm.

(i)   Make up the stain with 0.5 g of Sudan black B in 100 ml of 70% aqueous ethanol. Mix by shaking thoroughly and leave to stand overnight; filter before use.
(ii)   Prepare smears, fix by air-drying and stain.
(iii)   Cover the smear with Sudan black solution and stain for 30 min.
(iv)   Drain the excess stain and blot dry.
(v)   Rinse with 50% aqueous ethyl alcohol.
(vi)   Wash with water, dry in air and examine.

5.2.4 *Staining masked lipids*

Nile blue sulphate is used to detect acidic lipids as well as nuclei. Lipids stain pale pink; nuclei dark blue.

(i)   Make smears and fix with gentle heat.
(ii)   Stain with a saturated aqueous solution of Nile blue for 20 min.
(iii) Rinse with water.
(iv)   Bleach with 1% glacial acetic acid for 15−20 min, depending on the appearance of the specimen.
(v)   Wash, apply a coverslip and observe.

5.3 Immobilization of cells

It is recommended to make yeast cells adhere firmly to coverslips or slides before subjecting the specimens to complicated fixation and staining procedures. Generally, one of the following methods is convenient.

(i)   Prepare smears by spreading a yeast suspension as thinly as possible with a thin glass rod or the edge of a coverslip. Allow the smears to air-dry for a sufficient period of time. Do not speed the drying process by heating the microscope slide.
(ii)   Prepare a coverslip coated with a thin film of fresh egg white. Prepare egg-albumin-coated coverslips as follows. Stir the egg white with an equal volume of distilled water in a cylinder. When the floccules of insoluble globulins settle, decant the soluble albumin fraction. Apply a drop of the supernatant to the coverslip and smear evenly with the tip of a finger.
(iii) Allow yeast cells to adhere to the coverslip using Haupt's adhesive (dissolve 1 g of pure gelatin in 100 ml of distilled water at 30°C; when dissolved, add 2 g of phenol crystals and 15 ml of glycerol). Smear the adhesive on the slide, add the cells to be attached and allow to evaporate.

5.4 Fixing and staining procedures

The procedures covered in this section were introduced and refined for yeast cell observations by Robinow (e.g. ref. 3). Only nuclear staining methods will be considered in detail since they find wide application in many current research projects.

5.4.1 *Fixative fluids*

(i)   Acetic acid−formaldehyde−alcohol fixative: 50 ml of absolute or 70% ethanol, 5 ml of glacial acetic acid, 5 ml of formaldehyde. Allow a minimum time of 45 min for fixation.
(ii)   Modified Helly's fluid: 5 g of $HgCl_2$, 3 g of potassium dichromate, 100 ml of distilled water; before use add formaldehyde to 5%. Fix for 10−20 min, rinse several times with 70% ethyl alcohol and store the slides for 2 days at the most.
(iii) Schaudin's fluid: 20 ml of saturated solution of $HgCl_2$, 10 ml of absolute alcohol, 30 ml of distilled water, 1 ml of glacial acetic acid. Fix the specimen with the fluid for 10 min, transfer and wash with 70% ethyl alcohol.

5.4.2 *Staining nuclei of fixed cells*

The structural background of yeast nuclear division is not yet fully understood and some essential aspects are still inferred from genetic analysis.

Complementary staining of fixed nuclei with acid fuchsin (for nucleolus and mitotic spindles) and with HCl−Giemsa or HCl−acetate orcein (for chromosomal components) is needed to explain the cytological basis of mitosis and meiosis in different yeast species.

(i) *Acid fuchsin staining.*

(a)   Fix the cells with one of the above fixative fluids.
(b)   Rinse several times in 1% glacial acetic acid (beaker or Coplin jar).
(c)   Immerse for 1−5 min in 1:40 000 acid fuchsin in 1% acetic acid.
(d)   Examine in 1% acetic acid under the coverslip; use a green filter and optimal illumination.

(ii) *HCl−Giemsa staining.*

(a)   Fix the cells as above.
(b)   Immerse in 1% NaCl at 60°C for 1 h.
(c)   Hydrolyse by immersion for 8−10 min in 1 M HCl at 60°C (alternatively in 10% perchloric acid allowed to act overnight at room temperature).
(d)   Rinse with tap water at 37°C.
(e)   Stain for 1−2 h with Gurr's 'Improved Giemsa stain R 66' diluted 1:10 with Gurr's Giemsa (phosphate) buffer (pH 6.8).
(f)   Check the quality of staining with ×40 or ×70 *water-immersion* objective.
(g)   Differentiate overstained specimens with acidulated water in a Petri dish (40 ml of distilled water and a loopful of acetic acid). Experiment with the differentiation step in order to achieve the best results.
(h)   Mount the specimen in a solution containing two drops of Giemsa stain per 12 ml of Giemsa buffer.

(iii) *HCl−aceto-orcein staining.*

(a)   Fix with one of the fixatives described above.
(b)   Hydrolyse for 8−10 min with 1 M HCl at 60°C.
(c)   Rinse with water.
(d)   Immerse for a few minutes in 60% acetic acid.
(e)   Stain with aceto-orcein for 20−30 min (1% synthetic orcein in 60% acetic acid).
(f)   Examine the specimen in the staining solution under the coverslip.

6. FLUORESCENCE MICROSCOPY TECHNIQUES

Fluorescence microscopy differs from conventional light microscopy in the way the image is formed. UV or near-UV excitation causes yeast cells with specific fluorochromes to become luminescent. Two types of fluorescence can be studied: (i) primary fluorescence which is the natural fluorescence of substances present in cells (e.g. riboflavin), and (ii) secondary fluorescence which is induced by staining the cells with fluorescent dyes. Coupling fluorochromes to antibodies makes it possible to observe the location of antigens on the surface or within the cells.

A source of short-wavelength light (generally a mercury vapour lamp) is used in conjunction with thermal and excitatory filters inserted between the light source and the specimen, and with suppression filters interposed between the object and the eyepiece. The light emitted from the specimen has a higher wavelength than the absorbed light

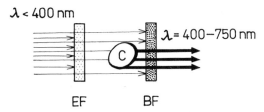

λ < 400 nm

λ = 400−750 nm

EF   BF

Figure 12. Principle of fluorescence microscopy: the selective transmission and suppression of short wavelengths by excitatory (EF) and barrier (BF) filters result in the emission of brilliant long-wave light by the specimen (C).

and can be monitored readily after suppressing the residual exciting light with a barrier (suppression) filter (*Figure 12*). The contrast obtained by this technique is high.

The last decade has seen an extraordinary refinement and specialization of techniques for fluorescence microscopy. Improvements have been mainly in the development of epi-illumination systems, including dichroic beam-splitters, matching excitation and barrier filters and special objectives.

Our initial studies involved transmitted illumination but now we obtain superior results with epi-illumination equipment. We have been using oil-immersion objectives ×40 and ×60 in conjunction with cardioid condensers of high numerical aperture. We prefer glycerol among the non-fluorescent immersion fluids. To obtain really good results, strict measures have to be taken to ensure cleanliness of all the optical parts of the microscope and of the glassware. Checking slides and coverslips for autofluorescence is also necessary.

6.1 Primary fluorescence

Autofluorescence is significant in many yeast species, the common pattern being a weak diffuse blue fluorescence of the cytoplasm and a yellow fluorescence of the granules. Checking for autofluorescence is a prerequisite for any experiments involving fluoro-chromes.

One way to reduce primary fluorescence is to use narrow-band cut-off suppression filters. Another possibility is to experiment with the conditions of cultivation of the yeast to be examined.

6.2 Fluorescent dyes

Classical fluorochromes acting as vital dyes are of permanent interest for yeast cytology. To demonstrate the differential fluorescence of cytoplasm and organelles under various conditions, primulin, berberin sulphate, aurophospin, coriphospin, thioflavine S, neutral red or acridine orange can be used and offer interesting results. The nature, colour and essential characteristics of these fluors have been reviewed (9) but the mechanism of their action is largely unknown.

Powerful DNA fluorochromes widely used at present for mapping nuclei in cyto-logically difficult organisms such as yeast, are stressed here since they have much to offer for the advancement of knowledge in yeast karyology. They also have potential in quantitating nuclear and mitochondrial DNA during yeast development.

Still another important group of fluorescent dyes are potential-sensitive fluors (10) used as probes with yeast cells that are too small to be impaled by microelectrodes. Methods involving the use of these probes are beyond the scope of this chapter.

In some cases, there is an advantage in using two or more fluorochromes simultaneously. In the case of contrast fluorochroming, excitation and emission is identical for both fluorochromes (e.g. mithramycin and acridine orange). Demonstration of complementary details can also be achieved through switching of various filter sets when excitation and emission are different with both fluors (e.g. 4′,6′-diamidino-2-phenylindole, DAPI and fluorescein).

The general procedure is given below.

(i)   Mount a small amount of yeast, previously washed by centrifugation, in a drop of the given fluorochrome solution on the slide. The staining period ranges from a few minutes to 15 min for individual dyes.

(ii)   Apply the coverslip.

(iii)   Make the preparation as thin as possible, by removing excess dye with a filter paper, to avoid a fluorescent background reducing the contrast of the specimen (important).

Alternatively, centrifuge the samples free of nutrient medium, resuspend the cells in the fluorochrome solution, centrifuge free of excess dye and use the cells in aqueous suspension for preparing specimens.

The main practical advantages of the techniques involving fluorochromes are:

(i)   simplicity of staining procedure;

(ii)   nearly instantaneous monitoring of samples in the course of experiments;

(iii)   minimum interference with normal cell physiology when using vital staining;

(iv)   very good resolution of features under study.

6.2.1 *Fluorochroming nuclei*

For the sake of completeness, staining of yeast nuclei with the fluorescent dye acridine orange is described first. This dye, however, does not stain DNA quantitatively and requires careful control of pH. Fluorescent DNA probes, mithramycin and DAPI bind relatively specifically to double-stranded DNA and exhibit an enhanced fluorescence in the association. These fluors are water-soluble and can be used as vital stains. With fixed preparations, their fluorescence reveals considerable detail of chromosome number and arrangement and is also proportional to the amount of DNA present. In some cases, selective counterstaining with acridine orange, Calcofluor or ethidium bromide to contrast nuclear structures against a differentially stained cytoplasm is advisable.

The acridine orange procedure produces a brilliant orange fluorescence of the cytoplasm and an intense green, yellow-green or yellow fluorescence of the nucleus (11).

(i)   Prepare the air-dried smears in the usual way.

(ii)   Rinse the slides in Carnoy's fixative (absolute alcohol:glacial acetic acid 3:1) for 30 min.

(iii)   Wash briefly in ethanol and air-dry.

(iv)   Immerse in 0.2 M acetate buffer (pH 4.5) for 15 min.

(v)   Stain with a solution compounded of the above buffer and 0.01 − 1 mg/ml of acridine orange for 20 − 120 min.

(vi) Rinse in buffer for 5–25 min.

(vii) Air-dry and observe.

When mithramycin is applied to the yeast culture, the nuclei fluoresce bright yellow-green. Discrimination of division figures is mostly obscured by the condensed chromatin (*Figure 11b*); unstained areas represent nucleoli. Generally, the method is designed for observing stages of nuclear division rather than the internal organization of the yeast nucleus (12).

The procedure for vegetative cells using mithramycin is as follows.

(i) Add a sample of culture to an equal volume of 50% aqueous ethanol solution containing mithramycin (0.4 mg/ml) and 30 mM $MgCl_2$, and stain for 15 min.

(ii) Counterstain briefly with acridine orange (1 μg/ml) if desired.

(iii) Centrifuge and observe as wet mount immediately or within 24 h.

The procedure for sporulating cells using mithramycin is given below.

(i) Fix sporulating cells in Carnoy's fixative for 1 h, then make slides by allowing a centrifuged sample of the fixed material to dry on a microscope slide.

(ii) Stain in 2 vol of aqueous 22.5 mM $MgCl_2$ containing mithramycin (0.6 mg/ml) for 20 min.

(iii) Counterstain briefly with acridine orange (1 μg/ml) if desired.

(iv) Apply a coverslip without rinsing and observe immediately or within 24 h if stored in the dark at 5°C.

With high magnification epifluorescence microscopy, DAPI techniques overcome the difficulty in cytological resolution of the yeast nucleus. Clear views of well-spread chromosomes and their dynamics during meiosis have been published (13,14). Chromosomes display a strong blue-white fluorescence; spots of weaker fluorescence localized in the cytoplasm are mitochondrial nucleoids (*Figure 11c*).

For living cells, the following steps should be carried out.

(i) Mix the yeast suspension with a drop of aqueous DAPI staining solution at a concentration ranging from 0.5 to 2 μg/ml.

(ii) Mount the stained specimen on a microscope slide, apply a coverslip and observe immediately.

DAPI can also be used in combination with ethidium bromide, emitting a pale orange fluorescence of the cytoplasm and a more pronounced orange colour of the nucleus. Double staining enhances the general fluorescence of the cells: the yeast suspension is first stained with DAPI for 5 min at 2°C and then post-stained with ethidium bromide (2.5 μg/ml) for 30 min (15).

Flattened protoplasts are most useful for counting meiotic chromosomes when observing fixed protoplasts.

(i) Prepare protoplasts according to Chapter 4, Section 11.1 or Chapter 7, Section 3.2.

(ii) Fix in Carnoy's fixative for at least 5 min.

(iii) Centrifuge and resuspend in Carnoy's fixative; mount a drop of the suspension on a slide.

(iv) Cool the slide to 0°C and heat in flame for 3 sec; allow to air-dry.

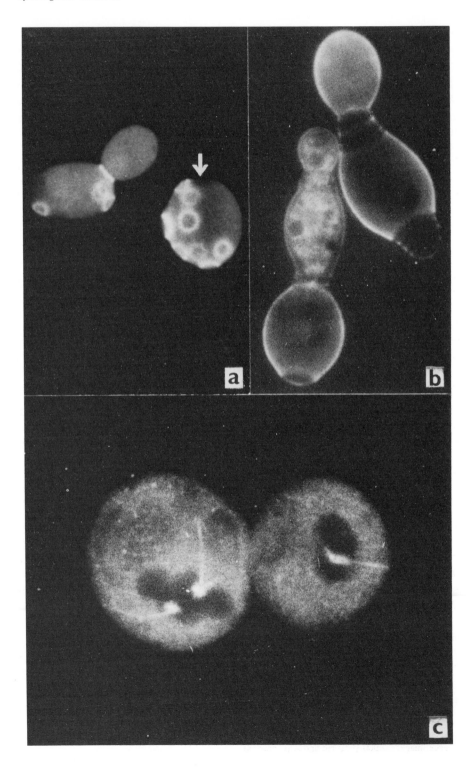

(v)   Immerse the preparation in staining solution composed of 1 μg/ml DAPI dissolved in the following buffer: 2 mM Tris$-$HCl, pH 7.6, 0.25 M sucrose, 1 mM EDTA, 1 mM MgCl$_2$, 0.1 mM ZnSO$_4$, 0.4 mM CaCl$_2$, 1.5% β-mercapto-ethanol.

(vi) Mount a coverslip and observe immediately or within 24 h if stored in the dark at 5°C.

6.2.2 *Fluorochroming mitochondria*

The existence of DNA in yeast mitochondria evokes such questions as how much mitochondrial DNA (mtDNA) is present per yeast cell? How much mtDNA is present per mitochondrion? What are the effects of mutations on the distribution of a mtDNA in the cell? Answers to these questions are greatly facilitated by the possibility of following mtDNA from the level of fluorescence microscopy to details seen in the electron microscope.

Williamson and Fennel (16) have described the fluorescent localization of mtDNA using DAPI. This probe specifically binds to mtDNA (*Figure 11c*) giving a pink-white fluorescence to the mitochondrial nucleoids.

(i)   Centrifuge cells free from nutrient medium.

(ii) Fix in 70% ethanol for 30 min.

(iii) Centrifuge, wash in distilled water and centrifuge again.

(iv) Apply the suspension to the slide and add aqueous staining solution (Section 6.2.1 above), but containing 0.1$-$0.5 μg/ml DAPI.

(v)   Apply a coverslip and observe immediately.

Mitochondria can be localized *in vivo* using rhodamine 123, berberine sulphate or coriphospin on concentrations ranging from 0.25 to 50.0 μg/ml prepared for vital staining as usual. Fluorochromed mitochondria appear yellow against the diffuse fluorescing cytoplasm.

6.2.3 *Fluorochroming the cell wall*

Primulin is one of the thiazole dyes; it is not to be confused with an anthocyanin of the same trivial name. The fluor is widely used in yeast cytology especially in cell wall studies. Its application includes determination of cell viability, characterization of the porous structure of the cell wall, quantification of bud scars and identification of regional cell wall growth at the population level as well as detection of the newly formed cell wall on the surface of regenerating protoplasts. Recently, the dye was found to be extremely useful in deciphering the mechanism of microfibril assembly.

Figure 13. Epifluorescence micrographs of primulin- and Calcofluor-stained cells. **(a)** Living cells of *Saccharomyces cerevisiae* are stained with primulin; note the presence of conspicuous bud scars and a birth scar (arrow), ×2500 (E.Streiblová). **(b)** Transitory stages from vital to post-mortal staining with Calcofluor: cell on the left is damaged and its cytoplasm displays a distinct fluorescence. The mother cell on the right side shows distinct multiple scars (17) giving rise to its apiculate shape. Accentuated fluorescence of the joined daughter cell (upper right) is due to prolonged light exposure rather than to the incorporation of new cell wall material, ×2500 (E.Streiblová). Epifluorescence micrograph of monoclonal antitubulin-stained cells of *S. uvarum* demonstrates the potential of immunofluorescence microscopy with yeast. **(c)** In MBC-treated cells spindle microtubules completely disappear with only residual cytoplasmic microtubules left in the fluorescence image, ×4000 (J.Hašek and J.Svobodová).

31

In living yeast cells, the cell wall fluoresces in various shades of green-yellow (*Figure 13a*) depending on cell wall texture and type of cell division (17).

(i)   Centrifuge the yeast sample free of nutrient medium.
(ii)   Mix a small amount of yeast suspension with a drop of primulin staining solution (0.1−0.5 μg/ml).
(iii) Apply a coverslip and observe.

6.3 Optical brighteners

Optical brighteners used in paper and textile treatments or added to detergents are efficiently absorbed by yeast cells. These colourless compounds are highly fluorescent when excited by UV to blue light. They are non-toxic at low concentrations and stable with respect to fluorescence when bound to cell structures. When added to the nutrient medium they can act as stable markers that impart fluorescence to the subsequent genera-tions of cells.

The following brighteners are of potential importance in yeast cytology: Tinopal 4 BMT (Geigy), Tinopal BOPT (Geigy), Photine LV (Hickson and Welch), Leucophor C (Sandoz) and Fluolite RP (ICI). Only the most popular brightener Calcofluor White RW (Cyanamid) that has essentially the same application as primulin will be considered in detail.

6.3.1 *Fluorescence emitted by Calcofluor*

Calcofluor White, primulin and Congo red are substances that display selective topo-optical properties when bound to one of the main components of the cell wall. Their affinity to the microfibrillar wall polymers is non-specific, although they have often been used incorrectly as a specific cytochemical test for chitin (see ref. 18 for details).

The flat linear molecules of Calcofluor, primulin and Congo red (*Figure 14*) display an intrinsic absorption anisotropy, on the basis of which one can characterize the spatial

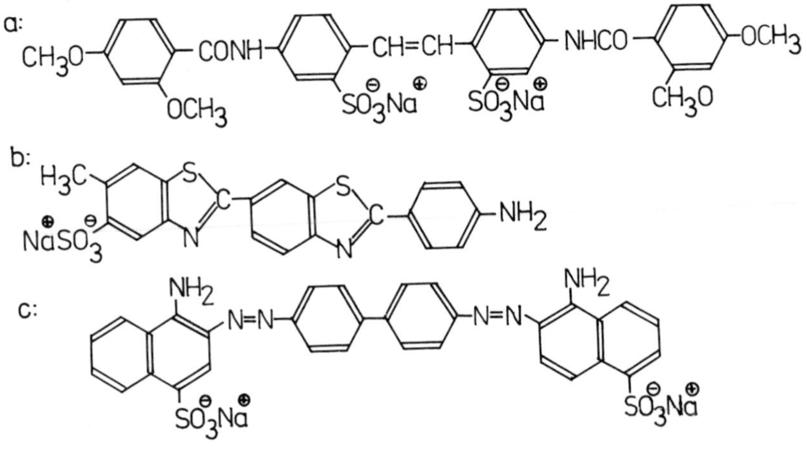

Figure 14. Dye molecules involved in the staining of yeast cell walls: **(a)** Calcofluor white; **(b)** primulin; **(c)** Congo red.

32

disposition of microfibrils in the native yeast cell wall. In a staining solution, the molecules of the above dyes form dipoles which attract one another and give rise to a polydisperse staining solution. This solution diffuses into the interfibrillar capillaries of the cell wall where the dye molecules are adsorbed in parallel to the microfibrillar axes. Our attempts at inducing polarized fluorescence of the yeast cell wall were not successful due to the low anisotropy of this structure.

(i)   Prepare a $0.1-0.5$ μg/ml staining solution of Calcofluor White.

(ii)   Mix a small amount of yeast suspension, centrifuged free of nutrient medium, with the staining solution on a slide.

(iii)   Apply a coverslip and observe.

Calcofluor and primulin are used routinely for distinguishing between living and non-living yeast cells, since the plasma membrane of dead cells is permeable to colloidal solutions of the dyes. In damaged or dead cells the dye enters the cytoplasm where it is covalently bound to the proteins of the cytoplasm, giving a powerful, bright yellow fluorescence.

Distinction between dead and living cells is usually simple. However, even partial damage of plasma membrane permeability can influence the interpretation of the fluorescent image. The following factors must be taken into account when staining with Calcofluor or primulin.

(i)   The cell under observation can be damaged by UV light, particularly during long photographic exposures. Both fluorochroming agents are of low toxicity but they enhance the photodynamic damage of the cell.

(ii)   The cell can be damaged mechanically by the pressure on the coverslip during specimen preparation.

(iii)   Even a gentle chemical extraction of the samples can be reflected in an increased permeability of the plasma membrane to the fluorochrome.

The above effects must be taken into consideration especially during observations of cells with polarized growth, the tips of which are extremely sensitive to external influences, for example buds, hyphal tips and germinating spores. Generalized fluorescence need not indicate an increased incorporation of polysaccharides into the cell wall but could reflect damage caused by the above factors (*Figure 13b*).

6.4 **Fluorescent antibodies**

The use of fluorescence microscopy is greatly extended by methods that employ fluorochrome-marked antibodies to identify appropriate antigens in biological materials. Antibodies produced in response to an antigen combine specifically with that antigen and their coupling with fluorescent dyes allows mapping of sites of antigen—antibody combination. The sites of immunocytochemical reaction are recognized by brilliant fluorescence in a dark-ground fluorescence microscope.

The two fluorochromes used most widely as labels are fluorescein and rhodamine B, both used in the isothiocyanate form which binds when coupled to antibodies without affecting their immunological specificity. Fluorescein isothiocyanate (FITC)-marked antibodies give a strong green-yellow fluorescence; excitation is maximal in the region of 550 nm. Rhodamine B-labelled antibodies display bright red-orange fluorescence with maximum excitation at about 640 nm. Rhodamine B can also be used as simple counterstain in experiments with fluorescein-bound antibodies. The major problem accompanying the use of both fluors is light-induced bleaching apparent as rapid fading of the emitted fluorescence, especially with epifluorescence illumination.

With yeast, the immunofluorescence methods were performed largely for the purpose of diagnosis of mycotic diseases and identifying pathogens such as *Candida albicans*. The investigations have so far been done predominantly with whole cells. The yeast cell wall is a traditional source of complex antigenic mixtures which tend to generate antisera with wide ranges of antibodies and often considerable cross-reactivity. These antisera when labelled with fluorescein have proved to be a powerful tool for locating regions of cell wall extension and growth in a variety of yeast species.

The main potential of immunofluorescence labelling for yeast cytology lies in marking particular intracellular proteins or organelles in order to identify, localize and understand their dynamics. The production of antibodies specific to cytoskeletal structures, particularly microtubules, has raised the possibility of examining their distribution in yeast cells. The application of appropriate techniques to yeast has, however, progressed slowly, primarily because the cell wall forms a barrier to antibody penetration.

In the recent past, modifications of animal cell immunofluorescence procedures have been used to produce immunofluorescence images in yeast (19−21). The critical steps of the procedure include stabilization of the cytoskeleton and penetration of antibodies into the cell. The procedure detailed below (Section 6.4.2) provides a starting point for the fluorescent detection of antigenic sites within yeast cells in general.

6.4.1 *Direct immunofluorescence staining*

In direct (one-step) methods, the antibody marked with a fluorescent dye (usually fluorescein) is applied directly to the specimen. It combines with the appropriate antigen on the surface or inside the cell and can then be localized by the fluorescence microscope (*Figure 15b*).

A rapid technique for the control of purity of baker's and brewer's yeast is based on the detection of non-fluorescent contaminants among fluorescent pure-culture cells. Observation under combined fluorescent and normal light permits the detection of one contaminant cell among 1000 pure-culture cells (22).

(i)   Centrifuge a portion of the yeast culture and resuspend it in a few drops of saline.
(ii)   Prepare smears and fix them by gentle heat (60°C).
(iii) Spread the labelled antiserum (22) over the fixed smear and place the slide in a moist chamber to avoid drying; a staining time of 30 min is usually adequate.
(iv)   Drain the slides free of the antiserum and wash thoroughly by two changes in phosphate-buffered saline (PBS) pH 7.
(v)   Rinse in buffered saline for 10 min.
(vi)   Dry the slides and mount in phosphate-buffered glycerol (pH 7; 9:1) under the coverslip; this pH is recommended as fluorescence tends to fade at acidic pH values.

6.4.2 *Indirect immunofluorescence staining*

In indirect (two-step) techniques, the preparation first reacts with an unlabelled antibody

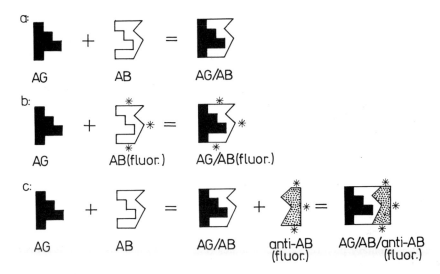

Figure 15. Methods of immunofluorescence: **(a)** combination of antigen (AG) with specific unlabelled antibody (AB) gives rise to a complex (AG/AB); **(b)** with direct methods, combination of antigen (AG) with fluorochrome-labelled antibody (AB fluor) produces a fluorescent complex (AG/AB fluor); **(c)** with indirect methods, the AG/AB complex reacts with fluorochrome-labelled antigamma globulin against the first (specific) antibody (anti-AB fluor).

(antiserum) which is allowed to bind to its appropriate antigen. In the second step, the antigen − antibody complex becomes coated and rendered visible with a second, fluorescent, antibody (conjugate) that is specific for the first antibody (*Figure 15c*).

Indirect immunofluorescence techniques are widely used, as they are readily controlled and very sensitive due to their inherent amplification effect. One must utilize reagents previously characterized as to their activity and specificity, which is generally achieved by using commercially available conjugates obtained from potent antisera and with discrete labelling.

Most yeast cells do not stain without application of wall-modifying treatment since fluorescent antibodies do not penetrate across the cell wall. The wall can be either removed by digestive enzymes (19,20) or by a short disintegration (Section 7.1.3). Application of digestive enzymes prior to fixation leading to formation of protoplasts (21) is used in the following method.

(i)   Prepare protoplasts according to Chapter 4, Section 11.1 or Chapter 7, Section 3.2.

(ii)   Add gently centrifuged protoplasts to osmotically stabilized medium containing 4% formaldehyde. The fixative is used in microtubule-stabilizing buffer (0.1 M potassium phosphate buffer, pH 6.5, 30 mM EDTA and 1 mM $MgCl_2$); pre-fix for 80 min.

(iii)   Fix for 10 min with 1% glutaraldehyde without washing.

(iv)   Centrifuge the fixed protoplasts free from the fixative mixture.

(v)   Wash twice with PBS, pH 7.0.

(vi)   Reduce the free aldehyde groups with $NaBH_4$ (0.5 mg/ml) for 20 min.

(vii)   Rinse in PBS three times, 1 min each.

(viii) Mount the protoplasts on coverslips and allow to air-dry. All subsequent steps involve gentle transfers of coverslips with adhering protoplasts.

(ix) Plunge the specimen into absolute alcohol for 10 min and into acetone for 30 sec, both at −20°C.

(x) Pre-incubate the specimen for 60 min with 1% bovine serum albumin (BSA) in PBS at 37°C.

(xi) Layer the coverslips with mouse anti-tubulin antibody and incubate for 60 min at 37°C (21).

(xii) Wash in BSA/PBS three times, 37°C.

(xiii) Layer the coverslips with anti-mouse γ-globulin conjugated with FITC diluted 10-fold with 1% BSA and incubate for 60 min at 37°C (21).

(xiv) Wash three times in PBS.

(xv) Post-stain the preparation with DAPI (Section 6.2.1) and mount in 50% glycerol with 0.1% *n*-propyl gallate (23) to reduce photobleaching.

Complementary pictures of microtubules (*Figure 13c*) and nuclei (*Figure 11c*) are obtained by using the appropriate filter sets.

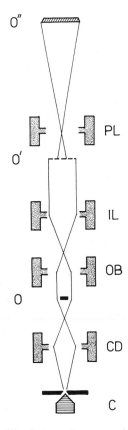

Figure 16. The basic optical system of the electron microscope: electron source — cathode (C), magnetic condenser lens (CD), specimen (O), magnetic objective lens (OB), intermediate lens (IL), intermediate image (O′), magnetic projection lens (PL), final image (O″) on viewing screen or photographic plate.

7. ELECTRON MICROSCOPIC TECHNIQUES

The electron microscope is the only magnifying instrument making it possible to study directly the subcellular and macromolecular architecture of the yeast cell. The effective wavelength of the electron beam used in the construction of the electron image can be as small as 0.0025 nm. The practical resolution limit of the modern electron microscope is better than 0.2 nm. Thus, significant development of specimen preparation techniques will be required before yeast cytologists can take advantage of the currently available levels of this resolution.

A beam of electrons produced in high vacuum by passing a high voltage current through a tungsten filament (cathode) is focused by a series of electromagnetic lenses, in a manner corresponding to that of the lenses of the light microscope. When the electron beam passes the specimen, various beam − specimen interactions occur, and different instruments (conventional transmission, scanning, scanning-transmission and high voltage electron microscope) gather ultrastructural information of various kinds. A further set of electromagnetic lenses projects the final image onto a viewing screen or a photographic plate (*Figure 16*).

The intrinsic contrast of yeast cells is too low for details of their subcellular architecture to be seen without contrast enhancement. Contrast in the electron microscope relies on differential scattering of electrons, which depends on selective combination of staining material of high atomic weight with appropriate regions of the specimen. Electron-dense areas of the electron image appear dark and areas with little dense material appear light when transformed into a visible pattern on the viewing screen. Thus, fixed thin sections are exposed to solutions of heavy metal salts such as lead hydroxide, acetate or citrate, or uranyl acetate; yeast cell surfaces and cytoplasmic and membrane fracture faces are studied by shadowing techniques using heavy metals such a chromium, platinum or gold − palladium; and, subtle particulate structures such as cell wall microfibrils and cytoplasmic skeletal elements are outlined by suspending them in a solution of heavy metal: uranyl acetate or phosphotungstic acid.

Preparatory procedures used with yeast for electron microscopy follow the current trends of refinement in techniques for other biological materials. However, the presence of a rigid cell wall and the relatively high density of the cytoplasm frequently pose particular problems in routine specimen preparation.

7.1 Ultrathin sectioning

Only thin section techniques of fixed material can demonstrate the detailed subcellular organization of individual yeast cells in addition to the basic features common to all cells.

Potassium permanganate is the only fixative that diffuses through the yeast cell wall. Other fixing agents, such as OsO_4 and glutaraldehyde, do not penetrate the intact yeast cells easily and show less detail than permanganate fixation.

A clearer definition of the yeast fine structure is currently achieved by removal of the yeast cell wall. This can be done by treatment of cells during or after fixation with snail enzymes (24,25) or by disrupting the cell wall of pre-fixed cells by mechanical means (26,27). A controlled stretching of the cytoplasm is also possible with the latter method; the enlargement of the cytoplasm minimizes the obscuring effect of densely spaced ribosomes and exposes the delicate organization of the yeast cytoskeleton.

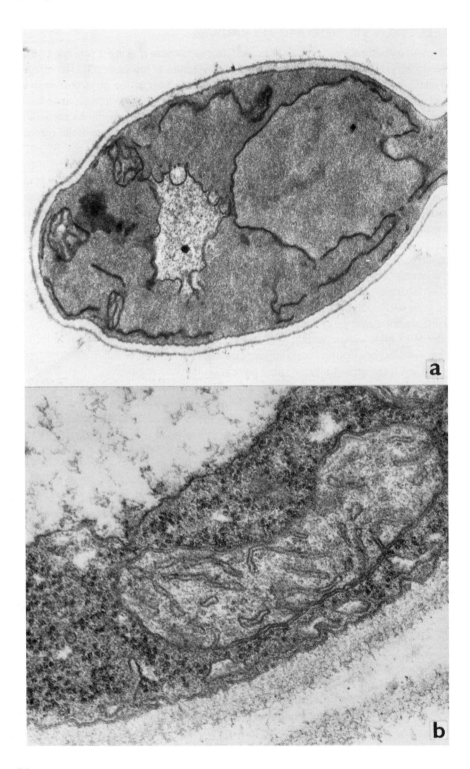

7.1.1 *Potassium permanganate*

Many ultrathin section studies have been carried out on yeast cells treated with this fixative. The procedure is simple and reliable when studying cell walls and membrane systems which appear as black lines against a homogeneous background. Cytoplasm and nucleoplasms have no clearly defined ultrastructural identity since ribosomes and most proteins are destroyed (*Figure 17a*). It is advisable to use permanganate at lower concentrations than usual, and at low temperature, to reduce its powerful oxidizing action. Uranyl acetate or nitrate, or lead citrate or oxide, applied in the dehydration step enhance the final contrast considerably (28).

(i)   Prepare a 1.5% potassium permanganate solution in distilled water; the solution must be fresh.

(ii)   Use a round-bottom centrifuge tube containing enough fixative to cover the sample easily. Transfer the yeast pellet to be examined into the fixative, plug the tube and keep for 30−60 min at 4°C.

(iii) Centrifuge the sample free of fixative and wash out by several changes of distilled water before dehydration.

7.1.2 *Osmium tetroxide*

Fixation with this reagent alone is not successful with whole yeast cells. At present, OsO_4 is widely used in a fixative combination following the glutaraldehyde step. Post-osmication contributes to the stabilization of proteins and reacts with unsaturated lipids and phospholipids. Osmium can also be regarded as an important factor for the enhancement of contrast in the specimen, since any structure in which this agent is deposited during fixation appears dense in the electron image.

7.1.3 *Glutaraldehyde and osmium tetroxide*

The most useful general combination for the current documentation of yeast ultrastructure is glutaraldehyde fixation followed by post-fixation with OsO_4. Partial removal of cell walls is required to ensure rapid and complete penetration of the fixatives. In general, contrast is further increased by exposing the material to solutions of stains before dehydration, during dehydration or after sectioning. The approach is useful in demonstrating many aspects of yeast morphology (*Figure 17b*), especially the cytoskeletal structures (*Figure 18*). Glutaraldehyde alone produces minimal denaturation and preserves many systems for the study of their enzyme activities.

Excellent results using glutaraldehyde in combination with other fixatives and stains were obtained by Robinow (24), Byers (25) and others. We prefer the method derived from the technique described by May (26).

(i)   Add glutaraldehyde directly to yeast cells in nutrient medium to give a final fixative concentration of 3% v/v.

Figure 17. Transmission electron micrographs demonstrating general aspects of the yeast ultrastructure in thin sections. (**a**) *Saccharomycodes ludwigii* fixed with potassium permanganate, ×30 000. (**b**) Fine structure of *Nadsonia elongata* that has been preserved successfully by careful control of the pH, osmotic pressure and ionic strength of the fixative; glutaraldehyde and OsO_4 fixation, uranyl staining, ×85 000 (M.Havelková).

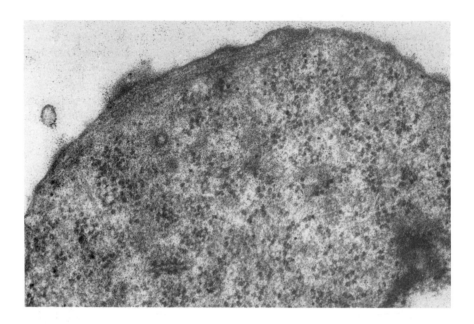

Figure 18. Thin section showing both essential components of the yeast cytoskeleton in *Schizosaccharomyces pombe* impaired in the *cdc*12 gene: the microtubule is positioned in the cytoplasm and the bundle composed of microfilaments is located beneath the plasma membrane; glutaraldehyde and OsO_4 fixation, uranyl and lead staining, ×38 000 (E.Streiblová).

(ii)   Fix at room temperature; microtubules tend to dissociate when fixed in cold solutions.

(iii)   After 10 min of initial fixation, harvest the cells quickly by centrifugation and resuspend in 0.7 M sorbitol for osmotic stabilization of the suspension.

(iv)   Rupture the pre-fixed cell suspension in a disintegrator using 0.5 mm diameter Ballotini beads for 10−20 sec, depending on the appearance of the specimen under a light microscope: the treatment results in stripping of cell walls from a substantial part of the cells which retain their original shape. Most of the success depends on the time of pre-fixation and on the disintegration step which should be estimated empirically.

(v)   Centrifuge the sample free of beads and broken cells; immediately add a 3% solution of glutaraldehyde in 0.1 M phosphate buffer (pH 7). Fix for 1 h including pre-fixation.

(vi)   Centrifuge free of glutaraldehyde and wash the specimen out in two 10-min changes of phosphate buffer.

(vii)   Add just enough of the 1% solution of OsO_4 in 0.05 M phosphate buffer (pH 7) to cover the pellet in a glass-stoppered vessel; keep at room temperature for 1 h. The solution should be prepared at least 1 day before by breaking the glass ampoule containing the crystals in the buffer solution within the bottle to be used for storage. Care is necessary in the use of the reagent.

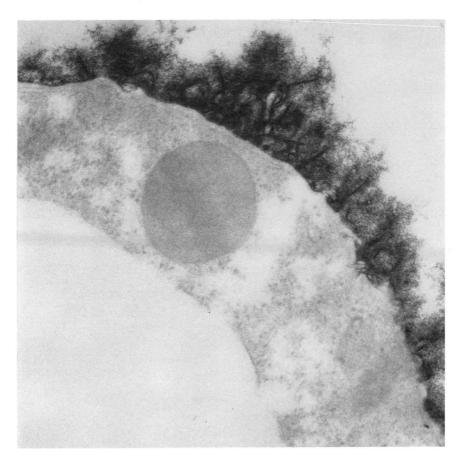

Figure 19. Thin section showing the surface of a regenerating protoplast of *Saccharomyces cerevisiae*: the densely stained coat consists of amorphous matrix and fine microfibrillar material; OsO_4 and glutaraldehyde fixation, ruthenium red staining to demonstrate carbohydrate-rich material, ×70 000 (M.Havelková).

(viii) Wash thoroughly, but not too long, in six changes of distilled water to remove phosphate.

(ix) Stain overnight with 0.5% uranyl acetate.

(x) Embed the yeast pellet in 2% agar and cut into small pieces after gelling.

Dehydration of agar cubes is achieved by passing the material through increasing concentrations of ethanol, finishing with absolute ethanol as usual. An alternative to ethanol is acetone. The samples are then embedded in embedding medium in the usual manner. We have achieved good results with Durcupan. Sections are cut with an ultra-microtome and post-stained with various stains such as lead citrate or lead oxide.

7.1.4 *Cytochemical and tracer techniques*

In recent years, ultrastructural techniques with sectioned material have extended the use of the electron microscope beyond the limits of pure morphology. Some of these

41

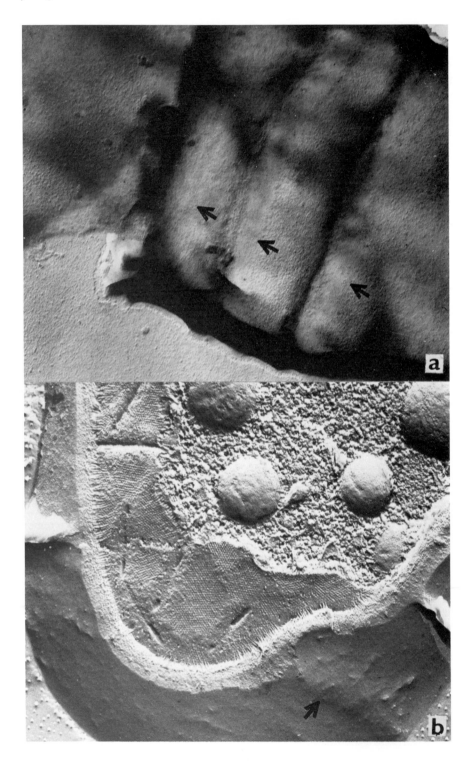

techniques have evolved from existing light cytochemistry procedures. An example is ruthenium red used in conjunction with OsO_4 impregnation to stain the coat of freshly prepared and regenerating yeast protoplasts (29). Adequate contrast to mark the assembling wall microfibrils is made available by the affinity of the bound dye for osmium (*Figure 19*). The Periodic – Schiff reaction to demonstrate polysaccharides by light microscopy can also be demonstrated by electron microscopy by the use of a silver proteinate compound instead of the coloured Schiff reagent (28).

Among the enzymic reactions that have been demonstrated successfully in yeast by precipitation of heavy metal salts at the reaction sites are those of acid and alkaline phosphatases.

Labels suitable for the localization of antigens at the electron-microscopic level are now available. Apart from immunoferritin methods, immunoperoxidase techniques appear to be very promising. Tracer techniques also include the use of colloid gold-labelled lectins which are currently popular tools in binding specifically to mannans and galactomannans of the yeast cell wall.

7.2 Shadowing techniques

These methods are designed for surface studies of particulate objects such as protoplasts, isolated organelles and complex macromolecules. The specimen is best examined in a suspension spread on formvar (polyvinyl formaldehyde)- or parlodion (nitrocellulose)-coated grids reinforced by a layer of evaporated carbon.

Shadow-casting involves the evaporation of metal (e.g. chromium, platinum, platinum – carbon, gold – palladium) under vacuum. The material falls at a pre-determined angle on the specimen. In this way, metal atoms are deposited at one side of the structure under examination and shadow is cast on the other side. If the angle of shadowing is known, the size and shape of the material can be calculated from measurement of the length of the shadow. The resolving power of the method is, depending on the metal used, up to $1.5 - 2$ nm. Shadowing has been successfully applied to isolated cell walls (*Figure 20a*) and to microfibrillar nets isolated from the surface of yeast protoplasts.

(i)   Place a small drop of aqueous suspension of the structures under examination (e.g. isolated cell walls) on formvar-coated grids.

(ii)   Fix by exposing the specimen to vapour from 2% OsO_4 for about 30 sec.

(iii)   Allow the preparation to settle for 5 min.

(iv)   Blot off excess liquid with a piece of filter paper and air-dry.

(v)   Place the grids in a vacuum chamber and coat the specimen with gold – palladium at an angle of $45°$.

(vi)   Observe under the transmission microscope.

Figure 20. Resolution of surface features using metal-coating of isolated cell walls and fracturing and etching of frozen specimen. (a) A clear resolution of three division scars (17) with adjacent scar bands (arrows) is obtained with isolated cell walls of *Schizosaccharomyces pombe* examined in transmission electron microscope with platinum coating, ×30 000. (b) A multiple scar (17) is seen on the cell wall of *Saccharomycodes ludwigii* (arrow); closely spaced particles and invaginations are apparent on the plasma membrane leaflet exposed by fracturing. Both structures, found on the plasma membrane, are thought to be a presentation of the living state; freeze-etching and fracturing, carbon – platinum coating, ×38 500 (E.Streiblová).

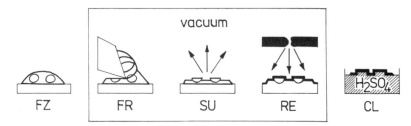

Figure 21. Steps in the freeze-fracturing and etching technique: freezing (FZ), fracturing (FR), sublimation (SU), shadowing (RE), cleaning of the specimen (CL).

Surface replicas are generally used to examine fine details of the cell wall on the whole yeast cells. A thin film of carbon is deposited under vacuum on the specimen. The carbon film replicating the yeast cell surface is stripped from the specimen by chemicals, cleaned, mounted on a Cu grid and shadowed obliquely with metal to accentuate minute details (30).

7.3 Freeze-fracturing and freeze-etching

These techniques involve the making of replicas of surfaces and fracture faces of the frozen yeast cell. Fixation and dehydration are not necessary, but the replicas are of cells which differ from the living state in that they are frozen. Generally, the fracture proceeds along structurally weak paths within frozen membranes and thus, after etching, three-dimensional images of the cell interior are obtained. The replicas are examined under the transmission electron microscope.

A yeast pellet is frozen on a support disc in liquid Freon at $-155°C$ and mounted on a pre-cooled specimen stage in a freeze-etch device. The essential steps of the procedure are carried out under high vacuum. Fracturing produces a series of breaks in the frozen cell interior in front of the cooled microtome knife edge. Etching involves sublimation of ice from the object; this gives a rough appearance to the cross-fractured cytoplasm and leaves fractured edges and surfaces of membranes standing slightly above it. In the subsequent step the cold specimen is coated with a mixture of platinum and carbon at an angle of $45°C$. Cleaning of the replica simply involves chemical removal of the cell material (chromic acid) and washing of the replica before picking it up and examining it in the electron microscope (*Figure 21*).

The first really successful freeze-etch device was used with yeast cells by Moor and Mühlethaler (31). It is commercially available from Balzers AG, Lichtenstein. Various simple freeze-fracture devices were developed elsewhere. The techniques are particularly applicable to membrane structure investigations and show great potential in combination with physical and biochemical methods. Freeze-fracture nomenclature in electron micrographs follows Branton *et al.* (32).

Of particular interest are the surfaces exposed by the process of mechanical fracturing of the frozen material. There is a tendency for the cells to fracture along membranes, thus exposing their inner surfaces for examination. New information may also be gained by allowing ice to sublime from whole yeast cells or protoplasts. In this manner, otherwise hidden details of yeast subcellular architecture are exposed which are not distorted

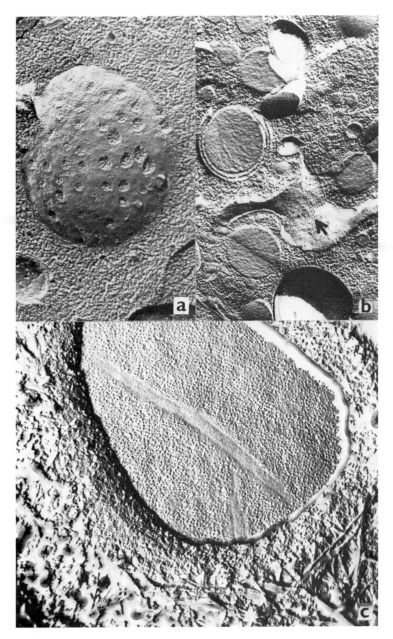

Figure 22. Cellular membrane systems and surface features in freeze-fractured and etched cells: **(a)** fractured nuclear membrane reveals conspicuous nuclear pores, ×40 000; **(b)** within the cytoplasm of wild-type yeast, endoplasmic reticulum is present in rather small amounts. Very prominent is the endomembranous system in septation mutants of *Schizosaccharomyces pombe*; notice the presence of lipid droplets and of vesicles discharged by the endoplasmic reticulum (arrow), ×58 000; **(c)** particles seen on the surface of the inner leaflet of the plasma membrane of *Schiz. pombe* corresponding to protein macromolecules traversing the full thickness of the membrane; deep etching illustrates features of the microfibrillar wall texture developing on the surface of regenerating protoplasts, ×30 000 (E.Streiblová).

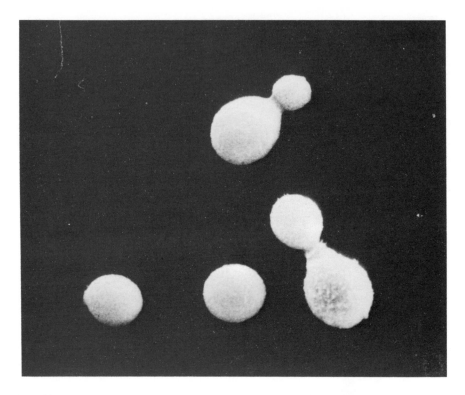

Figure 23. Scanning electron micrograph showing surface features of cells of *Cryptococcus* sp.; an additional external capsule is present; critical point drying, gold−palladium coating, ×5000 (O.Kofroňová).

by conventional fixation and dehydration procedures. Examples may be seen in *Figures 20b* and *22a−c*.

7.4 Scanning electron microscopy

The scanning electron microscope uses a combined approach based on the technology of electron microscopy and that of television electronics in the process of image formation. Although generally possessing lower resolution power than the conventional electron microscope (∼5−10 nm), this instrument provides a means for the examination of surface structures of complex biological units with a striking three-dimensional effect. The uses of this technique are relatively specific compared with other electron microscopic methods.

With yeast, this procedure is popular for observing structural relationships between chains of individual yeast cells and for interpreting shape characteristics of single cells (*Figure 23*). The technique also demonstrates very good preservation of freshly prepared and growing yeast protoplasts.

(i)   Prepare the specimen and use glutaraldehyde−OsO$_4$ in 0.1 M sodium cacodylate buffer (pH 7.1) according to Cole and Samson (5).

(ii)   Wash the specimen by ten changes of 0.1 M cacodylate buffer.

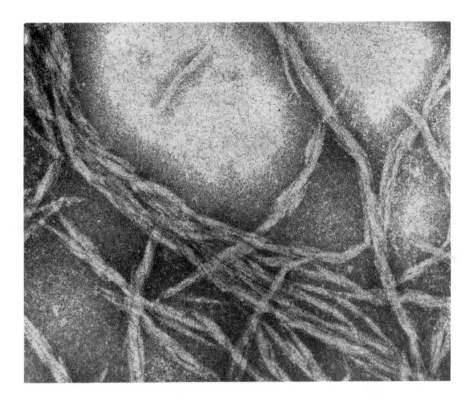

Figure 24. High-magnification micrograph showing the pattern of cell wall microfibril bundling as seen by negative staining with phosphotungstic acid, ×100 000 (M.Kopecká).

(iii)   Fix with 1% buffered OsO_4 for 2 h; this is considered to be very important for hardening the surface of soft cells.

(iv)   Wash thoroughly by 10 changes of cacodylate buffer.

(v)   Replace the water in the specimen sequentially; first using a graded ethanol series.

(vi)   Then use a graded series of amyl acetate in absolute ethanol, finishing with pure amyl acetate; this is designated as the intermediate liquid step.

(vii)   Use CO_2 as the so-called transitional liquid, having the critical point within the range of design of the critical-point apparatus for scanning microscopy.

(viii)   Carry out the critical-point drying procedure according to detailed instructions attached to the critical-point apparatus at one's disposal.

(ix)   Place the specimen on an omnirotating table and coat with a thin layer of conducting material (usually carbon or aluminium followed by gold—palladium).

(x)   Observe under the scanning microscope.

7.5 Negative staining

This is one of the most important, reliable and rapid methods of studying isolated subcellular structures and macromolecules. Generally, a thin layer of negative stain such as uranyl acetate or phosphotungstic acid is spread across the surface of the specimen. Direct examination by electron microscopy gives finely detailed images of subtle surface features of the material under study. The method is of particular interest for the future classification of yeast cytoskeletal fibres since it permits the visualization of filamentous protein aggregates, thus providing pictorial confirmation of their molecular shape as predicted by other methods. Kreger and Kopecká (33) used this technique to study ultrastructural details of polysaccharide microfibrils assembled on the surface of yeast protoplasts (*Figure 24*).

(i)   Suspend the particles to be examined in distilled water or in appropriate buffer.
(ii)   Transfer a small drop of the sample with a micropipette onto a formvar-coated grid.
(iii)   Allow the material to settle within 30 min and ensure that the specimen does not dry out (important).
(iv)   Remove excess liquid and add immediately a drop of 1−2% aqueous solution of phosphotungstic acid.
(v)   Drain off excessive negative stain, allow to air-dry and observe under the transmission electron microscope.

8. ACKNOWLEDGEMENTS

I am indebted to J.Fiala, J.Hašek, M.Havelková, E.Jelke, O.Kofroňová, M.Kopecká, L.Mueller, A.Svoboda and J.Svobodová for providing me with their microphotographs. My thanks are also due to J.Svobodová for help with line drawings.

9. REFERENCES

1. Pringle,J.R. and Hartwell,L.H. (1981) In *The Molecular Biology of the Yeast Saccharomyces. Life Span and Inheritance*. Strathern,J.R., Jones,E.W. and Broach,J.R. (eds), Cold Spring Harbor Laboratory Press, New York, p. 97.
2. Pringle,J.R. and Moor,J.R. (1975) In *Methods in Cell Biology*. Prescott,D.M. (ed.), Academic Press, New York, Vol. XI, p. 131.
3. Robinow,C.F. (1975) In *Methods in Cell Biology*. Prescott,D.M. (ed.), Academic Press, New York, Vol. XI, p. 2.
4. Cole,G.T. and Kendrick,W.B. (1968) *Mycologia,* **60**, 340.
5. Cole,G.T. and Samson,R.S. (1979) *Patterns of Development in Conidial Fungi*. Pitman, London.
6. Mueller,R. (1956) *Naturwissenschaften,* **43**, 428.
7. Poon,N.H. and Day,A.W. (1974) *Can. J. Microbiol.,* **20**, 739.
8. Kopecká,M. and Gabriel,M. (1977) *Arch. Microbiol.,* **119**, 305.
9. Drawert,H. (1968) *Vital Staining of Plant Cells and Tissues*. Springer, Vienna.
10. Waggoner,A. (1976) *J. Membr. Biol.,* **27**, 317.
11. Rustad,R.C. (1958) *Exp. Cell Res.,* **15**, 444.
12. Slater,M.L. (1976) *J. Bacteriol.,* **126**, 1339.
13. Williamson,D.H., Johnston,L.H., Fennel,D.J. and Simchen,G. (1983) *Exp. Cell Res.,* **145**, 209.
14. Kuroiwa,T., Kojima,H., Miyakawa,I. and Sando,N. (1984) *Exp. Cell Res.,* **153**, 259.
15. Toda,T., Yamamoto,M. and Yanagida,M. (1981) *J. Cell Sci.,* **52**, 271.
16. Williamson,D.H. and Fennel,D.J. (1975) In *Methods in Cell Biology*. Prescott,D.M. (ed.), Academic Press, New York, Vol. XII, p. 335.
17. Streiblová,E. (1971) In *Recent Advances in Microbiology*. Pérez-Miravete,A. and Peláez,D. (eds), Assoc. Mex. Microbiol., Mexico DF, p. 131.
18. Streiblová,E. (1984) In *The Microbial Cell Cycle*. Nurse,P. and Streiblová,E. (eds), CRC Press, Boca Raton, FL, p. 127.
19. Kilmartin,J.V and Adams,A.E.M. (1984) *J. Cell Biol.,* **98**, 922.
20. Adams,A.E.M. and Pringle,J.R. (1984) *J. Cell Biol.,* **98**, 934.
21. Hašek,J., Svobodová,J. and Streiblová,E. (1986) *Eur. J. Cell Biol.,* **41**, 150.
22. Kunz,C. and Klaushofer,H. (1961) *Appl. Microbiol.,* **9**, 469.

23. Giloh,H. and Sedat,J.W. (1982) *Science*, **217**, 1252.
24. Robinow,C.F. and Marak,J. (1966) *J. Cell Biol.*, **29**, 129.
25. Byers,B. (1981) In *The Molecular Biology of the Yeast Saccharomyces. Life Span and Inheritance.* Strathern,J.R., Jones,E.W. and Broach,J.R. (eds), Cold Spring Harbor Laboratory Press, New York, p. 59.
26. May,R. (1974) *Z. Allg. Mikrobiol.*, **14**, 161.
27. Streiblová,E., Hašek,J. and Jelke,E. (1984) *J. Cell Sci.*, **69**, 47.
28. Kopp,F. (1975) In *Methods in Cell Biology.* Prescott,D.M. (ed.), Academic Press, New York, Vol. XI, p. 23.
29. Havelková,M. (1981) *Folia Microbiol.*, **26**, 65.
30. Bradley,D.E. (1958) *Nature*, **181**, 875.
31. Moor,H. and Mühlethaler,K. (1963) *J. Cell Biol.*, **17**, 1.
32. Branton,D., Bullivant,S., Gilula,N.B., Karnovski,M.J., Moor,H., Mühlethaler,K., Northcote,D.H., Packer,L., Satir,B., Satir,P., Speth,V., Staehelin,L.A., Steere,R.L. and Weinstein,R.S. (1975) *Science*, **190**, 54.
33. Kreger,D.R. and Kopecká,M. (1976) *J. Gen. Microbiol.*, **92**, 207.

CHAPTER 3

Synchronous cultures and age fractionation

J.M.MITCHISON

1. INTRODUCTION

Certain aspects of the yeast cell cycle can be studied at the single cell level (1) and in normal asynchronous cultures, but when bulk measurements have to be made it is necessary to use synchronous cultures or age fractionation. There are many methods available and they all have advantages and difficulties. The reader is strongly advised to consider the full range of methods before making a final choice of the one most fitted to his needs and to his resources. The practical details are mostly given for the fission yeast *Schizosaccharomyces pombe* with which I have had the greatest experience but there are only small differences in handling and behaviour between this yeast and *Saccharomyces cerevisiae*.

2. SYNCHRONOUS CULTURES

2.1 Selection methods

The basic technique is to collect the cells from a normal asynchronous culture and separate off one age group — usually the smallest and youngest cells at the beginning of the cell cycle. These cells are then grown up separately as a synchronous culture. The yield is necessarily smaller than with induction synchrony since most of the collected cells are discarded. On the other hand, there is less distortion of the cycle and it is possible to make controls for the side effects of the selection procedure.

2.1.1 *Velocity separation in tubes*

This is a quick and easy method of making small synchronous cultures (2). A concentrated suspension of growing cells is layered on a gradient made in a centrifuge tube. It is then centrifuged for a short time and the cells are separated by size. The top layer of small cells is removed and resuspended to make the synchronous culture.

The details for *Schiz. pombe* are as follows.

(i)   Harvest about 10^{10} cells from a 4 litre exponential phase culture by filtering on a 32 cm Whatman No 50 filter in a holder similar to those used for membrane filters. This takes $10-15$ min.

(ii)   Scrape the paste of cells off the filter with a spatula and resuspend in medium to a vol of $4-5$ ml.

(iii)   Layer on top of an 80 ml linear gradient of $7.5-30\%$ lactose (made up in medium) produced in a 100 ml centrifuge tube by a conventional gradient former (3).

(iv) Centrifuge the tube in a swing-out head until the cell layer has moved about half-way down the tube. This takes approximately 5 min at 800 *g* at 25°C.

(v) Remove the top layer of cells (5−10 ml) by a syringe with a long needle and add to warm 'conditioned' medium (the supernatant from the harvesting) to make the synchronous culture.

The final yield is about 5×10^8 cells, about 5% of the cells loaded on the gradient. The subsequent progress of the synchronous culture is best followed with an electronic cell counter.

The following points relating to this method are worthy of comment.

(i) This is a simple method which requires little apparatus apart from the centrifuge and the gradient former.

(ii) Harvesting can be done with a membrane filter or a continuous flow centrifuge but these are more expensive.

(iii) Lactose is used for the gradient since it is not metabolized by *Schiz. pombe* but it is also possible to use sucrose, sorbitol or glycerol. Gradients with low osmotic pressure can be made with Ficoll, Percoll, protein (e.g. serum albumin), Renografin, Urografin or Ludox.

(iv) The method is easy to scale down by using smaller tubes which need shallower gradients (5−15% sucrose in 15 ml centrifuge tubes). The only easy way of scaling it up is to use several tubes.

(v) It is important to make control asynchronous cultures in order to test whether events seen in the synchronous cultures are genuinely related to the cell cycle or are artefacts produced by the perturbing effects of the collection or the selection on the gradient. These cultures are made in the same way as the synchronous cultures up to the moment before the top layer of cells is taken off the gradient. At that point the whole gradient is shaken up and a sample is diluted with medium and grown up in the same way as the synchronous cultures. This gives an asynchronous culture which serves as a control for most of the artefacts that may be produced. But it does not cover all of them since many of the cells will have been in higher concentrations of lactose than the top layer used for the synchronous cultures.

(vi) Heavy clumping of the cells stops efficient gradient separation. Light clumping is not a serious problem provided there is an appreciable proportion of single cells. The clumps go down to the bottom of the tube.

(vii) As with all methods of generating synchronous cultures, the degree of synchrony is not perfect but it is almost as good as that produced by the best alternative methods. *Figure 2* could have come from skilful tube selection.

(viii) This method can be used on other yeasts with only very minor changes in the details, though in the case of *Sacch. cerevisiae* it may be necessary to discard the very top of the cell layer which sometimes contains small unbudded cells that do not grow.

(ix) Apart from the low yield, the main disadvantage is that the synchronous cultures can be perturbed for several hours after they have been made. In some cases, for example DNA synthesis, the perturbations are minor but in others, such as enzyme activity (4) or amino acid incorporation [(5) and *Figure 3*], they can be severe. The control cultures described in (v) above are therefore vital. The reasons for the perturbations are not clear but it may be that the cells partially enter stationary phase when packed together during collection and on the gradient. Certainly the lengthy process of collection by batch centrifugation produces delay in the growth of the synchronous culture and should be avoided.

2.1.2 *Velocity separation in a zonal rotor*

This is essentially the same as the previous method except that it is scaled up by the use of a zonal rotor rather than tubes.

The details for *Schiz. pombe* and *Sacch. cerevisiae* are as follows.

(i)   Introduce a linear 15−40% gradient of sucrose (or lactose) by a peristaltic pump into a 1500 ml A XII zonal rotor (type A), made by Measuring Scientific Equipment Ltd (MSE), running at 600 r.p.m.

(ii)   Harvest about 10^{12} cells as described in Section 2.1.1 above and suspend in 40 ml of medium.

(iii)   Load these cells into the centre of the rotor with a 50 ml syringe followed by 30 ml of medium to displace the cells from the central core. Once loaded, accelerate the rotor to 2000 r.p.m. and follow the progress of the cells by eye.

(iv)   When the fastest-moving cells have travelled about 80% of the distance to the edge of the rotor, reduce the rotor speed to 600 r.p.m.

(v)   Remove the cells by pumping the highest density sucrose solution into the periphery of the rotor and collecting them from the centre. The small cells come out in the earlier fractions and are used to start the synchronous culture after addition of warm conditioned medium.

It takes about 15 min from loading to collecting the cells. Synchronous cultures can be started with as many as 5×10^{10} cells of *Sacch. cerevisiae* and 3.5×10^9 cells of *Schiz. pombe*. The exact limits of yield are uncertain but it is likely to be about 20 times greater with a 1500 ml zonal rotor than with a 80 ml tube. As with all selection techniques, the yield can be raised by increasing the size of the selected fraction but this will reduce the degree of synchrony.

The following points relating to the above method should be noted.

(i)   The outstanding advantage of this method is the high yield. Against it, is the cost of the rotor, centrifuge, pump and large gradient-former.

(ii)   Perturbations are also likely to be greater than with tubes because of the longer time spent in collecting the large number of cells and in running them through the gradient. Controls made by pumping out all the cells and resuspending them are therefore even more important.

(iii)   As with tubes, gradients can be made with materials other than sucrose and with different concentrations. Gradients of 10−25% sucrose have been used successfully with *Kluyveromyces lactis* and some haploid cultures of *Sacch. cerevisiae*.

(iv)   Other types of zonal rotor have been used, for example the MSE HS and the Sorvall TZ-28. In some cases these are smaller, which reduces the yield of selected cells, but they are easier to handle than the MSE Type A which is heavy and cumbersome to clean, assemble and mount.

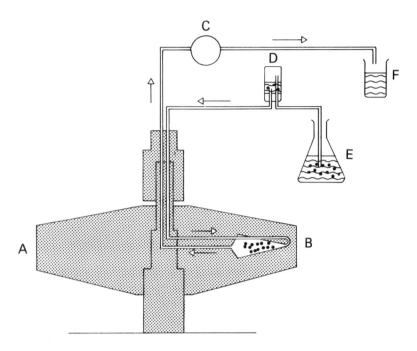

Figure 1. Diagram of elutriator rotor (Beckman JE-6): **A**, rotor; **B**, rotor cell; **C**, peristaltic pump; **D**, bubble trap; **E**, culture flask; **F**, effluent flask. Reproduced, with permission, from (5).

2.1.3 *Velocity separation in a continuous-flow centrifuge*

Lloyd *et al.* (6) have described a simple method of using a continuous-flow centrifuge to select small cells. The rotor speed and the inflow of growing cells from a normal culture are adjusted so that about 90% of the cells are retained in the rotor. The 10% of small cells that are carried away with the effluent are used to start a synchronous culture. In principle, the method is a good one since the small cells are only in the rotor for a few seconds and they are always in contact with the same medium. A control culture can be made by decreasing the rotor speed so that all the cells are carried over in the effluent. The authors claim success with *Schiz. pombe* and with *Candida utilis*. In this laboratory, however, we have not been able to get the method to work reliably with *Schiz. pombe* in spite of a good deal of effort. It might work better with other yeasts and would perhaps be worth trying if a suitable centrifuge were available.

2.1.4 *Velocity separation in an elutriating rotor*

This is the most recent method for selection synchrony, and probably the best one, but the yield is not high and it needs relatively expensive apparatus (5). The Beckman JE-6 elutriator rotor (*Figure 1*) works by a counter-current method in which the yeast in the 4 ml rotor cell is kept suspended by an inward flow of medium which can be pumped through the cell at up to 40 ml/min. Successive layers can be emptied out of the cell (the smaller yeast cells first) by lowering the rotor speed or increasing the pumping rate. The process can be watched with a stroboscope since the rotor cell is transparent.

Details for *Schiz. pombe* are as follows. The apparatus consists of a Beckman JE-6 rotor (complete with stroboscope and revolution counter), a Beckman J-21 centrifuge (slightly modified by the manufacturers for this rotor) and a variable speed peristaltic pump (e.g. Watson-Marlow 501 Z). *Figure 1* shows the piping arrangements in diagram form and the simple bubble trap which is needed on the inflow line. Any bubbles which get into the rotor cell cause turbulence and prevent efficient separation.

(i)   Load the rotor from an exponential phase culture at 32°C with a pumping speed of 22 ml/min.

(ii)   Run the rotor at about 3700 r.p.m. and take care that the yeast layer never rises above the broadest part of the rotor cell otherwise small cells will be carried away in the effluent. The refrigeration is switched off and the rotor temperature rises to about 30°C. The yeast cells continue to grow in the constant flow of warm medium.

(iii)   Collect some of the effluent-conditioned medium and keep it warm for subsequent dilution of the cells. Loading takes 10−40 min, depending on the cell density of the culture and the maximum loading is about 7×10^9 cells.

(iv)   To select the small cells for a synchronous culture, increase the pumping speed to 33 ml/min but keep the rotor speed unchanged.

(v)   When the layer of yeast cells has risen above the broadest part of the rotor cell, small cells begin to come into the effluent which should be collected in 10 ml samples. Some experience is needed here to judge the correct pump speed.

(vi)   Examine these samples under the microscope to check for a population of small cells, pool the best samples and, if necessary, dilute with the warm conditioned medium to give the synchronous culture. A good culture can contain up to 2 $\times 10^8$ cells (3% of the maximum loading).

(vii)   For an asynchronous control, change the inflow to the cell-free conditioned medium, increase the pumping speed and reduce the rotor speed so that all the cells go into the effluent.

(viii) Dilute to an appropriate cell density with conditioned medium.

After use, the rotor cell must be cleaned. It is also important to disinfect the rotor and tubing since bacteria can grow in parts of the complex circulation system. This can be autoclaved but it is simpler to use a disinfectant. Alcide (Life Science Laboratories Ltd, Sarum Road, Luton, Beds., UK) is effective if left in the rotor system overnight, and then washed out with 1 litre of distilled water.

Figure 2 shows a synchronous culture of *Schiz. pombe* made by this technique. The degree of synchrony is slightly better on average than that with other selection methods.

(i)   The advantage of this method is that it reduces the perturbations that can be produced during synchronization. *Figure 3a* and *b* show the perturbations in leucine incorporation produced in control cultures for tube separation. They were made in a slightly different way from that described in Section 2.1.1 above, without centrifugation. In contrast, the control elutriator cultures in *Figure 4c* and *d* have no perturbations after the first 30 min. It cannot, however, be assumed that perturbations are always absent and tests with control cultures are needed in each case. For example, a repeat of the experiments in *Figure 4c* and *d* with a different strain of *Schiz. pombe* (972) and a different minimal medium showed marked perturbations.

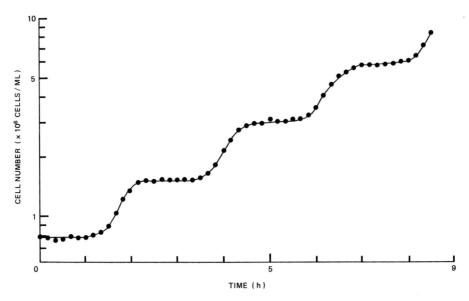

Figure 2. Increase in cell numbers in a synchronous culture of *Schiz. pombe* (strain 972 h⁻) prepared by elutriation and grown at 35°C in minimal medium (EMM3). Unpublished result by Dr B.Novak.

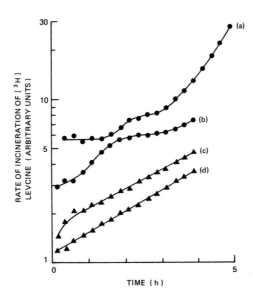

Figure 3. Incorporation of leucine pulses in control asynchronous cultures of *Schiz. pombe* (strain NCYC 132, ATCC 24751) grown at 32°C in minimal medium (EMM2). (a) and (b), controls for tube gradients. Growing cells were concentrated by filtration, suspended for 4 min in medium plus 20% (w/v) sucrose and then diluted in the original medium to a cell density of 2×10^6 cells/ml at time zero. Successive samples were labelled for 14 min with [³H]leucine (4−8 μCi/ml, 50−58 Ci/mM) and then treated with 5% trichloroacetic acid, filtered, washed and counted. (c) and (d), controls for the elutriator rotor. Growing cells were collected in the rotor [in (d) 6.5×10^9 cells were collected over a 40 min period]. After collection, they were pumped out and diluted in the original medium to a cell density of 2×10^6 cells/ml. Labelling was as above. Reproduced, with permission, from (5).

56

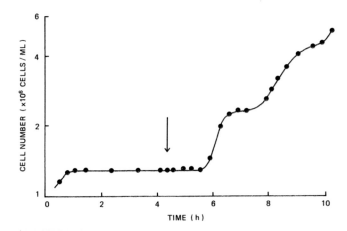

Figure 4. Induction synchrony of *cdc* 2.33 (*Schiz. pombe*). An asynchronous culture was shifted from 25 to 35°C at zero time and returned to 25°C after 4.5 h (arrow). Reproduced from (18).

(ii)   The method has been used successfully with *Sacch. cerevisiae* and *K. lactis* (7). As with other selection methods, the very smallest cells may have to be rejected.

(iii) The yield is about the same as that from an 80 ml tube gradient, but it can be increased by a factor of 5−10 by using the new high capacity rotor (JE-10X) which is available from Beckman. At the time of writing, this rotor is being improved by the manufacturers.

2.1.5 *Equilibrium separation*

Some yeasts have changes in buoyant density during the cell cycle so it is possible to separate cycle stages by equilibrium sedimentation in a gradient. This has been done for *Sacch. cerevisiae* with Renografin gradients (8) and dextrin gradients (9), and for *C. utilis* with dextrin gradients (10). The main advantage is that such gradients can be loaded more heavily than velocity gradients: 5×10^{10} cells can be loaded onto a 14 ml gradient (8). Much higher speeds are needed (giving up to 45 000 g) and temperature control is important. Perturbations have not been looked for but they are likely to be worse than in other methods if the loading is heavy. We have not been able to separate *Schiz. pombe* this way, possibly because the buoyant density changes through the cycle are much less marked than in *Sacch. cerevisiae* (11,12).

2.1.6 *Conclusions on selection methods*

Velocity separation in tubes is the simplest method and requires a minimum of apparatus. It is the method of choice provided the yield is adequate and provided an asynchronous control shows no lasting perturbations. Whatever property of the culture is being measured (e.g. enzyme activity) should show a smooth exponential increase in the control. If it does not, and there are serious perturbations, the best alternative is the elutriating rotor (either the Beckman JE-6 or the new high capacity JE-10X). It is likely to reduce the perturbations but may not eliminate them completely. If low yield is the problem rather than perturbations, then a zonal rotor may be the best solution. In all cases, it is important to make control cultures and to follow cell numbers as a routine both in these cultures and the synchronous ones.

Except at maximum growth rate, synchronous cultures of *Sacch. cerevisiae* lose synchrony faster than those of *Schiz. pombe*. This is because of the difference in generation times between mothers and buds (13).

2.2 Induction methods

These methods are quite different from selection. Instead of separating off a small fraction of a growing culture, the whole culture is induced into synchronous division. The yield is therefore much greater but there are problems about the extent to which the induction treatment has affected the cells. If the object is to deduce from the synchronous culture the events of the cycle in single cells in balanced growth, then the use of induction methods may be dangerous. But if the object is to have, for example, an enriched yield of dividing cells at the appropriate time, then induction synchrony can be very useful. Most of the methods are simple and do not need expensive apparatus.

2.2.1 Feeding and starving

This is one of the earliest techniques for inducing synchronous cultures of *Sacch. cerevisiae* and it remains one of the best (14,15). The following up-to-date version has been provided by Dr D.H.Williamson, National Institute for Medical Research, Mill Hill, London NW7 1AA.

The procedure works well with NCYC strain 239, a diploid of brewing origin. It has been used with other strains, notably diploids, but aα diploids which sporulate well tend to be refractory. In principle, stationary phase unbudded cells are subjected to alternating short periods of feeding and starvation such that they do not produce new buds. They thus remain in a 'resting' condition, and may be stored as such at 4°C in starvation medium. When inoculated into growth medium there is an initial lag period before the emergence of the first round of buds, which occurs with considerable synchrony. Subsequent divisions follow with a degree of synchrony which is still evident even after five division cycles (14).

The following media are required:

$\times 2$ MYGP: 2 g of glucose, 0.6 g of Difco yeast extract, 0.6 g of Difco malt extract, 1 g of Oxoid mycological peptone, 100 ml of water; pH 5–6.
$\times 1$ MYGP: half strength version of the above medium.
Starvation medium: 0.75 g of KCl, 0.26 g of $CaCl_2$, 0.5 g of $MgCl_2.6H_2O$, 1 litre of water.

(i)   Grow the cells with reciprocal shaking in 400 ml amounts of $\times 2$ MYGP for at least 10 days at 25°C. There should be less than 5% cells with young buds.
(ii)   Eliminate 'small' cells (amounting to ~40% of the population, distinguishable by their size and relatively dark appearance under phase-contrast) by repeated differential centrifugation in 15% mannitol. (This is rather laborious and can be shortened by banding the cells in a 20–40% sorbitol gradient in a swing-out head on a bench centrifuge for a few minutes. The lower band is greatly enriched for the larger cells.) The small cells should be reduced to less than 3–4% of the population by this treatment; see also step (v) below.

(iii) Suspend the large cells derived from one 400 ml starting culture in 100 ml of pre-warmed ×1 MYGP and incubate at 25°C for 40 min without shaking or aeration.

(iv) Centrifuge the cells, rapidly wash three times in pre-warmed starvation medium at 25°C, and incubate in starvation medium at this temperature with vigorous aeration for at least 2 h, preferably 6 h.

(v) Usually it is convenient to store the cells overnight at 4°C before repeating steps (iii) and (iv) twice more, but this storage is not essential. If the cells are stored in 100 ml medical flat bottles, or similar, residual small cells tend to remain suspended, and may be removed by deftly pouring off the supernatant. Repeated after each period of storage, this stratagem helps reduce the time required initially for step (ii).

(vi) Note: the small cells, which are essentially the daughter cells of the last round of cell division before entering stationary phase, must be removed, as they fail to be synchronized by this procedure. The procedure should yield $1-2 \times 10^{10}$ cells from a 400 ml culture. The population may be stored and used as required for $2-3$ weeks at 4°C; after more than a few days, a cycle of feeding and starving is desirable just before use.

The advantages are that the method is simple, the synchrony is good and the yield is high. The disadvantages are that it only works on some strains of budding yeast and that it takes 10 days. In addition, there is evidence that growth is unbalanced for the first cycle and it is safer to use the subsequent cycles in situations where the relationships between growth and division are critical.

2.2.2 *Block and release of the DNA-division cycle*

The principle here is to impose a block on the progress of cells through the dependent sequence of DNA synthesis, mitosis and division. The cells in the cultures move round the cell cycle until eventually all of them are held at the block point. This takes one generation time. The block is then released and a synchronous culture is produced.

This method has been used extensively with mammalian cells but is less popular with yeast, partly because of the shortage of efficient chemical blocking agents. It is important that the block should last for at least a cycle time and also that it should be easily reversible. Two good ways of imposing a block are described here but there are other ways known, for example, the use of hydroxyurea to block DNA synthesis (16).

(i) α-*Factor block.* α-Factor is a low molecular weight peptide which is excreted into the medium by haploid *Sacch. cerevisiae* of the α mating type. It has now been synthesized and can be bought from the Sigma Chemical Co. (Catalogue No. T7015). At 25 units/ml, it will block growing haploid cells of the opposite mating type (a) at the boundary between the G_1 and S phases. The cells show only a small decrease in viability during an 8.5 h treatment (17) and the block can be reversed by changing to fresh medium free of α-factor. During the block the cells continue to grow and they develop elongated processes after $3-4$ h ('shmooing').

(ii) *cdc mutants.* There is no mating pheromone in *Schiz. pombe* equivalent to α-factor but an alternative way of imposing a block is to use a temperature-sensitive cell division cycle (*cdc*) mutant. These mutants impose a block on the DNA division cycle when the temperature is raised from the permissive to the restrictive value. The best mutant to use in *Schiz. pombe* is *cdc* 2.33 which blocks just before mitosis. It also blocks at the G_1/S boundary but this has little effect on this use of the mutant since a large majority of the cells in a normal growing culture have passed this boundary and are in S and G_2 phases. *Figure 4* shows a synchronous culture produced by shifting up from the permissive temperature of 25°C to the restrictive temperature of 35°C for 4.5 h and then shifting down to 25°C. The first synchronous division starts at 1.3 h after the shift down and the second one about 2 h later. Although 35°C was used in this experiment, it is safer to use the slightly higher temperature of 36.5°C to reduce the risk of some cells leaking past the block.

The following points should be noted.

(i)   The most serious problem of these methods stems from the fact that the cells continue to grow during the period when division is blocked. As a result the cells at the first synchronous division are considerably larger than the normal division size and are also as variable in size as those in asynchronous culture. This distortion of the normal relationships between growth and the DNA division cycle makes it difficult to interpret the results when these relationships are important. In addition, an effect of large size at division is to shorten one or more of the subsequent cell cycles. This can be seen in *Figure 4* where the cycle time is about half of the normal cycle time of 4 h at 25°C.

(ii)   Synchrony is much shorter-lived than with a selection method. For example, compare *Figures 2* and *4*. It is, however, improved by raising the permissive temperature to 30°C.

(iii)   Cell morphology may be affected. Shmooing in the presence of α-factor has been mentioned above. In the case of *Schiz. pombe* synchronized by a *cdc* 2.33 shift, an appreciable number of cells containing three division septa appears after the first synchronous division.

(iv)   It is important with *cdc* mutants to carry out a control with the same temperature shift applied to wild-type cells.

(v)   It is, in principle, possible to use block and release with other *cdc* mutants of *Schiz. pombe* and *Sacch. cerevisiae* though the efficiency of the technique will depend on the tightness of the block and degree of recovery afterwards. Many of these mutants can be obtained from National Culture Collections, though it may in practice be easier to request them from workers in the field. They should always be screened before use for their *cdc* phenotype.

2.2.3 Repetitive heat shocks

Environmental changes of temperature or light applied once per cycle time have been used to induce synchrony in a variety of lower eukaryotes. This technique has been used with *Schiz. pombe* and produces an initial degree of synchrony about the same as with selection (19). The temperature of a normal asynchronous culture is raised from the growing temperature of 32°C to a sublethal temperature of 41°C for 30 min. It is then returned to 32°C for a further 130 min. This alternation of temperature is repeated 5−6 times, by which time the culture has become synchronous. Relatively large cultures can be handled and it is easy to do especially if the temperature control of a water bath can be automated. The dissociation of growth and division is not as marked as in the block and release methods but there is the problem that cell cycle events may be distorted by the heat shocks.

3. AGE FRACTIONATION

In principle, cells on a gradient or in an elutriator separate by size and therefore by age in the cell cycle. Successive samples down the gradient are fractions of the population of increasing age. These samples are assayed for a cell parameter, for example DNA content, and normalized to a per-cell value. A plot of these values against sample number gives the pattern of DNA synthesis through the cycle.

This method has been used on *Sacch. cerevisiae* both with a zonal rotor (20) and with an elutriator (21). It has also been used on *Sterigmatomyces halophilus* with equilibrium centrifugation on a zonal rotor (22). The details for *Sacch. cerevisiae* are exactly the same as those for selection in Sections 2.1.2 and 2.1.4 above, except for two points. First, successive samples are taken of the whole population in the centrifuge rotor and not one only. Secondly, the centrifuge is kept at 4°C so that the cells stop growing during the separation process. Cycloheximide (100 µg/ml) can also be added to stop protein synthesis. Cell number has to be measured in each sample for the normalizing and it is also desirable to measure some bulk parameter like mean cell volume or protein content for reasons which are apparent below.

There are several considerable advantages of age fractionation. The whole culture is used instead of only one small fraction as in selection, so the yield of cells is much greater. There is no distortion of the growth patterns in contrast to many of the induction methods. Finally the possibility of perturbations is much reduced since the separation of the fractions is done after the culture has been chilled.

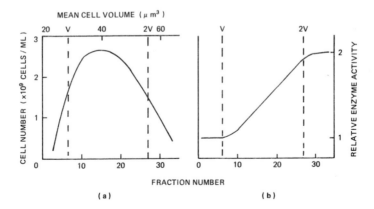

Figure 5. Hypothetical results of enzyme activity measured on fractions from a zonal or elutriating rotor. **(a)** Cell numbers and mean cell volume in the fractions. **(b)** Enzyme activity in the fractions after correction for cell numbers. V and 2V are on the Salmon−Poole criterion (see text). The cell cycle lies between the two solid ordinates if all fractions are included or between the dashed ordinates if only fractions between V and 2V are included.

A major problem in age fractionation is how to analyse the data and deduce the cell cycle pattern. *Figure 5* shows results for enzyme activity, cell volume and cell numbers in fractions from a rotor. They are hypothetical but are similar to some published results (e.g. 20). The volume covers a range of three times, rather than a range of twice, in the ideal cell cycle. This is because of the 'momentary variation' of cell volume at any given stage of the cycle. The problem is what should be done about the tails of the distribution where there are comparatively small numbers of very small and very large cells. If they are included in the analysis (as has usually been done in the past), the result is the periodic 'step' pattern of enzyme activity shown in the complete curve in *Figure 5b*. But this may overemphasize the contribution of the tails and Salmon and Poole (22) have produced a criterion for rejecting them. The best estimate of the birth volume V is that value where the area under the volume between V and 2V is greatest. All fractions with volumes less than V and greater than 2V are rejected, and the remaining fractions used after a conversion of volume to stage in the cycle using a known or assumed relation between these variables. *Figure 5b* shows that the effect of using this criterion is to convert the step pattern of enzyme activity through the cycle into a pattern of continuous increase. The method of analysis may therefore be critical in finding the true cell cycle pattern. The problem has not yet been studied sufficiently for there to be a definitive solution but the approach of Salmon and Poole seems the best one to adopt at the moment. The theoretical analysis by Fraser and Barnes (23) should also be studied.

A second problem is that the efficiency of separation of the cycle stages may drop with the larger cells. Discrimination of the S phase in the early fractions from an elutriator is good but the separation of the later stages is less efficient (24).

A third problem occurs with slower growth rates where buds are smaller than mother cells (13). The first fractions contain buds only whereas the later fractions contain both buds and mothers.

In the case of *Schiz. pombe*, it is difficult to get separation of the later stages. We have occasionally got good results [e.g. *Figure 4* in ref. (1)] but the technique is not reliable with this yeast.

A final point is that a cell cycle event is repeated two or three times in the two or three cycles of a growing synchronous culture. In age fractionation it only appears once and it may therefore carry less conviction, especially if it happens very early or very late in the cycle.

4. GENERAL CONCLUSIONS

Synchronous cultures prepared by selection give the most accurate representation of the cycle of the single cell in balanced growth. They also have the valuable feature that controls can be made which check the effects of the selection procedure. Their yield, however, is limited. If a large yield is needed, age fractionation can be tried though there are problems in converting the data into cell cycle timings. Feeding and starving is probably the best induction method and will work with large volumes. Block and release induction is mainly useful for enriching cultures with cells which are dividing or synthesizing DNA. The counsel of perfection is to try more than one method but this has seldom been done [though see (25)].

5. ACKNOWLEDGEMENTS

I am grateful to Dr D.H.Williamson for his contribution on feeding and starving synchronization, and to Dr J.Creanor for helpful criticism.

6. REFERENCES

1. Mitchison,J.M. and Carter,B.L.A. (1975) In *Methods in Cell Biology.* Prescott,D.M. (ed.), Academic Press Inc., New York, Vol. 11, p. 201.
2. Mitchison,J.M. and Vincent,W.S. (1965) *Nature*, **205**, 987.
3. Britten,R.J. and Roberts,R.B. (1960) *Science*, **131**, 32.
4. Mitchison,J.M. (1977) In *Cell Differentiation in Micro-organisms, Plants and Animals.* Nover,L. and Mothes,K. (eds), Gustav Fischer Verlag, Jena, p. 377.
5. Creanor,J. and Mitchison,J.M. (1979) *J. Gen. Microbiol.*, **112**, 385.
6. Lloyd,D., John,L., Edwards,C. and Chagla,A.H. (1975) *J. Gen. Microbiol.*, **88**, 153.
7. Creanor,J., Elliott,S.G., Bisset,Y.C. and Mitchison,J.M. (1983) *J. Cell Sci.*, **61**, 339.
8. Hartwell,L.H. (1970) *J. Bacteriol.*, **104**, 1280.
9. Wiemken,A., Matile,P. and Moore,H. (1970) *Arch. Mikrobiol.*, **70**, 89.
10. Nurse,P. and Wiemken,A. (1974) *J. Bacteriol.*, **117**, 1108.
11. Kubitschek,H.E. and Ward,R.A. (1985) *J. Bacteriol.*, **162**, 902.
12. Baldwin,W.W. and Kubitschek,H.E. (1984) *J. Bacteriol.*, **158**, 701.
13. Hartwell,L.H. and Unger,M.W. (1977) *J. Cell Biol.*, **75**, 422.
14. Williamson,D.H. and Scopes,A.W. (1962) *Nature*, **193**, 256.
15. Williamson,D.H. (1964) In *Synchrony in Cell Division and Growth.* Zeuthen,E. (ed.), Wiley, London, p. 589.
16. Slater,M.H. (1973) *J. Bacteriol.*, **113**, 263.
17. Iida,H. and Yahara,I. (1984) *J. Cell Biol.*, **98**, 1185.
18. Benitez,T., Nurse,P. and Mitchison,J.M. (1980) *J. Cell Sci.*, **46**, 399.
19. Kramhøft,B. and Zeuthen,E. (1971) *C.R. Trav. Lab. Carlsberg*, **38**, 351.
20. Sebastian,J., Carter,B.L.A. and Halvorson,H.O. (1971) *J. Bacteriol.*, **108**, 1045.
21. Gordon,J. and Elliott,S.G. (1977) *J. Bacteriol.*, **129**, 97.
22. Salmon,I. and Poole,R.K. (1983) *J. Gen. Microbiol.*, **129**, 2129.
23. Fraser,R.S.S. and Barnes,A. (1983) *J. Cell Sci.*, **62**, 187.
24. Ludwig,J.R., Foy,J.J., Elliott,S.G. and McLaughlin,C.S. (1982) *Mol. Cell Biol.*, **2**, 117.
25. Lorincz,A.T., Miller,M.J., Xuong,N.-H. and Geiduschek,E.P. (1982) *Mol. Cell Biol.*, **2**, 1532.

CHAPTER 4

Yeast genetics

J.F.T.SPENCER and DOROTHY M.SPENCER

1. INTRODUCTION

The methods in yeast genetics have undergone a major revolution since the discovery of restriction enzymes, the perfection of methods for transformation of yeasts as well as bacteria, the use of cloning for isolation of specific genes and related techniques. It might be thought that this would have made the method of 'classical' yeast genetics obsolete, since it is no longer necessary to rely on the methods of classical mating of selected haploid yeast strains, sporulating them and dissecting asci and characterizing the clones obtained. Even the term 'clone' has undergone a drastic change in meaning. The newer methods complement the older ones rather than replacing them, so that in combination, the two classes of technique become extremely powerful tools for elucidating the genetic structure and function of yeasts, not only of *Saccharomyces cerevisiae*, but of all yeast species. Techniques of 'classical' yeast genetics include mating, isolation of hybrids obtained in this way, dissection of asci to yield single-spore clones, characterization of the clones so obtained, mutation of both haploid and diploid strains and mapping of the genes thus made visible in the strain under investigation. Between the two divisions into 'classical' and modern yeast genetics lies the area occupied by protoplast fusion, an area fogged by clouds of mystery, poorly understood at present, not as precise in its operation as the introduction of cloned genes into a yeast cell by transformation with a plasmid. However, protoplast fusion is quick and cheap and probably the method of choice for improvement of industrial yeast strains where the introduction of precisely defined sequences of DNA is not so important. Its uses will be described later in this chapter.

2. GENERAL TECHNIQUES

The student of yeast genetics must possess a thorough competence in the basic techniques of microbiology. Failure to master these skills can only lead to misleading and spurious results, obtained from cultures, probably mixed, which are not what they seem, and which can only lead to self-deception by the investigator. There are a few techniques in cultivation and handling of yeast strains which, while based on general principles of microbiology, are widely used in yeast genetics, and which should be familiar to all investigators. The use of standardized methods, materials and media, where possible, makes the comparison of results in a rapidly growing field of great practical and fundamental importance much easier.

2.1 **Media**

Yeasts may be grown on many media, the requirements mainly being for a nitrogen source, a carbon source and any necessary vitamins and minerals. However, in yeast genetics as it now exists, the media have been standardized and reduced in number to a few, except for special purposes. Strains are generally maintained in the short term on yeast extract − peptone (YEP) − glucose agar (2%, 1%, 1%) plus 1−2% agar. Tests for sugar utilization and fermentation are normally done on Wickerham's Yeast − Nitrogen Base (YNB) (for composition see Appendix I), which may be made up from laboratory reagents, but which is more conveniently purchased from a supply house. This avoids the problem of growth inhibition by traces of impurities which may be toxic to the yeast, in reagents obtained from some companies.

2.2 **Methods**

2.2.1 *Tests for respiratory competence*

These tests can be done on YEP − glycerol or YNB − glycerol agar, and YNB − glycerol agar containing 3% ethanol to prevent sporulation is used to isolate prototrophic hybrids obtained by rare matings of prototrophic *petites* with auxotrophic respiratory-competent strains.

2.2.2 *Tests for auxotrophic requirements*

These tests are normally done on YNB (without amino acids) plus glucose, supplemented with all of the known requirements of the strains, except one (dropout medium). Double dropout media may sometimes be used. Clones are normally tested by replica plating to the different dropout media.

2.2.3 *Sporulation*

Saccharomyces spp. are normally sporulated on media containing acetate as the principal carbon source. In our hands, McClary's medium (1% potassium acetate, 0.25% yeast extract, 0.1% glucose) usually gives the best results, but raffinose − acetate solutions, sodium acetate solutions, and similar media are sometimes used. *Sacch. diastaticus* strains will often sporulate on yeast extract − starch medium.

2.2.4 *Streaking out of cultures*

The method for re-streaking bacterial and yeast cultures, as commonly taught, requires one plate for each colony thus purified, and yeast genetics requires the isolation of a rather larger number of pure clones than most types of microbial studies, with the possible exception of taxonomic and ecological investigations. Hence, one method of reducing the amount of material required is to divide each plate into at least two, and possibly four zones (in emergencies six) and streak a clone on each one (*Figure 1a*). If care is taken to begin with as light an inoculum as possible, and to raise the loop on its edge as soon as possible so as to deposit a minimum amount of material on the plate, then reasonably well-separated colonies can usually be obtained.

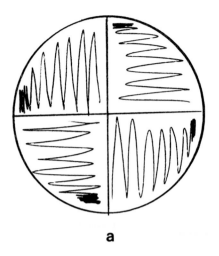

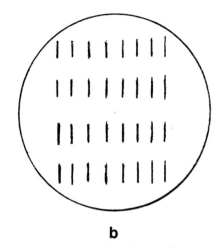

a   **b**

Figure 1. Patterns for re-streaking yeast colonies.

2.2.5 *Picking colonies*

Sterile toothpicks are convenient for investigations in yeast genetics, partly because of the number of strains involved. Especially when picking clones from single-spore isolates, obtained by microdissection, a template is useful, so that the clones can be arranged in groups of four, usually 32 per plate (*Figure 1b*). The toothpicks can either be discarded after use, or reclaimed, cleaned superficially if desired, and re-sterilized. The flat toothpicks used in North America are preferable, normally, to the round, pointed ones common in Europe: the blunt end is normally used for picking up the colonies. Sometimes the pointed ends are used in isolating viruses from plaques.

2.2.6 *Replica plating*

This is a very useful technique after ascus dissection in determining the reaction of clones towards particular sugars as sole carbon sources, or in determination of fermentative ability, and determination of auxotrophic requirements, using dropout media. Three general methods can be used.

(i)   Hand-held, using a velvet stretched over a disc, and formed into a handle at the back (*Figure 2a*). The velvet is used as a kind of stamp and is satisfactory for users with a steady hand. This method tends to smear the colonies more than with the other methods.

(ii)   Block method using velvet. The sterile velvet is stretched over a block of any convenient material and held in place with a rubber band or, preferably, a ring of the proper size. The plates are inverted over the velvet (*Figure 2b*).

(iii)   Block method using filter paper. In this method, the velvet is stretched over the block as in (ii), three pieces of 7 cm filter paper are placed on the velvet to absorb excess moisture, and a piece of 12 cm paper (Whatman No. 1 or equivalent) is placed on top of these and held in place with a band or ring. The plates are

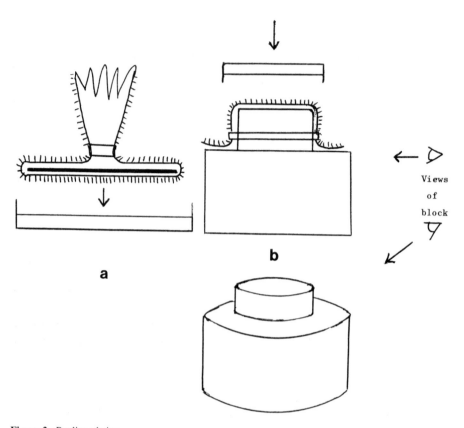

Figure 2. Replica plating.

inverted over the paper as in *Figure 2b*. It is not usually necessary to sterilize the filter paper before use, if a freshly-opened box is available. Otherwise the papers can be sterilized by conventional means. This method usually gives the most clearly-defined replicas, with the least smearing.

In all three methods, the plates are marked on the bottom with an index mark, the master plate is inverted over the sterile velvet or paper, pressed down firmly and tapped lightly, and care taken during removal to make sure that the agar does not come off on the velvet. The remaining plates are handled in the same way, keeping the index mark in the same orientation as the master plate. Normally, all the plates of a group are incubated together.

3. DISSECTION OF ASCI

3.1 Micromanipulators

Several such instruments are available, ranging from rather elaborate and exact micromanipulators such as the Leitz, to very simple ones which can, if necessary, be assembled from parts available from several supply houses. All have in common a mechanism to move the needle tip in three dimensions by relatively fine screw adjustments, each controlled separately. The Leitz instrument has both coarse and fine adjustments; the less sophisticated units have only a single adjustment in each dimension, but they are both relatively slow-acting but accurate. The Leitz instrument can be used for microinjection of animal cells and similar exacting tasks; the others, mainly for work on separation of relatively durable objects, not sensitive to somewhat rougher handling.

Another class of instrument, represented by the de Fonbrune and Singer micro-manipulators, uses a type of control stick for movement of the needle in two dimensions, and a screw adjustment for the vertical movement. Both have been used extensively for selection and separation of microbial cells, and the Singer instrument was designed specially for the dissection of yeast asci. Both require practice before the operator can dissect yeast asci and isolate single-spore clones successfully, but the technique is not difficult to acquire. The de Fonbrune micromanipulator uses aneroid chambers and air lines to control the movements of the needle, while the Singer instrument uses a purely mechanical linkage.

A third type of micromanipulator clamps to the microscope stage, and the chamber holder is moved by accurate machine screws so that it is possible to return exactly to the location of any spore that is put down. However, the same effect can be achieved by a consistent system of marking the agar during dissection. The instrument described by Sherman (2) is of this type, and is also relatively cheap.

In all of the micromanipulators mentioned, most of the movements are done using the stage adjustments. Once the needle tip is located in the microscope field, only the vertical movement of the micromanipulator is used in actual dissection.

3.2 **Ancillary equipment**

Dissection chambers, Pyrex (borosilicate) dissecting needles, needle holders for spare needles, dissection agar, microspatulas, small loops (0.5 mm), tweezers, enzymes for removal of ascus walls, and other equipment for use in a normal microbiology laboratory, are required.

3.2.1 *Dissection chambers*

Most dissection chambers described in the literature are rather small, which makes their use awkward, especially for the beginner, and may lead to breakage of needles. We use a chamber made from a 37 × 77 mm (1½ × 3″) microscope slide, with a strip of metal approximately 10 × 210 mm, cemented to it, with one end left open, so that a second slide can be set easily on top of the chamber. This allows the low-power objective to focus on the lower side of an agar slab carrying the ascus suspension, which is on the underside of the cover.

3.2.2 *Microscope*

An ordinary student microscope is generally adequate for microdissection of yeast asci. The eyepieces should be ×15 or ×20 if these are available, but normal ×10 eyepieces can be used. If the stage adjustments can be attached so that the operator can move the chamber with the free hand without reaching past the needle, work will be considerably easier.

3.2.3 *Needles*

These are flat-ended, bent at right-angles near the tip, and having a diameter such that they can just be seen with the naked eye. The actual diameter is not critical, as the needles are many times the diameter of the yeast ascospores. Slightly larger needles are often better for picking up vegetative cells which do not stick to the needle as readily. The needles are made from 2 or 3 mm borosilicate glass rod, for the Singer and other instruments. The needle holder for the de Fonbrune instrument will only take a smaller size of rod, which is seldom available in microbiological laboratories except on special order. Drawing out the needles requires a microburner, as it is seldom possible to obtain a low enough flame with an ordinary burner to make the fine end. Satisfactory needles, which will also fit the de Fonbrune instrument, can be made from thin-walled borosilicate capillary tubing. The small hole in the end of the needle does not interfere with picking up spores or cells, and may occasionally be an advantage.

The needles are made in two stages. In the first, a section of the rod is pulled out to a diameter of approximately 0.25−0.5 mm. In the second, the thin section is heated in the flame of the microburner, and pulled out to the desired thickness (easily visible to the naked eye, as noted above) and bent over at right-angles. This step is the most difficult part, and the operator will have to work out his own method. Usually it is better to hold the part to be bent well above the flame. The needle is pulled out to its final thickness at the same time as the bend is made.

After obtaining a satisfactory shape and thickness, the needle is trimmed to the desired length (a little less than 0.5 cm) with a razor blade or very sharp knife, while holding the needle against a wooden block, preferably. The end of the needle should be examined under the microscope, to see if there is a spike left on one side. Any such spikes should be removed by further trimming. The needles may be conveniently stored in a beaker lined with a strip of corrugated cardboard, the needles being slipped into the corrugations.

3.2.4 *Other materials and apparatus*

These include a screw-capped jar of alcohol the right size to hold two or three spare slides, and tweezers for handling the slides. A microspatula can be used for cutting out and handling the slabs of agar for dissecting, and a template for cutting the slabs to the desired size is desirable. Six slabs can usually be cut from one plate of dissection agar. For making the spore digest, sterile tubes and a normal and small loop are required. Finally, Petri dishes of YEP−glucose agar should be available for transferring the slabs after the dissections are complete, along with a plate of dissection agar. The latter consists of YEP−glucose medium containing 2−2.5% agar for added stiffness and strength, though normal YEP−glucose agar can be used, and is dispensed into screw-capped bottles or other containers in 10-ml portions. These are melted just before use, and a plate poured, 10 ml giving a slab of the best thickness.

3.2.5 *Enzymes*

Glusulase (Helicase; snail enzyme; sterile preparation), Novozyme, Zymolyase, and several other commercial enzymes are available and are satisfactory for dissolution of ascus walls. If live snails are available, it is possible to prepare the enzyme by killing the snails with chloroform, breaking open the shell and the body and extracting the juice from the crop, which is on the left side of the body, with a Pasteur pipette. This crude preparation is not sterile, and since it contains numerous bacteria and even some yeasts, must be filter-sterilized. A less drastic method of obtaining enzyme is from common mushrooms, which can be cut up and homogenized in a Waring Blendor. The supernatant is active enough to be used for removal of most ascus walls. Several species of micro-organisms, including *Arthrobacter* and *Oerskovia* spp., also produce active enzymes, and some of them are used in preparation of the commercial enzymes.

3.3 Procedures

3.3.1 *Preparation of the spore suspension*

(i) Dilute the enzyme into sterile water (1:4 or higher dilution, up to 1:40, depending on the activity of the enzyme and the susceptibility of the yeast strain), and dispense into tubes, 0.2 ml/tube.

(ii) Add a loopful of spore suspension, if liquid, or a small amount of sporulated culture from solid sporulation medium. Avoid too much spore preparation; if the culture sporulates well, a rather dilute suspension is easier to handle and one can find well-separated asci which can be picked up readily. Incubate the spore suspension in the enzyme for 10−30 min or more, depending on the resistance of the ascus walls to dissolution.

(iii) While the spore suspension is being incubated, melt a tube of dissection agar (see above) and pour a plate. It will probably harden by the time the ascus walls have been dissolved.

3.3.2 *Preparation of the dissection chamber*

(i) Wipe out the dissection chamber with alcohol. Remove a slide from the alcohol jar and flame off. Place it on top of the chamber.

(ii) Flame the microspatula in alcohol, place the plate of cooled dissection agar on the template and cut out a slab of agar and place it on the flamed slide on the dissection chamber, approximately centred.

(iii) Cut a thin strip from the main slab and leave in place beside it (see *Figure 3*).

(iv) Using the small loop (~0.5 mm diameter), streak a loopful of digested spore suspension *along the outside edge* of the secondary strip. Leave a little space along the inner edge of this strip, for working space when picking up asci.

(v) Invert the slide over the dissecting chamber so that the slab with the spores is below the slide.

(vi) Mount the chamber in the slide holder of the microscope, so that the open end is toward the micromanipulator (depending on whether the operator is right- or left-handed). Use the low-power objective and ×20 eyepieces if available.

(vii) Mount the needle in the micromanipulator and adjust the height approximately, then slide the instrument into place so that the end of the needle is just below the corner of the thin strip, at the end where the first asci are to be dissected.

(viii) Centre the controls, then locate the end of the needle in the centre of the microscope field. It will appear as a dark circle, several times larger than the spores or cells. Lower the needle slightly, to be still visible, but out of harm's way during the initial search for the spores.

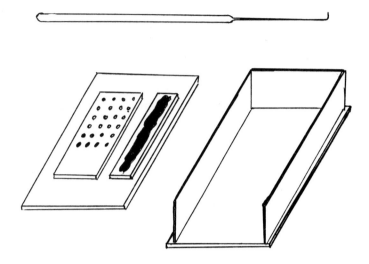

Figure 3. Needle, agar slab and dissecting chamber.

3.3.3 *Dissection of asci*

(i)   First find the spores. Move the chamber, using the stage controls, until the edge of the streak is located. Then look for an ascus, reasonably well-separated from the rest of the spores and cells. It will probably appear as a diamond of four spores (see Chapter 1, *Figure 1*). (A 'family' of four cells may look like an ascus.) If the ascus wall is still visible, it will be difficult or impossible to separate the spores. It may be impossible to pick up the ascus, as naked spores adhere to the needle more readily than the cells.

(ii)   Having located an ascus, free of interfering spores or cells, touch the agar surface with the end of the needle, close beside the desired ascus. Lift the needle away from the agar, using the fine controls. The ascus should stick to the needle. If it does not, repeat the process until it does.

(iii)   Move the chamber, using the stage controls, until the area on the main slab where it is desired to set down the spores is in the field. Touch the end of the needle to the agar surface and remove it; the spores should remain on the agar. If not, repeat until they do.

(iv)   Move the chamber again so that the edge of the strip closest to the spore streak is in the field, lower the needle and move the chamber back and forth to mark the agar, so that the spores can be located again if necessary.

(v)   Return the chamber to the position where the isolated ascus is in the field. Touch the agar surface beside the spores, very gently, and tap the bench top or the base of the instrument (also very gently) to vibrate the needle and disturb the spores. They should separate readily.

(vi)   When one or more spores separate, mark the agar again, and pick up the remaining spores, singly or all at once if convenient. Move the chamber approximately 2 mm transversely and set down the spores again. Leave one spore behind, pick up the others and move the chamber again, until all four of the spores have been set down in a line across the slab.

(vii) Return the edge of the slab to the original mark, then move the chamber along the slab approximately 3 mm and mark the edge of the agar again. Return to the streak of spores, find another ascus, pick it up and return to the main slab at the mark. Dissect this ascus as before.

(viii) When enough asci have been dissected, slide the micromanipulator away from the microscope and make sure that the needle is out of harm's way. Remove the chamber from the microscope, flame the microspatula in alcohol again, remove the slide from the chamber and turn it over, and remove the main slab from the slide and place it on the surface of a plate of YEP−glucose agar and incubate. Discard the small strip with the spore streak. Small colonies should appear on the slab in 2−3 days, and can be picked to YEP−glucose agar for further study.

(ix) Put away the needles immediately, where they are well protected from breakage.

3.4 Dissection directly on the surface of the agar in a Petri dish

This can be done either with the Singer instrument, or with the Berkeley type of micromanipulator mounted on the microscope stage.

3.4.1 *Singer instrument*

A dissecting microscope with approximately 100 cm clearance between the objective and the stage is required, with a magnification of at least ×50 and preferably ×100 or more. The dissection is usually done with a microloop rather than a flat-ended needle, though this has the disadvantage of requiring the services of a glassblower, or at least the availability of a microforge and some expertise in its use. A flat-ended needle, particularly one made from thin-walled capillary tubing as described above, may also be used satisfactorily in this method, provided the end is drawn out far enough for the tip of the needle to reach the surface of the agar. Since the dissection is done on the surface of an open plate, the procedure must be carried out in a clean area, free of air currents.

(i) Prepare the spore digest as described above. Carry out dissection on the surface of a plate of YEP−glucose agar. The concentration of the agar may be increased to 2% if desired.

(iii) Place a small loopful of the spore digest on the agar, at the centre of the plate, and spread with the loop to give an area of about 1 cm. Place the plate on the microscope stage and focus the microscope on the spore digest. Mark the plate according to *Figure 4*. Remember to keep the lid of the plate sterile and available.

(iii) Mount the microloop or needle in the micromanipulator, adjust it to approximately the correct height, and raise the micromanipulator so that the loop can be moved to the centre of the plate.

(iv) Find an ascus, pick it up with the microloop, and move the plate so that the loop is over one end of the first mark. Set down the ascus and dissect, placing one of the spores at each cross-mark.

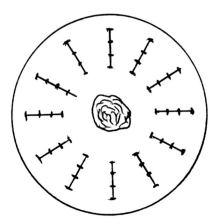

Figure 4. Dissecting on open plates—Queen Mary College method.

(v)   When the dissections are finished, remove the micromanipulator and microloop, cut out the area with the spore digest and discard, replace the cover on the Petri dish, replace the microloop and incubate the plate with the spores at 25−30°C until colonies appear.

3.4.2 *Berkeley instrument*

This modification is quite different, as it uses the stage-mounted instrument devised in Berkeley in the late 1950s. For the present technique, a plate holder which will hold the Petri dish in an inverted position has been constructed, and the plate can be moved through a range of 9−10 cm laterally. Dissection is with a flat-ended needle which is long enough (~2 cm) to reach the surface of the agar when the plate is mounted in the holder.

(i)   Prepare the spore digest as before, and streak along a diameter of the dish.
(ii)   Mount the plate in the holder and mount the needle in the micromanipulator. Locate the needle tip in the microscope field, using the low-power objective, preferably long-focus, and proceed with the dissection. Set the spores out at right-angles to the streak, and well separated from it. If desired, the groups of spores can be set out on both sides of the streak, so that a considerable number can be grown on a single plate.
(iii)   When the dissections are completed, remove the plate from the holder, cut out the area carrying the spore digest and discard (a miniature 'cookie cutter' can be made for this), replace the cover on the Petri dish, replace the needle in its storage jar, and incubate the plate as before, until colonies appear.

3.4.3 *Goldsmiths' instrument*

This modification also uses a flat-ended needle like that for the Berkeley instrument, and a plate-holder which can be constructed by any competent machinist, of lucite (Perspex) and a pair of glass slides, as in *Figure 5*. The plate for dissection

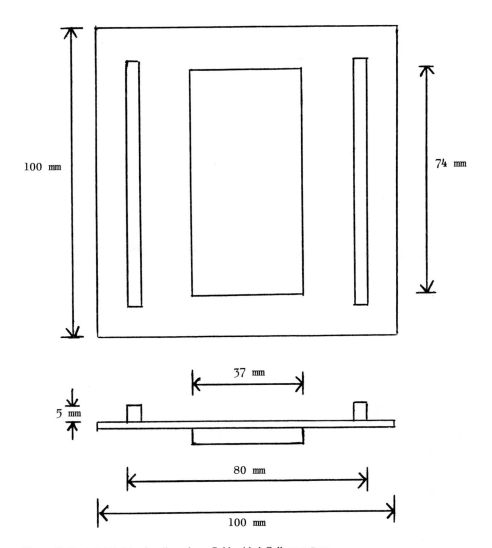

Figure 5. Petri dish holder for dissection—Goldsmiths' College pattern.

(YEP−glucose agar as before) should be relatively thin (8−10 ml of agar), to allow better focusing and visibility.

(i)   Prepare the spore digest as before, and spread on a small area as in the procedure for the Singer instrument.

(ii)   Mount the plate on the holder, using Blu-Tac or putty (or chewing gum) to stick it to the rails.

(iii)   Find the spore digest with the low-power objective and focus on an ascus near the edge.

(iv)   Mount the needle in the micromanipulator (any available instrument) and locate the needle tip in the microscope field.

75

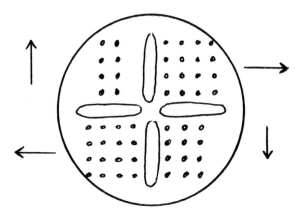

Figure 6. Dissection pattern using Goldsmiths' College plate holder.

(v)   Mark the surface of the agar along a transverse radius of the plate (*Figure 6*). This method of marking is necessary because in most microscopes the range of movement of the stage is insufficient to allow dissection along a complete diameter as in the Berkeley procedure.

(vi)   Return to the spore digest, pick out an ascus, move it to a suitable point on the mark, set it down, dissect and set out the spores at right-angles to the mark. Continue until all the space along the mark is used up, then turn the plate through 90° and repeat the procedure.

(vii)   Remove the plate from the holder and, with a sterile microspatula, cut out the area with the spore digest and discard. Replace the cover on the plate and incubate. Put away the needle as before.

The Berkeley procedure was designed specifically to allow detection of post-meiotic segregation, since the colonies arising from the spores can be replica-plated directly to selective media for detection of sectored colonies, which are not detectable if the clones are picked to other media before replica plating (3). However, the other procedures for dissection directly to an open plate can also be used for this purpose.

4. MATING

4.1 Mass-mating techniques

4.1.1 Drop overlay method

This method can be used for obtaining hybrids between auxotrophic haploid strains for investigating such phenomena as recombination, gene conversion and similar events. The strains should have at least one set of complementing requirements. If the strains both have the same requirements at any locus, the medium must be supplemented with the appropriate amino acids or other compounds.

(i)   Grow the strains in YEP−glucose overnight, and dilute an aliquot of each 1:10.

(ii)   Place two drops of one strain on the surface of a plate of YNB, supplemented if necessary as above, but lacking the auxotrophic requirements for the desired loci, on different locations. Allow the drops to dry.

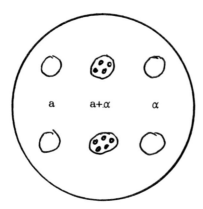

Figure 7. Mating of haploid strains—drop overlay pattern.

(iii) Place two drops of the other suspension on the agar, one of them on the dried drop of the previous culture, and the second in a completely different location (see *Figure 7*).

(iv) Incubate the plate at 25−30°C for a few days, until microcolonies appear in the area where the two cultures overlap. Pick these colonies and re-streak on the same YNB−glucose agar as the original.

(v) Test the diploids thus obtained for sporulation and dissect and test the clones obtained, if desired.

(vi) Note that markers other than auxotrophic requirements can be used. Mutants resistant to inorganic ions (Cu, Mn, etc.) and to antibiotics such as chloramphenicol, erythromycin or oligomycin, which are all readily available, can all be used, as long as they are stable and dominant.

4.1.2 *Mixing of prototrophic haploid strains, either in liquid or on solid medium*

This method is probably the only one possible with strains that carry no observable markers. The cultures are grown for 24−48 h on YEP−glucose medium and then mixed in the same medium, either liquid or solid. Mix the strains as early as possible in the day, and observe the cultures at intervals of about 1 h. Look for the presence of zygotes in the mating mixture. If zygotes are observed, isolate them with the micromanipulator as described previously and grow them for a few days, then test for sporulation, or for the presence of other characters, if any, that can be detected readily.

4.1.3 *Mating of homothallic strains*

This also requires the use of the micromanipulator if hybrids which are the actual products of the matings are to be isolated.

(i) Sporulate and dissect the heterothallic strain or strains as described previously, but onto acetate agar instead of normal dissection agar YEP−glucose (4).

(ii) Incubate the strains until microcolonies appear.

(iii) Mix the strains on a YEP−glucose plate (mass mating) and observe for zygote formation. Isolate the zygotes with the micromanipulator. This method is more certain if one of the strains, preferably not the homothallic one, carries a detectable marker, either auxotrophic or one for resistance to some toxic compound, so that the hybrid can be checked by sporulation and dissection of the asci.

4.1.4 *Rare-mating methods*

These methods were developed (5), for making hybrids involving prototrophic industrial yeast strains which have lost the ability to sporulate. The method depends on the occasional appearance of mating strains in the prototrophic culture, which arise from rare mating-type switching in diploids or polyploids (6), and these mating strains then mate with the autotrophic strain.

(i) Make a *petite* mutant from the above strain, either by treatment with acriflavine or by isolation of a spontaneous mutant. Ethidium bromide-induced *petites* are not satisfactory for this method, unless the intention is to dispense with resistance to antibiotics as a marker. This is feasible if another type of marker is available. Crosses with strains of *Sacch. diastaticus* can be made readily in this way, and the hybrids isolated on the basis of the ability to use starch.

(iii) Use a mating haploid or diploid, auxotrophic for preferably two or three markers at least. Grow both strains in YEP−glucose broth for 48 h. Mix aliquots of the two strains in fresh YEP−glucose broth, and either incubate in still culture for 5−6 days, or mix and centrifuge or membrane-filter, so that the cells are held in close contact for up to 2−3 h, though 30 min may be sufficient for some strains.

(iii) Recover the cells in any appropriate way, and wash at least once with sterile distilled water.

(iv) Spread the cells on plates of YNB−glycerol agar, containing 3% ethanol to inhibit sporulation. Isolate any colonies arising and re-streak on agar of the same composition. Be patient enough to allow slow-growing hybrids to appear.

(v)   Check for the reappearance of the marker(s) from the *petite* parental strain, and continue with the testing of the hybrid.

(vi) Note that a similar system can sometimes be used for the construction of hybrids of industrial yeast strains by protoplast fusion.

4.2 Spore−spore, spore−cell and cell−cell pairing

This method can be used to obtain hybrids very rapidly, as the spores and cells are selected from a culture at the most vigorous stage, set out on a plate or slab of dissection agar and paired by the micromanipulator. After pairing, the cells are observed at frequent intervals to determine whether fusion and zygote formation have occurred. Only haploid cells can be mated in this way, of course. Cells arising from dissection of asci from homothallic strains onto acetate agar can be paired with other cells or spores, in the hope that they will be mating haploids, but close observation is even more essential in this case to find and isolate the zygotes.

5. SPORULATION

Two processes associated with meiosis and spore formation, recombination and post-meiotic segregation are extremely important in the study of yeast genetics. This makes a knowledge of the processes involved in spore formation essential in such investigation and the ability to induce sporulation in yeasts is of first importance. In *Sacch. cerevisiae*, and probably in other yeast species that form spores, there are generally two major stages in sporulation: commitment and spore formation proper. If vegetative cells are transferred to sporulation medium and sampled at intervals, cells which are replaced in growth medium during the first few hours will revert to vegetative growth. After a certain time, cells which are transferred to growth medium continue the steps in the sporulation process and eventually form asci containing normal spores.

Some yeast species form spores readily on any medium: some require special media either enriched or impoverished in some constituent(s). *Sacch. cerevisiae* falls into the latter class and, in particular, requires a non-fermentable carbon source, a restricted nitrogen supply and an optimum level of aeration. The most effective medium for most strains of *Sacch. cerevisiae* is probably McClary's medium (7) (see Section 2.3). The yeasts are usually grown for approximately 48 h on a rich pre-sporulation medium, and then transferred to McClary's medium. The yeasts may be sporulated in either solid or liquid medium. The presence of potassium rather than sodium in the medium, as well as traces of glucose, appears to give improved sporulation. *Sacch. cerevisiae* strains will usually sporulate on medium containing sodium acetate and are sometimes sporulated in a medium containing raffinose. Some strains of *Sacch. diastaticus* will sporulate on yeast extract − starch medium.

5.1 Sporulation on solid medium

The yeast cultures are grown on a pre-sporulation medium, as previously described, and inoculated on to the surface of a plate or slant of sporulation medium. As well as McClary's medium, Kleyn's medium (0.25% Bacto-tryptose, 0.062% glucose, 0.062% NaCl, 0.5% sodium acetate and 2.0% agar, is desired), gypsum blocks or carrot or potato wedges have been used. McClary's medium is normally most satisfactory. The inoculated medium is incubated for 2−6 days at 25 or 30°C and examined at intervals for the presence of asci. Some strains sporulate better at 20−25°C than at 30°C. The spores can be used for some weeks, but survival is probably less than in liquid medium (see below).

5.2 Sporulation in liquid medium

This is preferred where a uniform suspension of asci may be required over a period of weeks. Usually it is most convenient to sporulate a small volume (2 ml) of culture in a standard culture tube. The culture is pre-grown for 48 h in YEP−glucose broth, and 0.2 ml of this culture is inoculated into 2 ml of McClary's medium in a 16 × 150 mm culture tube. The tube is usually incubated 2−3 days on a shaker or roller drum, operated vigorously enough to give adequate aeration and observed at intervals. When well-developed asci are found, the culture is washed once or twice with sterile distilled water, and the supernatant removed. The tube, with the asci suspended in the remaining few drops of water, is sealed with parafilm and stored at refrigerator temperature. The spores remain viable for several weeks.

6. TREATMENT OF CLONES OBTAINED FROM DISSECTION OF ASCI

After single-spore clones have been obtained, they are further tested to determine post-meiotic segregation, auxotrophic requirements (and hence recombination) and the

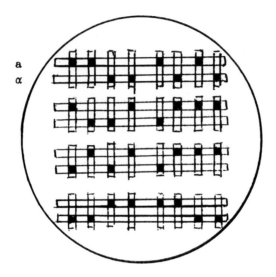

a
α

Figure 8. Cross-stamping pattern for determination of mating type.

appearance of any other characteristics as a result of hybrid formation. Determination of auxotrophic requirements and other nutritional requirements of the clones can be done by replica plating to dropout media (for auxotrophic requirements) or other specialized media, but mating type is best determined by cross-stamping, especially where the clones are known to contain auxotrophic requirements which can be utilized to detect the occurrence of prototrophic diploids arising from mating with tester strains of known requirements, usually for adenine and of known mating type.

(i)   The chief requirement is a set of tester strains, such as the adenine series, of both **a** and α mating types. Streak these strains in a single band across a plate of YEP−glucose agar [two strains, **a** and α, can be streaked on one plate (*Figure 8*)], not far from the diameter. Sterilize wooden strips, long enough to reach across eight streaks of single-spore clones, and keep in a suitable container. Streak the clones to be tested in sets of four, on a YEP−glucose agar plate, using a template to ensure that the streaks are close enough to be covered by the strip. Immediately after streaking, remove a wooden strip from the container by sterile forceps, and dip lightly into the streak of the selected tester strain, so that the whole edge is covered. Remove excess yeast by dabbing the edge of the strip on the uninoculated part of the plate of tester culture. Press the strip down on the upper part of the first set of eight streaks (see *Figure 8*) so that a line through them is inoculated. Wipe the first strip with tissue (Kleenex) and discard into another container for re-sterilization.

(ii)   Take a second strip and repeat the process until all of the streaked clones have been inoculated across the upper part with the **a** mating type tester strains. Then repeat for all of the strips with α mating type tester, this time inoculating the lower part of the streaked clones. After the cross-stamping is completed, re-autoclave the used wooden strips as soon as possible to avoid confusion with sterile strips, and to be sure of having enough strips on hand for future exercises.

(iii)   Incubate the cross-stamped plate overnight, or until the cultures have grown up sufficiently. Replica plate the cultures to a plate of minimal agar with glucose as sole carbon source. The prototrophic diploids will appear as small rectangles where the streaks cross the stamp, which allows the mating behaviour of the clones to be determined according to the mating type of the tester strain.

Note that in 'normal' diploids the mating type will segregate 2:2. However, some strains do not behave according to this rule, and in particular some industrial strains of *Sacch. cerevisiae* seem to produce many α mating type clones. In some strains, segregating a lethal mutation, the nature of this phenomenon is equivocal, as to whether the lethal character is associated with the **a** mating type and only the α clones survive, but recently we have observed aberrant behaviour in some hybrids, in which spore viability is high, and all four clones from any one ascus are of the α mating type. So far, we have no explanation for this phenomenon.

Mating types of prototrophic haploids cannot readily be determined by this method. However, cross-stamping is a convenient way of making a number of crosses for this purpose. After the stamping is completed, the cells at the points of intersection can be examined microscopically for the presence of zygotes. This method, though accurate where the strains mate readily, is laborious.

7. RANDOM (MASS) SPORE ISOLATION

In cases where a micromanipulator is not available, mass isolation of spores can be used. The ascus suspension may be mixed with powdered glass and ground for up to 10 min with a motor-driven pestle, to destroy the vegetative cells and free the ascospores. Alternatively, treat with enzyme to dissolve the ascus walls, as before, and separate the spores from the remaining vegetative cells. The first method unavoidably damages some of the spores. In the second method, the spore suspension may be sonicated lightly to break up clumps and tetrads of spores before further treatment. The spores are then separated from any remaining vegetative cells as follows.

7.1 **Partitioning in mineral oil**

(i)   Shake the spore suspension with sterile mineral oil and centrifuge the preparation briefly to separate the phases (and, if necessary, break any emulsion formed). The spores tend to aggregate in the oil layer.

(ii)   Repeat the process three or four times, until the spore suspension is free of vegetative cells.

(iii)   Add 15% gelatin solution and shake the suspension thoroughly.

(iv)   Spread aliquots on plates of YEP−glucose agar.

(v)   After incubation, replica plate the colonies to dropout media and other selective media, and score against the master plate (3).

7.2 **Partitioning in other biphasic mixtures**

Solutions of polyethylene glycol (PEG 6000) and dextran will also form biphasic systems, and the spore suspensions obtained by grinding with powdered glass or enzymatic treatment can be separated in this system, the spores occurring in one phase and the asci in the other. This system can also be used for separating *petite* mutants from respiratory-competent strains.

7.3 **Heat treatment**

(i) Hold a water suspension of material from a sporulating culture at $54-60\,^{\circ}$C for $10-15$ min, depending on the strain, and streak the resulting ascus suspension on plates of YEP−glucose medium. The time required to kill all the vegetative cells should be determined previously.

(ii) For recovery of the asci, remove aliquots of the culture at intervals of $1-2$ min, up to $10-15$ min after the time necessary to kill the vegetative cells, and plate on the desired agar.

(iii) Pick a number of the smaller colonies for testing. The larger colonies may arise from vegetative cells which have survived the heat treatment (9).

7.4 **Recovery of asci by ether killing**

This is another method which can be used for elimination of the vegetative cells from a sporulated culture.

(i) Add diethyl ether, 1:1, to the sporulated, chilled culture in a McCartney or similar screw-capped bottle, close tightly and agitate on a roller drum at $0-4\,^{\circ}$C for $10-15$ min. The time required for complete killing of the vegetative cells should be determined previously.

(ii) Separate the layers and plate the layer containing the asci on YEP−glucose agar or other complete medium (10).

This method and that described in Section 7.3 are particularly suited for isolation of single-spore clones from strains where the rate of sporulation and also spore viability are low, as in many industrial strains, since usually there is only one viable spore per ascus, and it is not necessary to disrupt the asci and free the spores. The method may not yield mating haploid clones, as numerous industrial yeasts are homothallic and self-diploidize readily. However, the spores may be plated on to acetate agar immediately after the ether treatment, so that microcolonies, delayed in self-diploidization, may be obtained and mated immediately.

7.5 **Isolation by sporulation of protoplasts**

This method may be used for either random spore isolation or for production of tetrads for isolation of clones by micromanipulation. If a strain which gives a very high yield of protoplasts is used, then the percentage of vegetative cells present is very low and can be ignored.

(i) Grow the yeast cultures to the mid-exponential stage.

(ii) Recover cells from (usually) 2 ml and convert them to protoplasts (University of Washington procedure; see Section 11.1).

(iii) Recover the protoplasts and wash once or twice with the solution of osmotically-stabilized (1.2 M sorbitol or 0.6 M KCl) solution used in protoplasting and then transfer to sporulation medium, also osmotically stabilized. McClary's medium with 0.6 M KCl can be tried.

(iv) After sporulation occurs, dilute the culture with water, either before or after recovery of the protoplasts and ascospores by light centrifugation, which bursts the remaining protoplasts of vegetative cells, leaving only the tetrads of ascospores. These can be sonicated very lightly, to break up the tetrads, and then spread on YEP−glucose agar for germination and growth of the spores, after which the colonies can be replica plated to dropout or other selective media. Otherwise, the asci can be dissected as given in Section 3 on use of the micro-manipulator, except that treatment with enzyme to dissolve the ascus wall is not required.

7.6 Separation of yeast asci and vegetative cells on renografin gradients

This system can be used for normally-sporulating yeast strains, and also for those with poor sporulation and very low spore viability.

(i)   Lay 3 ml of 35% renografin on top of 3 ml of 43% renografin solution in a centrifuge tube. Pipette the sporulated culture (washed and sonicated lightly) on top, and cover with a layer of 3 ml of 20% renografin.

(ii)   Spin the tube in a bench-top centrifuge at approximately 2500 r.p.m. for 15 min, and remove the layer containing the vegetative cells, at the interface, by Pasteur pipette. The asci should be in the pellet.

(iii) If there are still vegetative cells in the ascus suspension, repeat the procedure once or twice, using only the 43% and 35% layers of renografin solution.

Note that this method also can be used to isolate single-spore clones from cultures having low spore viability, as in most cases only one spore per ascus is viable.

7.7 Separation of vegetative cells into sporulating and non-sporulating fractions

This can also be done on renografin step gradients.

(i)   Grow the cells to stationary phase (~48 h) in YEP−glucose medium. Recover the cells and wash them, resuspend in water and sonicate lightly.

(ii)   Spin them down again and resuspend in 20% renografin.

(iii) Layer 2 ml of this on to 10 ml of 35% renografin in a 30 ml chlorex (plastic) centrifuge tube, and centrifuge as described above.

(iv)   Remove the layer of cells at the interface by a Pasteur pipette (fraction I). These cells should be large and vacuolated, and should sporulate on the appropriate medium. Fraction II (pellet) consists of smaller, non-vacuolated vegetative cells which do not sporulate.

(v)   If the two fractions are not uniform, re-run on a similar gradient.

(vi)   The cells in fraction I can be washed and placed in liquid sporulation medium, or spread on solid sporulation agar.

8. MUTAGENESIS

The study of yeast genetics depends to a considerable degree on the availability of mutants as markers for the different chromosomes, since it is very difficult to visualize the chromosomes by staining as in some other fungi, various plants and animals, and in particular, *Drosophila melanogaster*. Thus, where *Sacch. cerevisiae* is concerned, cytogenetics virtually does not exist. The techniques for obtaining a variety of mutants in yeast have been worked out in considerable detail, and have been used for everything from the study of recombination and linkage to the elucidation of mechanisms of secretion of extracellular enzymes and dissection of the glycolytic cycle.

Recessive mutations such as to amino acid and purine base auxotrophy, and to canavanine resistance, are normally induced in haploid strains, for the obvious reason that in auxotrophic mutants they are complemented by the gene product from the other strand of DNA in the diploid state, and are not expressed. Since the frequency of occurrence of these mutants is very low, a recessive mutation induced in a diploid or polyploid strain would be very laborious to detect, though in theory this is possible. Dominant mutations can, however, be detected in strains of any ploidy if desired, and consist mainly of mutations to resistance to various toxic metals and to some antibiotics. These are of considerable potential value in constructing hybrids of industrial yeast strains. Resistance to compounds such as 2-deoxyglucose, which is also dominant, can be induced in yeast strains used for ethanol production, and in baking yeast strains, and are of considerable interest as the resistant strains are normally also resistant to glucose repression.

The general requirements for obtaining mutants in yeast are relatively simple and consist of a suitable yeast strain, a mutagenic agent and some form of selection system for identifying and isolating the mutants. The mutagenic agents fall into two groups; physical and chemical.

8.1 Physical mutagenic agents

8.1.1 *Ultraviolet light and X-rays*

These are among the most commonly used agents for inducing mutations in yeasts. They have the great advantage especially in the case of UV-light, of cheapness, relative harmlessness to the operator and complete absence of carcinogenic residues which could pose a disposal problem after the procedure. The only requirements for either are a source of radiation (X-ray tube or UV light source), the yeast strain, and the necessary ancillary equipment and media for growing and testing the yeast strains after mutagenesis. For X-rays, the requirements for use of the irradiation equipment are more stringent, to protect the operator and any chance passers-by.

8.1.2 *UV irradiation in liquid medium*

(i)   Grow the desired culture in YEP−glucose broth for 24−48 h and dilute an aliquot 1:10.

(ii) Sonicate the diluted suspension lightly (10−15 sec at low power) to break up clumps and separate mature daughter cells from the mother cells, and then put it into a sterile Petri dish and irradiate.

(iii) A relatively sophisticated apparatus for irradiation with UV light has an automatically-controlled shutter which can be set to times from a few seconds to 15−30 min, and a shaker to keep the culture agitated. All that is necessary is to place the Petri dish of suspension on the shaker table, set the timer for the desired exposure and remove the cover from the Petri dish, then start the apparatus and, when the exposure is finished, replace the lid on the dish.

(iv) After irradiation, dilute the culture appropriately and spread aliquots on YEP− glucose agar plates.

(v)   After the colonies have appeared (3−5 days of incubation), plate replica to minimal medium, for selection of auxotrophic mutants, or to other selective media for isolation of resistance mutants and others.

(vi)   Pick cultures which fail to grow on minimal medium from the master plates and re-streak to purify, and test on dropout media to determine the nature of their requirements.

(vii) Pick those growing on medium containing inhibitors of some kind directly from the selective plates and re-streak on these media.

All of these operations, and those for mutagenesis by X-irradiation, should be done under dim yellow light, which avoids any complications brought about by exposure to visible light.

If such equipment for irradiation with UV-light is not available, then the culture can be irradiated using any UV source that is available. In such a case, take extra precautions to avoid exposure of the eyes to the light. The plate will have to be agitated by hand and control of the time of irradiation is not as precise.

The irradiation time is generally selected to give a considerable degree of killing of the cells, approximately 90%. The dilutions must be selected with this in mind. However, different yeast strains, and especially different species, differ widely in their sensitivity to UV light.

8.1.3 *Irradiation on solid medium*

(i)   Dilute the culture before irradiation to the point where a suitable number of cells can be spread on YEP−glucose agar plates before irradiation, allowing for 90% killing and allow the plates to dry.

(ii)   Irradiate the plates with the desired dose and incubate and isolate the mutant colonies as before.

This method has the advantage that where mutants resistant to a given compound are desired, the cells can be irradiated directly on the selective media.

8.1.4 *Induction of mitotic recombination with UV irradiation*

Diploid yeasts of any species can probably be induced to undergo mitotic recombination by low doses (1000−2000 ergs/cm^2) of UV light. This allows recessive mutations occurring on one of the chromosomes of a pair to be expressed. The method has been used to show that *Candida albicans* is probably a diploid, and that many strains carry naturally-occurring mutations to auxotrophy. It is possible that the same phenomenon may occur in industrial strains of *Sacch. cerevisiae*.

8.1.5 *Irradiation with X-rays*

The same procedure as for UV irradiation is followed except that the cultures are irradiated in the appropriate apparatus. It is best to find a qualified operator and assist only to the extent of handling the cultures.

8.1.6 *Irradiation with heavy ions: He, Ne, C, etc.*

This requires access to an accelerator, usually with special equipment for handling the cultures. Follow the instructions for handling the cultures, and leave the operation of the machine to the crew.

8.2 Obtaining mutations by drying

This method is probably not applicable for general use, as only a limited range of mutations have been reported in this way, adenine-requiring mutants apparently predominating. The yeast cultures, 18−20 h old, are suspended in a medium containing 5% skim milk and 0.5% glutamate, and either freeze-dried in tubes or vacuum-dried on membrane filters for 2−2.5 h, in the dark or under dim yellow light, after which they are resuspended in synthetic complete medium, diluted appropriately and plated on complete medium. The colonies appearing are tested for their nutritional requirements as before.

There is some evidence that menadione (vitamin K_3), which enhances the toxic effect of oxygen on yeast, may also enhance the rate of mutagenesis (11).

8.3 Chemical mutagens

Induction of mutations by chemical mutagens is done by exposing the cells to the mutagen in water solution for the appropriate time, neutralizing the mutagen, recovering the cells and plating suitable dilutions on complete medium as before, and then isolating and purifying the desired mutant clones.

8.3.1 *Classes of chemical mutagens*

The various classes of chemical mutagens include the following.

(i)   Base analogues, such as 5-bromouracil (5-BU), 5-deoxyuridine (5-BUdR), and 2-aminopurine, which cause replication errors when incorporated into DNA, in the enol form, so that G is incorporated instead of A. 5-BU can induce reversions of mutations of this type.

(ii)   Deaminating agents. Nitrous acid (NA) deaminates A to hypoxanthine, and hydroxylamine converts C to a derivative that pairs with adenine instead of guanine, producing a GC to AT transition, not reversible by NA.

(iii)   Alkylating agents [ethylethane sulphonate (EES), ethylmethane sulphonate (EMS), methylmethane sulphonate (MMS), nitrosoguanidine, ICR-170, and numerous other such agents] also produce transitions. Agents such as 4-nitro-quinoline-1-oxide, which also induces other types of mutations, and Mn^{2+} induce transitions as well.

(iv)   Acridine derivatives and ethidium bromide. These agents cause frameshift mutations, which can be reversed by acridine, or by a second frameshift, but not by mutagenic agents from the previous groups. Acridines and ethidium bromide cause mutations especially in mitochondrial DNA (mtDNA) (see Chapter 9). Mn^{2+} also causes numerous mutations in mtDNA.

(v)   Cross-linking agents such as nitrous acid, bifunctional alkylating agents and irradiation, may cause deletions of varying lengths in the DNA. Deletions are the most stable form of mutation, not being subject to reversion by other agents.

In addition to mutations induced by various chemical and physical agents, spontaneous mutations also arise, which include frameshifts, transversions and deletions. These arise primarily through enzymatic imperfections during DNA replication or recombination and repair. The following examples are given of procedures for induction of mutations by chemical mutagens.

8.3.2 *Nuclear mutations*

N-Methyl-*N'*-nitro-*N*-nitrosoguanidine (MNNG), EMS, MMS and NA are suitable mutagens.

The following materials are required.

(i)   For all procedures: yeast cultures, grown to stationary phase in YEP−glucose broth. Normally haploid strains should be used, and are essential if recessive mutations (e.g. auxotrophy or canavanine resistance) are desired. For dominant mutations (e.g. resistance to heavy metals), diploid strains may be used if desired.

(ii)   Other reagents: 0.05 M phosphate buffer, pH 7.1; 5% $Na_2S_2O_3$, for terminating the reactions with MNNG, EMS or MMS.

(iii) Media: complete (YEP−glucose) and minimal media. For characterization of mutant strains, use dropout media or other selective media.

(iv)   Water blanks, sterile tubes and pipettes, and other laboratory supplies as required.

The procedure is as follows.

(i)   Wash the cells in sterile phosphate buffer, pH 7.1.

(ii)   Resuspend the cells, approximately 10^7 cells/ml, in buffer containing the desired mutagen at the specified concentration.

(iii) Incubate the suspension at $25-30°C$ for $10-40$ min, the time depending on the mutagen and on the yeast strain.

(iv)   Dilute the cell suspension into 5% $Na_2S_2O_3$ to terminate the reaction (MNNG, EMS, MMS) or into 0.1 M phosphate buffer for NA.

(v)   Make appropriate dilutions of the suspension and plate aliquots on to complete medium. Incubate for $4-5$ days and select the desired mutants by replica plating to selective medium (YEP−glucose, for instance). Isolate auxotrophic mutants by comparison of the latter plates with the ones of complete medium, and select those which do not grow. Purify these strains and test them on dropout media to determine their nutritional requirements. They can then be crossed to strains carrying known mutations for these requirements to identify the gene(s) responsible; that is, if the diploid resulting from any given cross is still auxotrophic for the requirement under test, the mutations are allelic.

(vi)   The culture can be enriched in the desired mutants by transferring the mutagenized cells to a minimal medium containing an inhibitor (e.g. nystatin) which kills growing cells but not non-growing ones, so that when the culture is transferred to a complete medium lacking the inhibitor, the wild-type cells have mostly been killed, and the mutant cells constitute a greater percentage of the populations.

(vii) Mutations to resistance to inhibitors, heavy metals and other similar substances, can also be enriched by a similar procedure, by transferring the mutagenized culture to a medium containing the inhibitor under test, possibly making two or three serial transfers, and isolating the mutant strain(s) on solid medium, also containing the inhibitor. It may be desirable to grow the culture for a generation or two in complete medium without inhibitor, to fix the mutation in the cells, before beginning the transfer to selective medium.

8.3.3 *Petite mitochondrial mutations*

The frequency of spontaneous *petite* mutants in *Sacch. cerevisiae* is relatively high, and may reach at least $1-2\%$ in a few strains. Although the colonies of *petite* mutants can be identified by staining with an overlay of tetrazolium chloride, in practice it is simpler to pick small colonies to complete media having glucose and glycerol as sole carbon sources, and to assume that colonies growing on glucose but not on glycerol medium are *petites*.

For induction of *petites* by starvation, grow the cultures in a weak medium (e.g. Schopfer's medium, which contains, per litre, 30 g of glucose, 1 g of asparagine, 1.5 g of KH_2PO_4 and 0.5 g of $MgSO_4.7H_2O$) for $1-2$ weeks and plate out on YEP−glucose agar. Pick the small colonies.

In some strains, incubation of the culture at elevated temperatures will increase the frequency of *petite* mutants, which can be isolated as previously described.

Suitable chemical agents include acriflavine, ethidium bromide and Mn^{2+}. Acriflavine and ethidium bromide are the mutagens of choice. Acriflavine acts on growing cells and ethidium bromide on both growing and non-growing cells. The latter compound is more drastic in its action, eventually destroying all or nearly all of the mtDNA in the cells. *Petites* induced with acriflavine may contain enough non-functional mtDNA to be visible (fluorescent in UV light) in cells stained with 4′,6-diamidino-2-phenylindole and it may recombine with incoming mtDNA from other yeasts during hybridization. Mn^{2+} also induces numerous *petites*, but is more commonly used to induce mitochondrial mutations to antibiotic resistance. Mutations to antibiotic resistance, encoded on mtDNA, may be retained on the segment of mtDNA which is maintained and reduplicated during *petite* formation, and may again be expressed in the resulting diploid when the *petite* is mated with a sensitive *grande* strain.

The following materials are required for acriflavine induction.

(i)   Tubes of YEP−gluose broth, 5 ml/tube.
(ii)   Acriflavine solution, aqueous, 0.4%.
(iii)   Standard loop.
(iv)   Plates of YEP−glucose agar and (later) YEP−glycerol agar.
(v)   Fresh slants of the desired yeast cultures.

The following procedure should be carried out.

(i)   Add a loopful of the acriflavine solution to each of the YEP−glucose broth tubes required, and mix.
(ii)   Inoculate the tubes with a small inoculum from the slants.
(iii)   Incubate the cultures for $2-3$ days at 30°C, without agitation, in the dark.
(iv)   Streak the cultures on YEP−glucose medium and incubate for $3-4$ days, or until the colonies are large enough that the *petite colonie* mutation can be distinguished.
(v)   Pick small colonies to YEP−glucose and YEP−glycerol media, and retain those which do not grow on glycerol as sole carbon source.

8.3.4 *Antibiotic-resistant mitochondrial mutants*

These mutants are resistant to high levels of such antibiotics as chloramphenicol, erythromycin, paromomycin, oligomycin and numerous others, which affect mitochondrial function in sensitive strains. They may be spontaneous or induced by mutagens such as Mn^{2+}.

Spontaneous mutations may be obtained by spreading a heavy suspension of cells from a 48-h culture of the desired strain on plates of YEP−glycerol containing the antibiotic. Normal concentrations of the more commonly used antibiotics are: chloramphenicol and erythromycin, 1.5−2.5 mg/ml, oligomycin, 2.5 µg/ml. Mutants resistant to daunomycin, chlorimipramine, trimethoprim, adriamycin and many similar compounds may also be obtained. Some of these compounds are also carcinogenic to humans and other mammals and must be handled according to recognized procedures for such chemicals.

Colonies growing on the antibiotic media are picked and purified by streaking on the same media.

Alternatively, mutation can be induced by Mn^{2+} (12), using the following materials.

(i) YEP−glucose broth, in tubes, 5 ml/tube.
(ii) $MnCl_2$ solution, 79 mg/ml.
(iii) Plates containing the desired antibiotic (YEP−glycerol, + chloramphenicol, erythromycin, oligomycin, etc): not required until growth appears in the tubes.

Carry out the following procedure.

(i) Add 0.1 ml of the $MnCl_2$ solution to each of the tubes. Some strains are more sensitive to the ion and some are less, so that the concentration must be adjusted to obtain mutagenesis without completely inhibiting growth.
(ii) Inoculate the tubes with a small inoculum of the desired strains.
(iii) Incubate the tubes on a shaker or roller drum (for aeration) until growth is moderately heavy (7−10 days in some cases).
(iv) Spread approximately 0.5 ml of the culture on each of several plates of the antibiotic medium. Washing the cells is unnecessary. Let the plates dry, invert and incubate at 25−30°C until colonies appear, which may take 2 or 3 weeks depending on the strain.
(v) Pick colonies and re-streak on the same medium. In theory, one colony/plate is sufficient, but it is obvious from the colonial morphology and rate of growth, especially in industrial strains, that more than one mutant type is induced in each broth culture. Mn^{2+} induces *petites* at relatively high frequencies as well as mutations to antibiotic resistance.

9. ISOLATION OF PARTICULAR CLASSES OF MUTANTS

9.1 **Mutator mutants**

These mutants show an increased rate of spontaneous mutation, 6−20 times normal, at various loci. Mutator mutations may occur in both *petite* and *grande* strains. Von Borstel (13) has devised a very sophisticated method for the isolation and study of mutants of this type.

9.1.1 *Detection of mutator mutants using strains auxotrophic for lysine*

(i)   Grow the strains in normal medium (YEP−glucose, for instance) and wash the cells.

(ii) Spread $10^6 - 10^7$ cells per plate on synthetic complete medium containing 20 μg/ml of lysine and on lysine omission medium.

(iii) Incubate for several days. The yeast uses up the lysine in the medium in this time and only revertants to lysine independence continue to grow. Strains containing a mutator mutation show a much higher frequency of lysine-prototrophic colonies on the plates.

9.1.2 *Determination of revertant frequency*

This is a somewhat specialized procedure and requires some equipment not normally available in the microbial genetics laboratory, in particular the well system. Some form of multi-well plate, having wells holding approximately 2 ml each, capable of being used for aseptic culture, is required. Normally 1000−2000 wells/yeast strain are necessary. The requirements are:

(i)   Flask of limiting medium (e.g. 1 μg/ml lysine, 0.5 μg/ml uracil), 1.5 l, all other ingredients at normal levels according to the strain.

(ii) Yeast strain, grown in normal medium, as for Section 9.1.1.

(iii) Multi-well plates, for example, Sterilin micro-titre plates.

(iv) Automatic pipettor, to deliver 1 ml of medium/well.

(v)   Magnetic stirrer. The flask of medium should contain a sterile stirring bar.

The procedure is as follows.

(i)   Wash the yeast cells. Inoculate the flask of limiting medium with approximately 5×10^3 cells/ml, place on the magnetic stirrer and allow the system to equilibrate.

(ii) Connect the automatic pipettor, check for accuracy of delivery and fill the wells using aseptic techniques.

(iii) Plate aliquots on solid medium to determine the numbers of revertants in the original inoculum (see Section 9.1.1).

(iv) Seal boxes of plates and incubate at room temperature (24−25°C). Avoid agitating the wells.

(v)   After incubation, count the number of wells having no revertant colonies and record. Count also the total number of revertant colonies per strain.

(vi) Determine the total number of cells by haemocytometer (2 wells/plate).

(vii) Plate one revertant colony or well on complete, synthetic complete, minimal and omission medium (without the other auxotrophic requirements) to determine whether the reversions were at, for example, the *lys1* locus (growth on lysine omission medium only) or were due to the presence of a super-suppressor.

(viii) Calculate the mutation rate from the formula

$$e^{-m} = \frac{N_0}{N}$$

where N is the total number of compartments in the experiment, N_0 is the number of compartments without revertants, and m is the average number of mutational events/compartment. Also, m_b is the average number of mutants/well, as determined by direct plating, in the original inoculum, and m_g is the corrected number of mutational events, so that $m_g = m - m_b$, so that the mutational events/cell/generation, $M = m_g/2C$ where C is the number of cells/ well after growth has ceased in the limiting medium (the number of cell generations in the history of a culture is approximately twice the final number of cells).

(ix)   If the mutants are to be grouped into categories, the mutation rate is $M_i = f_i M$, where M_i is the mutation rate (per cell per generation), and f_i is the fraction of mutants tested which were found in the ith category.

The above procedure is for determination of the spontaneous mutation rate, without mutagen treatment. Other mutator strains may be induced by treatment with EMS, MMS, UV or other agents. The procedure is approximately the same as for detection of mutator strains as outlined in Section 9.1.1 above, the cell suspensions being spotted (one drop each) on the omission medium, and those with more prototrophic papillae per drop are further tested by re-spreading on omission medium for the presence of mutator genes. Finally, the actual mutation rate is determined as given.

9.2 Temperature-sensitive mutants

These mutants have been widely used in studies of the cell division cycle, and have numerous other applications in cases where the effect of switching off cell growth is under investigation. Cells are treated with the desired mutagen, the suspension is diluted and spread on complete medium. After incubation, the colonies are replica plated to pairs of plates of complete medium and one plate of each pair is held at the restrictive temperature (normally 36°C). Colonies which do not grow are isolated from the other plate and purified. They may be tested by staining with a nuclear stain (Giemsa) to determine the point in the cell division cycle at which growth has been halted. Further characterization of the mutant strains can be carried out as described. Osmotic-remedial mutants can be isolated by replica plating to a third plate containing elevated concentrations of KCl (1−3 M) and incubating at the restriction temperature with the second plate.

Cold-sensitive mutants may be isolated in a similar way.

9.3 *kar* mutants (defective in nuclear fusion)

Nuclear fusion does not take place in these mutants, either during normal mating or in protoplast fusion. They are very useful in transferring cytoplasmic elements (cytoduction) such as mitochondria and killer virus-like particles (VLPs), without affecting the nuclear genome, and in transfer of single nuclear chromosomes. It is simpler to buy or borrow most mutants, but they may be constructed according to the following procedure.

The following are required to carry out the procedure.

(i)   Yeast strains carrying several nuclear mutations, including mutations to resistance to canavanine and/or cycloheximide. Strain JC1 (*MATα his4 ade2 can1 nys^R rho^−*) and GF4836-8C (*MATa leu1 thr1 RHO^+*) have been used (14).

(ii)   Apparatus and reagents for mutagenesis. Media required are minimal medium, and glycerol minimal medium supplemented with histidine (0.3 mM) and adenine (0.15 mM), buffered at pH 6.5, plus canavanine (60 mg/l) and nystatin (2 mg/l) added after autoclaving.

Carry out the following steps.

(i)   Mutagenize strain JC1 according to methods used previously and cross to the other strain by mass mating. Spread mating mixture on YEP−glucose plates and incubate for 20−50 h at 30°C.

(ii)   Resuspend the cells from the plates in 1 ml of water, dilute 1:10 and spread 0.2 ml/plate on selective media.

(iii)   Isolate colonies from the plates and test. These will mostly be *MATα his⁻ ade⁻ canᴿnysᴿ (RHO⁺)*, having the nucleus from JC1 and the mitochondria from GF4836-8C, since the parental strains will not grow on this medium.

(iv)   Back-cross the strains to GF4836-8C and determine the frequency of heteroplasmon formation, using a cross of the original strains as a control.

9.4 Cell wall mutants

9.4.1 *Mutants having cell walls with mannans of altered chemical structure (15)*

These mutants have been used to investigate the biogenesis and chemical structure of the cell wall mannans of several yeast species. The method depends upon repeated treatment of the mutagenized yeast cells with antisera raised either against particular chemical structures, such as the pentasaccharide side chains of *Kluyveromyces lactis* cell wall mannan, after adsorption of the serum with cells of *Sacch. cerevisiae*. The mutagenized cells are allowed to grow for 2 days, and are then washed and agglutinated with the antiserum. The agglutinated cells are discarded (wild-type), and those remaining in the supernatant are recovered and plated out. After two or three repetitions of the procedure, the cells are plated out and colonies isolated and tested for agglutination by the antiserum.

For isolation of mannoprotein mutants by fluorescence-activated cell sorting, the mutagenized cells are grown up in YEP−glucose, washed and treated wtih fluorescein-labelled wheat-germ agglutinin. After labelling, the cells are sorted in a fluorescence-activated flow cytometer, and those showing less than 10% of the maximum fluorescence are retained for further testing with fluorescence-labelled antigens and specific antisera.

9.4.2 *'Fragile' mutants; Venkov's procedure (16)*

These mutants lyse at normal osmotic tensions, and can be used for obtaining cell components by less drastic treatments such as ballistic disruption, and at less cost than by enzymatic removal of the walls (for large quantities of yeast) prior to lysis of protoplasts. The cells are mutagenized according to standard procedures, and plated on osmotically-stabilized medium, for example containing 1.2 M sorbitol or 0.6 M KCl. After growth, they are replica plated to normal YEP−glucose agar and any colonies growing on the first medium and not on the second are isolated for further testing. Note that fragile mutants obtained in this way may be unstable and revert rapidly and that they may have increased sensitivity to temperature, antibiotics and other drugs.

9.4.3 *Mutants having cell walls more easily digested by lytic enzymes (17)*

These mutants were originally isolated for use as food yeasts, on the assumption that they would be more easily digestible.

The cells are mutagenized according to normal procedures. The mutations are fixed by growth in YEP−glucose broth, and diluted and plated (several plates). The cultures are incubated for 3−4 days, and the colonies are replica plated to YEP−glucose agar containing Helicase (snail enzyme, β-glucuronidase, etc). Colonies growing on YEP−glucose but not on medium containing Helicase are isolated and tested for lysis by pepsin, lipase and trypsin.

9.5 Antibiotic-resistant mutants; 'kamikaze' strains (18)

'Kamikaze' strain BL15, a temperature-sensitive strain which grows at 32°C but dies at 42°C, is used for isolation of mutants which are sensitive to such antibiotics as cycloheximide, emetine, fusidic acid, amecitin and others which inhibit protein synthesizing systems *in vitro*, since normal yeast strains are generally impermeable to these compounds. Thus, the first step in obtaining mutants which are drug resistant because of alterations in the protein synthesizing system is the isolation of yeast strains which are sensitive to these antibiotics. Use of the 'kamikaze' strain eliminates the need for screening large numbers of colonies and the concomitant use of large quantities of high-priced antibiotics.

(i)   The usual equipment and supplies, plus YEP−glucose plates containing the desired antibiotics, are required.

(ii)   Mutagenize the culture, dilute 1:5 into YEP−glucose medium and incubate overnight.

(iii) Dilute to approximately 4×10^7 cells/ml in YEP−glucose together with the antibiotic (for instance, emetine, 100 μg/ml) and incubate 1 h at 32°C.

(iv)   Raise the temperature to 42°C and hold there for 7 h. This reduces the viable count to 7×10^5 cells/ml.

(v)   Wash the cells with sterile water to remove the drug and plate on YEP−glucose plates.

(vi)   Replica plate to antibiotic-containing plates to detect those colonies which are sensitive to the drug.

Mutants sensitive to trimethoprim may be isolated using a similar procedure, on minimal medium containing 300 μg/ml of trimethoprim. Trimethoprim sensitivity results from a single-site mutation, but emetine sensitivity, from a double mutation.

9.6 *PEP4* mutants (19)

These mutations cause deficiencies in carboxypeptidase Y, proteinase A, RNase(s) and the repressible, non-specific alkaline phosphatase, and are recessive as well as pleiotropic. The *PEP4* gene product appears to be required for the maturation of several inactive precursors of vacuolar hydrolases, and therefore the mutation may be useful in investigating mechanisms for the maturation of vacuolar and other yeast enzymes.

(i)   Mutagenize the yeast strain in the usual manner, and dilute the treated cell suspension, plate on YEP−glucose medium and incubate for 2−3 days.

(ii)   When the colonies are reasonably well developed, detect *PEP4* mutants by pouring an overlay of agar containing *N*-acetylphenylalanine-β-naphthyl ester (APE).

(iii)   Treat the plates with fast garnet (GBC), a diazonium salt which gives a red colour reaction with the β-naphthol produced by the enzymatic reaction.

(iv)   Detect proteinase B-deficient mutants by treating the plates with hide powder azure reagent (19), these mutants failing to release the blue colour from the preparation.

There is a long phenotypic lag in expression of the mutant phenotype. The character segregates 2:2 on crossing with a wild-type strain and subsequent sporulation of the diploid and dissection of the asci. The mutant phenotype gives a yellow colour in the test, while the wild-type is red.

9.7 Membrane mutants

9.7.1 *Inositol-requiring mutants*

Inositol is a precursor of a number of membrane phospholipids which play an essential part in yeast metabolism and growth. Deficient mutants are made by standard methods of mutagenizing the yeast and plating on inositol-containing media, after which the colonies are replica plated or picked to inositol-deficient media. Fatty acid-requiring mutants may be isolated in the same way.

9.7.2 *Inositol-secreting mutants*

These mutants can be isolated in the same way, except that after the mutagenized cells are plated on complete medium (YEP−glucose) and incubated, the colonies are replica plated to inositol-deficient medium having a lawn of an inositol-requiring yeast strain which also carries an *ade1* or *ade2* mutation, spread on it. Inositol-excreting mutants are white and surrounded by a red halo of cells of the test strain.

Among the mutants isolated are those constitutive for inositol-1-phosphate synthetase, and less commonly, non-constitutive mutants in which the synthesis of phosphatidyl-choline is affected.

9.7.3 *Sterol-requiring mutants*

These have been isolated by selecting for mutants which were resistant to nystatin, after mutagenizing the yeast with UV light, at normal temperature, but which were temperature-sensitive at 36°C, or by selecting for mutants resistant to nystatin in the presence of cholesterol at normal temperatures (22°C) (20). The concentration of ergosterol or cholesterol used was 40 mg/l and of nystatin, 15−28 mg/l.

Mutations in the pathway for biosynthesis of sterols can also be obtained by selecting for strains requiring heme in the medium for growth. Ergosterol or cholesterol is then substituted in the medium for α-aminolaevulinic acid. The strains grow poorly at first, but spontaneous mutations in the biosynthetic pathway for sterols arise after several generations (21).

9.8 Mutants auxotrophic for 2′-deoxythymidine 5′-monophosphate (dTMP)

These mutants may be used for the investigation of DNA replication by using the incorporation of such analogues as 5-BUdR 5-monophosphate as markers. However, normal wild-type yeast cells are not permeable to thymidine monophosphate, and so the first step in obtaining mutants auxotrophic for dTMP is to obtain mutants which are permeable to this compound (22).

9.8.1 *Obtaining mutants permeable to dTMP*

(i)   Grow the yeast strain (wild-type) to exponential phase and wash in phosphate buffer.

(ii)   Plate at approximately 2×10^7 cells/plate, on minimum medium containing either 2% glucose (SD medium) or 3% glycerol (SG medium) as sole carbon source, plus sulphanilamide (5 mg/ml) and aminopterin (100 μg/ml), vitamin-free casamino acids (0.15%), adenine (30 μg/ml) and dTMP (100 and 10 μg/ml), and incubate at 34°C for 4 days.

(iii)   Pick emergent colonies and discard *petites* ($\sim$90%).

(iv)   Grow up *grande* colonies, mostly from the SG plates, and plate on SG medium containing aminopterin, sulphanilamide and dTMP (10 μg/ml). Colonies on these plates should be highly permeable to dTMP.

Mutants requiring dTMP are obtained by mutagenizing the cells obtained in the previous procedure with EMS, and diluting and plating according to standard procedures on YEP−glucose medium containing 100 μg/ml dTMP or SD medium containing 25 μg/ml of dTMP. The emergent colonies are replica plated to YEP−glucose and SD plates with and without dTMP and strains auxotrophic for dTMP are isolated.

Note: the frequency of dTMP-requiring mutants is low, amounting to 4−11 colonies per 11 000 colonies scored. At least one mutation in the *PHO80* gene, controlling expression of phosphatase, is allelic to one or more mutations conferring permeability to dTMP.

9.9 Glycolytic cycle mutants

These include mutants having altered hexokinase, alcohol dehydrogenase (ADH), pyruvate carboxylase, phosphoenolpyruvate carboxykinase (PEPCK), phosphofructokinase, mutants affecting galactose utilization and phosphogluconate dehydrogenase mutants. In general, they are isolated by methods based on their failure to grow on glucose or other hexoses, and by growth on ethanol, glycerol or other non-fermentable substrates. Isolation of some of these mutants is described here.

9.9.1 *Alcohol dehydrogenase mutants (23)*

These mutants are selected for their resistance to allyl alcohol. Cytoplasmic *petite* yeast strains are mutagenized with EMS, nitrosoguanidine or other mutagenic agents, and approximately 10^7 cells are plated on YEP−glucose medium containing 20 mM of allyl alcohol. Colonies growing on this medium are spread on plates containing the next higher (40 mM) concentration, until strains growing in the presence of 100 mM of allyl alcohol are obtained.

Alternatively, mutants may be isolated by continuous culture. The culture of the desired *petite* strain is established in a turbidostat, with a doubling time of approximately 2 h, without aeration. Allyl alcohol is then added to the feed reservoir, at the lowest concentration (10−12 mM). After the growth rate is restored to the original level, the concentration of allyl alcohol in the feed is increased to the next level, and growth is allowed to re-establish itself once more. When the concentration of allyl alcohol has reached the desired level, the fermentor is sampled and strains resistant to allyl alcohol are isolated. These strains are grown and tested for altered forms of ADH by making extracts of a small quantity of cells, vortexing the cell cream with glass beads in a culture tube, and determining the enzyme in the extract electrophoretically by SDS−PAGE (Chapter 5).

9.9.2 *Pyruvate carboxylase mutants (24)*

Since these mutations affect the anaplerotic reactions which supply intermediates to the Krebs cycle reactions, they may illuminate the interface between reactions occurring in the mitochondria and those in the cytoplasm. The mutation to pyruvate carboxylase deficiency is a recessive single-gene nuclear mutation. The mutant has an aspartic acid requirement, like biotin-deficient mutants, and pyruvate carboxylase is a biotin-requiring enzyme.

The mutants are constructed in an otherwise wild-type strain, because of the nutritional complexity of the *pyc* mutant. The yeast strain (haploid, wild-type) is mutagenized with EMS, in YEP−glucose broth. The cells are recovered and washed in 6% $Na_2S_2O_3$ and then in phosphate buffer, the suspension is diluted appropriately and aliquots to give 100−200 colonies/plate are plated on YNB−glucose agar, to eliminate amino acid auxotrophs. The colonies appearing are replica plated to YEP−ethanol and YNB−ethanol. Cultures growing on the former but not on the latter are isolated and mated to a wild-type haploid strain and segregants from this cross, unable to grow on YNB−ethanol, are isolated. These segregants are again crossed to wild-type strains and segregants isolated, until strains giving high spore germination (>95%) and 2:2 segregation of this character are obtained. This mutant is then crossed to an ADH-negative strain, to show that it has intact Krebs cycle and oxidative phosphorylation systems. If these pathways are absent, ADH-negative strains do not survive.

9.9.3 *Mutants defective in phosphoenolpyruvate carboxykinase (PEPCK) (25)*

These mutants are selected from cultures forming colonies on media containing glycerol, but not ethanol, as sole carbon source. The desired yeast strain is mutagenized with EMS and the population is enriched in mutants by growing in medium containing ethanol as sole carbon source plus nystatin, 3 μg/ml, to kill growing cells. The cells are diluted suitably and plated on YEP−glucose agar, and after incubation for 3−4 days at 25−30°C, replica plated to YEP−glycerol and YEP−ethanol media. Colonies growing on glycerol but not on ethanol are selected, purified and grown to amounts sufficient to yield 100 mg wet weight of cells. These are broken by vortexing with glass beads in 0.5 ml of imidazole buffer and the supernatant recovered, and the extract is tested for PEPCK levels, using an assay medium containing (in 0.5 ml) 25 μmol imidazole, 25 μmol $NaHCO_3$, 1 μmol, $MnCl_2$, 1 μmol ADP, 0.25 μmol NADH, 1 μmol phosphoenolpyruvate, 80 μkat malate dehydrogenase, 40 μkat hexokinase, and cell extract (26).

Mutants lacking fructose-1,6-bisphosphatase are not isolated by this procedure.

9.9.4 *Secretory mutants (27)*

Secretory mutants are defined as those in which an enzyme such as invertase or acid phosphatase is formed but not exported to its normal location, in the periplasmic space or external to the cell. The presence of the mutation is confirmed by breaking the cells as described above and testing the extract for invertase activity. Secretory mutants contain invertase in the cytoplasm, but not in the periplasmic space. The first secretory mutant was isolated from a collection of temperature-sensitive mutants. This mutant, *sec1*, failed to grow at 37°C but accumulated relatively high levels of intracellular invertase at this temperature, as compared with the wild-type strain. Acid phosphatase also accumulated at the non-permissive temperature. The enzymes accumulated in the cytoplasm inside large numbers of secretory vesicles, whose numbers decreased concomitantly with secretion of the enzymes, when the cells were returned to the permissive temperature (25°C). Although the degree of enrichment reaches approximately 33% in the densest 1−2% of the mutagenized cells, only 220 *sec* mutants were obtained from a total of 18 500 colonies from the dense fraction, so the method is somewhat laborious.

An unusual type of mutant, showing failure to excrete invertase, is a *petite* of a distiller's yeast, which grows on sucrose in the *grande* form, but fails to do so when converted to a *petite* (28). This strain contains a cytoplasmic invertase as mentioned in the preceding paragraph. However, the cells were not investigated for the presence of increased numbers of secretory vesicles.

Similar mutants have been isolated which do not excrete acid phosphatase. Likewise mutants which do not excrete the yeast mating hormones or killer toxin are known. Schekman (29) has investigated the nature of secretory mutants in detail.

9.9.5 *Transport mutants (30)*

Transport of materials into the yeast cell is normally by either facilitated diffusion or active transport, involving some form of membrane carrier (permease). Mutants showing failure in the transport system affect transport of carbohydrates and other carbon sources, amino acids and other nitrogenous compounds, vitamins and inorganic ions. Mutations giving rise to defects in transport, in particular of nitrogenous compounds, can be divided into those mutations which give rise to recessive phenotypes for resistance to various inhibitors of growth, an inhibitor very often being an analogue of the compound whose transport has been impaired (methionine and L-ethionine; uracil and 5-fluorouracil: basic amino acids and L-canavanine; adenine and 4-aminopyrazolopyrimidine; thiamine and pyrithiamine, to name a few examples). The other group of mutations consists of those which inactivate a transport system for some nutrient, but which do not confer resistance to any known inhibitor. Mutations affecting transport of sugars fall generally into the second group.

As in the case cited above for *sec* mutants, the *petite* mutation can affect sugar transport. Evans and Wilkie (31) showed that some yeast strains failed to grow on galactose and other sugars when converted to the *petite* form. Evans showed that these strains took up galactose, for instance, when they were permeabilized with dimethyl sulphoxide (DMSO), so the phenomenon occurred at the cell surface, and probably a galactose permease, located in the cytoplasmic membrane, was affected.

10. GENE MAPPING

Mapping of genes in yeast and assigning them to a particular chromosome consists of assigning the genes to the different linkage groups which have been established. The yeast strain carrying the unmapped gene is crossed to various strains carrying mutations which have been mapped to known positions on the different chromosomes, and determining whether or not linkage exists between the unknown gene and the markers used. Mapping is done by tetrad analysis or random spore analysis, trisomic analysis for assigning a gene to a particular chromosome, mitotic mapping after induction of mitotic crossing-over with UV light or X-irradiation, or treatment with mutagens, mitotic gene conversion, chromosome loss and chromosome transfer. Fine-structure mapping, for determining the location of intragenic mutations, is done by similar methods and by deletion mapping (for review, see ref. 32).

10.1 **Tetrad analysis**

Tetrad analysis is used to determine the distance between two markers on the same chromosome, and depends on the number of cross-overs between the markers, which in turn can be determined from the types of ascus present as shown by the segregation of the markers in the clones obtained on dissection. If there are no cross-overs, the asci will all be parental ditype (PD; AB AB ab ab), where A and B are the two genes in question. One cross-over will yield tetratype asci (T; AB Ab aB ab) only, while multiple cross-overs will yield three types of asci, including non-parental ditype (NPD; Ab Ab aB aB). The map distance is defined as the average number of cross-overs per chromatid, multiplied by 100, and can be calculated from the equation X = 100/2[(T + 6NPD)/(PD + NPD + T), which gives the map distance between the two genes to within a few percent. However, if the map distance is greater than 35−40 cM (centimorgans), this equation overestimates the true map distance due to the increasing number of triple and higher-order cross-overs in the region.

10.2 **Determination of centromere linkage**

This is related to second division segregation of the parental alleles in meiosis. Second division segregation frequency is zero for genes close to the centromere and increases to 67% for genes located at some distance (40−60 cM) from the centromere. A gene is therefore considered to be centromere-linked if the frequency of second division segregation is significantly less than 67%.

Centromere linkage in diploid strains forming unordered asci is determined by including in the cross a gene such as *trp1*, *pet8* or *met14*, for instance, which is closely linked to the centromere (second division segregation <1%). The frequency of T-type asci in the sporulated diploid is determined, which corresponds to the second division segregation frequency of that gene. The map distance of the gene from the centromere is half the second division segregation frequency, for low values of second division segregation, corresponding to approximately 10 cM.

There are two other methods of determining linkage to a centromere, that is use of linear asci and use of tetraploid hybrids. Centromere-linked genes in linear asci usually segregate almost exclusively in a linear array (AaAa), while genes which are not linked to a centromere segregate in other arrays (AAaa, AaaA, aAAa) as well (33).

However, the asci have to be dissected in such a way that the order of the spores in the ascus is preserved, which is a difficult technique and is seldom used at present. The order of the spores is lost if the ascus wall is digested with enzymes. In addition, few strains of *Saccharomyces* form significant numbers of linear asci. Strains obtained by intergeneric protoplast fusion sometimes form asci of this type. The other method requires inclusion of the unmapped gene in a tetraploid hybrid in a duplex state (A/A/a/a), in which case three types of asci are formed, Aa,Aa,Aa,Aa; AA,Aa,Aa,aa; and AA,AA,aa,aa. The ratio of the three classes is 2/3:0:1/3 for centromere-linked genes, and 4/9:4/9:1/9 for non-centromere-linked genes. This method also is seldom used at present.

10.3 Trisomic analysis for assignment of a gene to a particular chromosome

If a known chromosome is disomic, and the strain carrying it is crossed with a haploid carrying an unmapped gene, and the resulting diploid is sporulated and dissected, then if the unmapped gene is on the disomic chromosome, the segregation for it will be 4:0, 3:1 and 2:2. If the unmapped gene is on any other chromosome, the segregation will be 2:2 only. One or more known markers on the trisomic chromosome should be in the duplex condition (A/A/a).

If the chromosome bearing the unmapped gene is disomic, the strain is crossed to strains carrying markers on all chromosomes, or on as many as possible. The unmapped gene will be on the chromosome bearing markers which segregate with the unmapped gene as described previously for a trisomic chromosome.

If multiply disomic strains are available, these can be used in locating an unmapped gene. The strain carrying the unmapped gene can be crossed to these strains and chromosomes showing 2:2 segregation for the unmapped gene and other markers can be excluded. Usually up to four or five chromosomes can be eliminated per cross (34).

10.4 Mapping with super-triploids

Super-triploids can also be used in mapping genes to particular chromosomes. Such triploids may be constructed with at least one marker on each chromosome in the duplex (A/A/a) condition. Random spores are isolated after sporulation and mated with a haploid carrying the unmapped gene. If several strains constructed in this way are analysed genetically, 4.2 chromosomes per cross can be eliminated as sites for the unmapped gene (35).

Disomic strains can be isolated by recovering them as rare meiotic products of diploid strains, showing aberrant phenotypes.

10.5 Mapping using mitotic crossing-over

If a diploid yeast is heterozygous for several known markers and an unmapped gene, mitotic crossing-over will result in a sectored colony, wherever a marker is distal to a cross-over event. If an unmapped gene occurs in the same sector as a known marker, the two genes are on the same arm of the chromosome. Genes controlling colour, colonial morphology, resistance to toxic agents and auxotrophy for various nutritional requirements can be mapped to specific chromosomes by this method. Mitotic crossing-over can be induced by low doses of UV and other mutagenic agents, though

UV is probably the most convenient. The presence of mutations such as *rad18* and *chl1* also increase the frequency of mitotic crossing-over.

10.6 Mapping by mitotic chromosome loss

Chromosome loss in a diploid strain carrying several known markers and an unmapped gene may result in expression of the recessive markers. If the unmapped gene is expressed at the same time as any of the known markers, it is probably located on the same chromosome as the known marker. Diploids homozygous for the recessive mutations *cdc6*, *cdc14* and *chl1* show increased rates of chromosome loss, and the mutations *spo11* and *rad52* also induce increased rates of loss. Mitotic recombination and tetrad analysis can then be used to confirm the assignment of the gene to a particular chromosome and map it to a definite site on the chromosome.

Chromosome loss may also be induced by treatment of the cells with *p*-fluorophenylalanine, $CdCl_2$, $CoCl_2$, acriflavine and methyl benzimidazol-2-yl, a derivative of benomyl. The latter compound induces chromosome loss without increasing rates of mitotic recombination or mutation.

10.7 Fine-structure (intragenic) mapping

Deletion mapping is normally the method of choice for fine-structure mapping (36), and the interested investigator is referred to this publication.

10.8 Strategies for mapping

10.8.1 *Method 1*

(i) Determination of centromere linkage. Cross the strain carrying the unmapped gene to a strain having a centromere-linked marker such as *trp1*, and determine the frequency of tetratypes in relation to this gene.

(ii) If the unmapped gene is centromere-linked (frequency of tetratypes $<2/3$), cross the original strain to a set of tester strains carrying centromere-linked markers, and look for linkage to one of the markers. If there is no linkage, the gene is on a new chromosome (unlikely in a *Sacch. cerevisiae*) or the centromere of chromosome XVII. Otherwise, this step permits assignment of the gene to a particular chromosome.

(iii) Do mitotic cross-over analysis (Section 10.5) to assign the gene to a particular arm of the chromosome.

(iv) Cross the original strain to a strain carrying, preferably, several known markers on the relevant arm of the chromosome, and do tetrad analysis to localize the gene to a specific site in relation to the other markers and the centromere.

10.8.2 *Method 2*

This method requires a set of strains carrying 66 markers spaced approximately 50 cM apart over the entire yeast genome. A set of nine strains carrying appropriate markers is available from the Yeast Genetics Stock Center, Donner Laboratory, Department of Genetics, University of California, Berkeley, CA 94720, USA.

(i) Cross the strain carrying the unmapped gene to the tester strains. Do tetrad analysis on the hybrids obtained.

(ii)   Examination of four or five crosses should enable the mapping of most genes. Genes which cannot be mapped by this method are probably located distal to the markers or are in some other unmarked region.

10.8.3 *Notes*

(i)   Mapping of genes in *Sacch. cerevisiae* and other yeast species should be undertaken only after the investigator has gained some experience in genetic manipulation of yeasts, including use of the micromanipulator.

(ii)   *Sacch. cerevisiae* is unique in possessing 17 known chromosomes. Most other yeast species investigated so far using orthoganol-field-alteration gel electrophoresis (OFAGE) or field-inversion gel electrophoresis (FIGE) have shown from one to five chromosomal bands which could be separated electrophoretically. Thus, many more genes are found at relatively great distances from the centromere, which should be taken into consideration in mapping.

11. PROTOPLAST FUSION

Protoplast fusion can be found to obtain both intra- and interspecific hybrids of yeasts which, for a variety of reasons, do not mate in the normal manner. It has been used to introduce the gene(s) for the ability to metabolize starch directly, into industrial yeast strains which do not normally utilize this compound. Hybrids between *Sacch. cerevisiae* and *Kluyveromyces lactis*, resembling the former species in most of its morphological and biochemical characteristics, but with the ability to metabolize lactose from the latter, have been obtained. Various organelles such as mitochondria, and the VLP which produce the *Saccharomyces* killer toxin, have been transferred between yeast strains by protoplast fusion. Its uses are not limited to obtaining hybrids between yeasts, as it was originally developed for fusion of plant cell protoplasts (37,38), and has been used for construction of improved *Aspergillus* strains for use in soy sauce manufacture. At the moment it is often the method of choice for construction of new brewing and other industrial yeast strains, since it is simple, cheap and rapid, and requires only a suitable selection system for isolation of the desired hybrids.

The mechanism of transfer of genetic information during protoplast fusion is not well understood. At first, it was hoped that the method could be used for construction of interspecific and intergeneric (plant) hybrids, in which the resulting product would be, in effect, a blend of the two parental strains, each of which would contribute an approximately equal part of the total genome. However, it became evident that one parental nucleus rapidly became dominant and most of the genome of the subordinate nucleus was degraded, ejected from the final nucleus or rejected in some undetermined way, so that only a small part of the latter genome was expressed in the genome of the hybrid. A few isoenzymes, for instance, could be detected in tissue regrown from such fusion hybrids (39). A similar process apparently takes place in intergeneric fusants of yeast, whether the hybrids are selected with the aid of a *petite* mutant as one parent (40), or whether nuclear markers are used in isolation of the fusants (41). In either case, one nucleus becomes dominant and the hybrid resembles the dominant parent morphologically and biochemically, but characteristics controlled by genes contributed by the subordinate parent are obviously present in the offspring, as in the hybrids obtained in the fusion product from the cross *Sacch. cerevisiae* × *Zygosaccharomyces rouxii*, which yields baking yeast strains which have a better performance in sweet doughs with an elevated osmotic tension (42). We have suggested that a mechanism similar to single-chromosome transfer, as occurs in matings where one strain carries the *karl-1* mutation and nuclear fusion does not take place, might operate in intergeneric fusions and information transfer, but our recent results throw doubt upon this idea. We fused protoplasts from *Hansenula wingei* (*H. canadensis*) with protoplasts from a diploid, homozygous for several auxotrophic requirements, and obtained a sporulating diploid as we had expected, but in which all of the homozygous autotrophic markers had been simultaneously complemented (J.F.T.Spencer, D.M.Spencer, N.Reynolds and R.Cromie, unpublished data). Further investigation is required before an explanation of this phenomenon can be attempted.

The situation is different in fusions between strains of *Saccharomyces*, for instance, where the genomes are essentially compatible. Fusion between two haploid strains of the same mating type yields a diploid in which any auxotrophic requirements are complemented as in a diploid obtained by the normal mating process, with the exception that only one mating type is present. Nuclear fusion is stimulated if the protoplasts are exposed to mating hormone (α-factor) before fusion. The situation is more complex, in the case of fusions between non-mating and/or non-sporulating industrial yeast strains, where the processes associated with nuclear fusion before or after regeneration of the cell wall are obscure.

Transfer of cytoplasmic organelles such as mitochondria or of killer VLPs is most easily accomplished by use of a strain carrying the *karl-1* mutation, which delays or inhibits nuclear fusion. If a *petite* is made of a *karl-1* strain, carrying the killer VLP, it can be fused wth the industrial strain and diploids of the other strain can be selected after fusion, which can be tested for the presence of the killer character. Use of a *karl-1* strain which requires adenine (red colonies) aids in selecting the colonies derived from the industrial strain. If it is desired to transfer mitochondria instead of or as well as killer VLPs, then it is possible to make a *karl-1 petite* and fuse it with the strain selected as donor, select red respiratory-competent (RC) colonies, make a *petite* of the desired recipient strain, fuse it with the RC *karl-1* strain, and select white RC colonies. These will have the nucleus of the recipient strain and the mitochondria transferred from the donor strain to the *petite* of the *karl-1* strain in the previous step. If a *karl-1* strain carrying the killer VLP, previously constructed, is used as such a vector, the killer character can be introduced at the same time. Other plasmids, introduced into the donor strain by transformation, can be transferred to recipient strains in the same way.

11.1 **Formation of protoplasts**

We have used the University of Washington procedure, though other similar ones are equally effective (see Chapter 6, Section 3.2). All depend on pre-treatment of the cells (exponential growth phase) with an agent such as mercaptoethanol, glycollic acid or dithiothreitol, to break disulphide bonds in the compounds making up the cell wall, and removal of the wall enzymatically in an osmotically stabilized buffer.

11.1.1 *Materials and equipment*

The yeast culture should be grown to approximately 10^8 cells/ml, in approximately mid-log (exponential) growth phase. Cells in the stationary phase are difficult or impossible to protoplast. Pre-treatment buffer, containing 0.02 M EDTA, 0.02 M Tris buffer, pH 7.8 and 0.1 M β-mercaptoethanol, requires 1.2 M sorbitol or 0.6 M KCl as osmotic stabilizer. Protoplasting solution, citrate phosphate buffer, contains equal parts of 0.016 M sodium citrate and 0.08 M KH_2PO_4, pH 5.8. Enzyme for digestion of cell walls is snail enzyme (Helicase), Zymolyase, Novozyme or other enzyme having β-glucuronidase activity. Sterile centrifuge tubes and culture tubes, pipettes and other standard laboratory equipment are also required.

11.1.2 *Procedure*

(i)   Spin down 2 ml of cells. If the cell density is less than indicated above, adjust the culture volume used accordingly.

(ii)   Add 4 ml of pre-treatment solution, resuspend cells and incubate for $10-15$ min at 35°C. Washing is not required at this stage. The time of incubation may be adjusted according to the strain of yeast used.

(iii)   After incubation, spin down the cells and wash once with sterile 1.2 M sorbitol or 0.6 M KCl solution. Spin down again and discard the supernatant.

(iv)   Resuspend the cells in 2 ml of protoplasting solution and add $0.1-0.2$ ml of snail enzyme, or Novozyme to make a final concentration of 1 mg/ml.

(v)   Incubate at 35°C until protoplasting is complete. This may require anything from 5 min to 1 h, depending on the yeast strain used.

(vi)   Observe the progress of protoplasting under the microscope. The form of the cells changes and becomes spherical, and protoplasting can be confirmed by adding a drop of water at the edge of the coverslip. Protoplasts will burst.

Dithiothreitol, thioglycollic acid, L-cysteine or other -SH reagents can be substituted for mercaptoethanol.

11.2 **Fusion**

There are a number of procedures for fusion, nearly all depending on the use of PEG as a fusogenic agent, though electrofusion can also be used successfully for yeast protoplasts (43). The equipment for this method is rather expensive, however. Fusion in the presence of PEG gives enhanced yields of fusants if the PEG solution ($\sim 35\%$) contains 15% DMSO and can be carried out in bulk solution. A refinement of the procedure, devised by Kao (44) uses fusion at the interface between a 30% PEG$-$10% sucrose solution and a glucose$-$CaCl$_2$ solution of elevated pH (pH 10). Kao claimed $10-30\%$ fusants and $50-80\%$ viability, for plant protoplasts. It has not been tried for yeasts.

11.2.1 *Regeneration in solid medium*
Requirements for the method are as follows.

(i)   Petri dishes of OSY medium (YEP$-$glucose agar plus 0.6 M KCl), or other stabilized medium such as YNB$-$glucose or YNB$-$glycerol, depending on the requirements of the experiment.

(ii)   Water agar, containing 0.4 M $CaCl_2$. This is made up by mixing molten 3% water agar with attempered (45°C) 0.8 M $CaCl_2$ solution and holding in a water bath at this temperature. Autoclaving agar with this concentration of

CaCl$_2$ destroys its setting ability. The agar is dispensed in individual culture tubes and held in the molten state as above until use.

(iii)   PEG solution, 30−35%. PEG can most readily be dissolved in a microwave oven. Klinner (45) found that addition of 15% DMSO increased the yield of fusants by up to 10-fold.

(iv)   Dilution blanks of PEG solution or 0.6 M KCl.

(v)   Water bath set at 35°C, sterile tubes, pipettes, loops and other equipment.

The following procedure should be carried out.

(i)   Mix the protoplasts from the two different cultures, obtained in the previous section, in a centrifuge tube, centrifuge to pack the protoplasts and remove the supernatant.

(ii)   Add 2 ml of the PEG solution and mix with the protoplasts, gently, until they are resuspended. The protoplasts will aggregate into clumps of different sizes. Incubate up to 30−60 min, depending on the strains.

(iii)   Dilute the fusion mixture into the blanks, to give ×10, ×100 and ×1000 dilutions.

(iv)   Add 0.1 and 0.5 ml of the dilutions to tubes of the CaCl$_2$ agar, mix immediately and pour as an overlay on the OSY plates. The plates should be pre-warmed to prevent too rapid hardening of the agar. Do not add fused protoplasts to a series of tubes of molten agar, but pour each tube as soon as the protoplasts are mixed in, otherwise the protoplasts will probably be killed.

(v)   Incubate the plates at 25−30°C until colonies appear. Pick or replica plate to selective media to isolate the desired hybrids.

Regeneration may also be done in 12% gelatin containing 1% yeast extract, and 2% glucose plus 0.1−0.3% agar as a hardening agent. The concentration of agar is adjusted so that it melts at approximately 35°C. The advantage of this method is that higher concentrations of fused protoplasts can be regenerated, and the gelatin can then be melted and the cells recovered by centrifugation. The cells can then be transferred to selective liquid medium for enrichment of the culture in the desired hybrids (see below).

11.2.2 *Regeneration in liquid medium*

The materials and equipment are as above except that instead of plates of solid medium (OSY), liquid YEP−glucose medium, containing 35% PEG, is dispensed in shallow layers in Petri dishes. CaCl$_2$ agar and, probably, dilution blanks are not required.

(i)   Carry out fusion as in the previous section.

(ii)   After incubation in the fusion mixture, dispense aliquots into Petri dishes and incubate these as above until colonies are visible on the bottom of the dishes. Regeneration is usually well advanced in 24−48 h.

(iii)   Suspend the cells in the mixture, dilute in 0.6 M KCl as a precaution, and transfer to selective medium (normally liquid) for enrichment in the desired hybrids. Incubate these flasks until the desired strains have grown. The cells may be

recovered, washed and transferred through fresh medium for further enrichment if desired, but rigorous precautions against contamination must be taken.
(iv)   Plate out the resulting cultures and carry out final isolation and purification of the hybrids by standard methods.

11.2.3 *Fusion at a liquid—liquid interface; Kao method (44)*
Use the following equipment.

(i)   Sterile coverslips, 60×15 mm plastic disposable Petri dishes, Pasteur pipettes, micropipetter with disposable tips, silicone 200 liquid, centrifuge tubes and other standard equipment for microbiological investigations.
(ii)   Yeast protoplasts for fusion.
(iii) PEG solution, consisting of 100 ml of water, 30 g of purified PEG, 17 g of sucrose, 150 mg of $CaCl_2.2H_2O$ and 10 mg of KH_2PO_4.
(iv)   High pH solution consisting of 100 ml of water adjusted to pH 10 with KOH, 5.4 g of glucose, 294 mg of $CaCl_2.2H_2O$ and 66.3 mg of CAPS (cyclohexylaminopropane sulphonic acid, $pKa = 10.4$) (Solution K1825).
(v)   Kao solution 2 (44), containing 9 g of glucose, 100 mg of $CaCl_2.2H_2O$ and 10 mg of KH_2PO_4 in 100 ml of water, adjusted to pH 5.5 with KOH.
       Test this solution to see if the protoplasts burst. If so, increase the concentration of $CaCl_2$ or add the minimum concentration of KCl to preserve the protoplasts intact.
(vi)   Protoplast culture medium. Use YNB—glucose medium, stabilized with 0.6 M KCl.

Carry out the following steps.

(i)   Place a drop of sterile silicone 200 liquid in the centre of a small Petri dish, and place a sterile coverslip on it.
(ii)   Pipette approximately 300 μl of the PEG solution on it. Prepare 6—10 dishes in this way.
(iii) *Slowly* add 150 μl of protoplast mixture, about 5—10% v/v, in Kao solution 2 to the centre of the PEG layer. Incubate for 10 min.
(iv)   Add two 0.5 ml aliquots of solution K1825 at 5 min intervals, to the PEG-protoplast mixture.
(v)   After 5 min more, add 3 ml of protoplast culture medium. Incubate for another 30—60 min.
(vi)   With a Pasteur pipette, transfer the mixture to a centrifuge tube and sediment the protoplasts by centrifuging for 5—6 min at low speed. Wash the protoplast mixture twice more with stabilized culture medium and transfer to regeneration medium as described above. Isolate the colonies of hybrids by whichever method is suitable.

Kao obtained approximately 30% heterokaryocytes by this method, and 50—80% of the heterokaryocytes and parental protoplasts were viable. Yeast protoplasts are evidently more sensitive to osmotic lysis than protoplasts from plant cells, and in consequence, it may be necessary to increase the concentration of osmotic stabilizer in the different solutions. After fusion, as noted above, the fused protoplasts can be regenerated in any way that is suitable, in liquid or solid medium.

12. REFERENCES

1. Hieda,K. and Ito,T. (1973) Annex 1973−5 to Bull. *Int. Inst. Refrigeration*, Paris, p. 71.
2. Sherman,F. (1973) *Appl. Microbiol.*, **26**, 829.
3. Fogel,S., Mortimer,R.K. and Lusnak,K. (1983) In *Yeast Genetics: Fundamental and Applied Aspects*. Spencer,J.E.T., Spencer,D.M. and Smith,A.R.W. (eds), Springer-Verlag, New York, p. 65.
4. Palleroni,N.J. (1961) *Phyton*, **16**, 117.
5. Spencer,J.F.T. and Spencer,D.M. (1977) *J. Inst. Brewing*, **83**, 287.
6. Gunge,N. and Nakatomi,Y. (1972) *Genetics*, **70**, 41.
7. McClary,D.O., Nulty,W.L. and Miller,G.R. (1959) *J. Bacteriol.*, **73**, 362.
8. Fowell,R.R. (1969) In *The Yeasts, Vol. 1*. Rose,A.H. and Harrison,J.S. (eds), Academic Press, New York, p. 303.
9. Wickerham,L.J. and Burton,K.A. (1954) *J. Bacteriol.*, **67**, 303.
10. Dawes,I.W. and Hardie,I.D. (1974) *Mol. Gen. Genet.*, **131**, 281.
11. Chaput,M., Brygier,J., Lion,Y. and Sels,A. (1983) *Biochimie*, **65**, 501.
12. Putrament, A., Baranowska,H., Ejchart,A. and Jachymczk,W. (1977) *Mol. Gen. Genet.*, **126**, 357.
13. Von Borstel,R.C., Cain,K.T. and Steinberg,C.M. (1971) *Genetics*, **69**, 17.
14. Conde,J. and Fink,G.R. (1976) *Proc. Natl. Acad. Sci. USA*, **73**, 3651.
15. Ballou,L., Cohen,R.E. and Ballou,C.E. (1980) *J. Biol. Chem.*, **255**, 5986.
16. Venkov,P.V., Hadjiolov,A.A., Battaner,E. and Schlessinger,D. (1974) *Biochem. Biophys. Res. Commun.*, **56**, 599.
17. Mehta,R.D. and Gregory,K.F. (1981) *Appl. Environ. Microbiol.*, **41**, 992.
18. Littlewood,B.S. (1972) *Genetics*, **71**, 305.
19. Zubenko,G.S., Park,F.J. and Jones,E.W. (1982) *Genetics*, **102**, 679.
20. Karst,F. and Lacroute,F. (1977) *Mol. Gen. Genet.*, **154**, 269.
21. Taylor,F.R. and Parks,L.W. (1980) *Biochem. Biophys. Res. Commun.*, **95**, 1437.
22. Little,J.G. and Haynes,R.H. (1979) *Mol. Gen. Genet.*, **168**, 141.
23. Ciriacy,M. (1975) *Mutat. Res.*, **29**, 315.
24. Wills,C. and Melham,T. (1985) *Arch. Biochem. Biophys.*, **236**, 782.
25. Perea,J. and Gancedo,C. (1982) *Arch. Microbiol.*, **132**, 141.
26. Hansen,R.J., Hinze,H. and Holzer,H. (1976) *Anal. Biochem.*, **74**, 576.
27. Novick,P. and Schekman,R. (1979) *Proc. Natl. Acad. Sci. USA*, **76**, 1858.
28. Spencer,J.F.T., Spencer,D.M. and Miller,R. (1983) *Curr. Genet.*, **7**, 47.
29. Novick,P., Field,C. and Schekman,R. (1980) *Cell*, **21**, 205.
30. Cooper,T.G. (1981) In *The Molecular Biology of the Yeast Saccharomyces, Metabolism and Gene Expression*. Strathern,J.N., Jones,E.W. and Broach,J.R. (eds), Cold Spring Harbor Laboratory Press, New York, p. 399.
31. Evans,I.H. and Wilkie,D. (1976) *Genet. Res.*, **27**, 89.
32. Mortimer,R.K. and Hawthorne,D.C. (1981) In *The Molecular Biology of the Yeast Saccharomyces. Vol. 1. Life Cycle and Inheritance*. Strathern,J.N., Jones,E.W. and Broach,J.R. (eds), Cold Spring Harbor Laboratory Press, New York, p. 11.
33. Hawthorne,D.C. (1955) *Genetics*, **40**, 511.
34. Mortimer,R.K. and Hawthorne,D.C. (1973) *Genetics*, **74**, 33.
35. Wickner,R. (1979) *Genetics*, **92**, 803.
36. Fink,G. and Styles,C. (1974) *Genetics*, **77**, 231.
37. Kao,K.N. and Michayluk,M.R. (1974) *Planta*, **115**, 355.
38. Wallin,A., Glimelius,K. and Eriksson,T. (1974) *Z. Pflanzenphysiol.*, **74**, 64.
39. Wetter,L.R. (1977) *Mol. Gen. Genet.*, **150**, 231.
40. Spencer,J.F.T. and Spencer,D.M. (1981) *Curr. Genet.*, **4**, 177.
41. Groves,D.P. and Oliver,S.G. (1984) *Curr. Genet.*, **8**, 49.
42. Spencer,J.F.T., Bizeau,C., Reynolds,N. and Spencer,D.M. (1985) *Curr. Genet.*, **9**, 649.
43. Zimmermann,U., Scheurich,P., Pilwat,G. and Benz,R. (1981) *Angew. Chem.*, **93**, 332.
44. Kao,K.N. (1986) *J. Plant Physiol.*, in press.
45. Klinner,U., Becher,D. and Böttcher,F. (1980) *Wissenschaft. Z. Ernst-Moritz-Arndt Univ., Greifswald, DDR: Math.-Naturwiss. Reihe*, **29**, 55.
46. Svoboda,A. (1978) *J. Gen. Microbiol.*, **109**, 169.

CHAPTER 5

Yeast genetics, molecular aspects

JOHN R.JOHNSTON

1. INTRODUCTION

Much of the current genetic manipulation of yeast at a molecular level has stemmed from the development of the general methodology of recombinant DNA technology in the 1970's, which is now familiar to a whole generation of biologists. Thus the first isolation of yeast genes by their expression in *Escherichia coli* was accomplished around a decade ago (1,2). The advent of genetic transformation of yeast by the use of recombinant plasmids (3,4) effectively led to a scientific explosion in the cloning of yeast genes and subsequent studies with those cloned DNA sequences.

This chapter describes the most basic of methods developed for the molecular genetic analysis and manipulation of the yeast *Saccharomyces cerevisiae* and closely-related species. No attempt is made at comprehensive coverage of many of the more advanced techniques, most of which are detailed in several laboratory manuals (5−8). Nor, in general, is discussion extended to several other yeasts, such as *Schizosaccharomyces pombe, Kluyveromyces lactis* and *Pichia pastoris*, which are common subjects of genetic study.

2. DNA ISOLATION

The cloning of genes requires the isolation of both donor strain (genomic) DNA and preparation of DNA of the appropriate vector, either plasmid or viral, such as λ phage. Since most DNA *in vivo* is present in association with RNA and proteins, it is necessary first to isolate crude complexes of nucleoproteins from cells and then purify and separate DNA from proteins and RNA.

2.1 Genomic (high molecular weight) DNA

The preparation of high molecular weight DNA of very high purity (*Table 1*) is based upon the method of Cryer *et al.* (9). If DNA of adequate purity for restriction analysis and for transformation is sufficient, then the method (*Table 2*) based upon that of Struhl *et al.* (10) may be used.

2.2 Plasmid DNA

The vast majority of plasmids used in the cloning of yeast genes are hybrid DNA molecules or 'shuttle' vectors, that is, they combine both yeast and bacterial DNA and can replicate both in yeast and *E. coli*. The various types of these plasmids have been comprehensively reviewed (7,11−13). They include those designated as YIp (yeast

Table 1. Preparation of high molecular weight DNA.

1. Grow 1 litre of cells in YEP−glucose broth on shaker (30°C) to late log phase (~5 × 10⁷ cells/ml). Harvest by centrifugation (6000 g for 5 min at 4°C).
2. Resuspend in 50 ml of distilled water and wash by centrifugation.
3. Wash in 50 mM EDTA, pH 7.5. Resuspend in 40 ml of sorbitol (SEC) buffer.
4. These cells may be divided into aliquots, for example 8−10 ml, centrifuged and stored at −20°C. Step 3 may be continued at total volume or at aliquot volume, appropriately scaled.
5. For ~10 g of wet weight of cells (thick paste), resuspend the cells in 40 ml of SEC buffer. Add 0.5 ml of β-mercaptoethanol, incubate for 15 min at 30°C or room temperature.
6. Centrifuge, resuspend in 40 ml of SEC buffer, add 2−5 ml of Zymolyase 60 000 (Miles Laboratories) (1 mg/ml freshly-prepared in TEN buffer) and incubate at 30°C (shaker or occasional shaking) for 1 h. (Alternatives for steps 5 and 6 are incubation with β-mercaptoethanol and enzyme simultaneously and substitution for Zymolyase of Glusulase, Helicase, Novozyme 234 or Lyticase. However, the extent of DNase activity of these enzyme preparations should be examined).

   The 1 h period may be shortened for certain strains but may require to be prolonged, occasionally considerably, for strains more resistant to spheroplasting. The rate of spheroplasting can be assayed by mixing a drop of culture suspension in 1 ml of water and adding a drop of 10% SDS solution. Clearing of the suspension indicates a high level of spheroplast conversion.
7. Centrifuge the spheroplasts more gently than cells (3000 g for 5 min at 4°C) and wash by centrifugation in SEC buffer. Spheroplasts may be stored at −20°C for several days.
8. Resuspend the spheroplasts in 18 ml of TE buffer, add 1−2 mg of freshly-prepared solution of either proteinase K or pronase (in TEN buffer), add 2 ml of 10% SDS and incubate at 37°C for 3 h with occasional gentle shaking.
9. Heat this lysate at 65°C for 30 min and cool to room temperature.
10. Add an equal volume of phenol/chloroform (1:1) and mix gently until a homogenous white emulsion is formed.
11. Separate the phases by centrifugation at 13 000 g for 10 min at 4°C and transfer the upper (aqueous) phase by wide-bore pipette to a clean, large centrifuge tube or bottle. Repeat steps 10 and 11 with chloroform/n-amyl alcohol (24:1).
12. Add 2 vol of cold absolute alcohol (−20°C) over the aqueous phase, leave on ice for 10−15 min with occasional swirling (or leave overnight at −20°C).
13. Spool the white precipitate onto a glass rod, dissolve in 18 ml of 1/10 diluted SSC and add 2 ml of 10 × SSC.
14. Add 0.1 ml of 10 mg/ml boiled RNase A buffer and incubate at 37°C for 1 h.
15. Repeat steps 10−12, spool DNA into 2 ml of TEN buffer and store over 0.1 ml of chloroform at 4°C. For highest purity, free from RNA, it may also be necessary to repeat step 14.
16. Estimate concentration and purity by A_{260}/A_{280} ratio.

YEP−glucose = yeast extract−peptone−glucose medium: see Appendix I.
SEC, SP, SSC, TE, TEN: see Appendix IV.

integrating plasmid), YEp (yeast episomal plasmid), YRp (yeast replicating plasmid) and YCp (yeast centromeric plasmid). For maximal expression in yeast of genes inserted into plasmids, several 'expression vectors' have been constructed. These include particular yeast regulatory sequences to optimise transcription and translation of the inserted gene, which in many cases is heterologous (of non-yeast origin), such as human.

Diagrams of a commonly used cloning vector, YEp13 (14) and a recently developed expression vector (15) are shown in *Figure 1*.

The procedure for preparation of larger (mg) quantities of highly purified plasmid DNA is detailed in *Table 3*. Such DNA preparations are required for extensive restriction analysis, nick-translation preceding DNA strand hybridizations and DNA sequencing. Often it is necessary to isolate plasmid DNA from several or numerous recombinant

Table 2. Total yeast DNA (limited purity).

1. Grow the cells (shaker) in 20 ml of YEP−glucose at 30°C to $2−5 \times 10^7$/ml (late log phase). Harvest in one or more 5 ml aliquots by centrifugation and resuspend the cells in 1 ml of 1 M sorbitol.
2. Spin down the cells in a microcentrifuge tube (0.5−1 ml) and resuspend in 0.5 ml of SP buffer.
3. Add 2 µl of β-mercaptoethanol. Mix and stand for 10 min.
4. Add 50 units of Glusulase (or 0.1 ml of Zymolyase, 0.5 mg/ml), incubate, with occasional shaking, at 30°C for 30 min. Some strains may require a longer period.
5. Centrifuge, wash in 1 M sorbitol and resuspend in 0.5 ml of 50 mM EDTA, pH 8.5.
6. Add 20 µl of 10% SDS (1 µl of diethyloxydiformate may also be added at this point).
7. Heat for 15 min at 70°C in a water bath.
8. Transfer the tube to ice, add 100 µl of cold 5 M potassium acetate solution, gently mix and stand in ice for 1 h.
9. Centrifuge (13 000 g) for 10 min, transfer supernatant to a fresh microcentrifuge tube and add ethanol with gentle mixing to fill the tube.
10. Centrifuge for 10 min, drain off liquid and vacuum-dry briefly.
11. Dissolve the DNA in 0.5 ml of 1 mM EDTA, 10 mM Tris−HCl, pH 7.5 and 10 µg/ml boiled RNase A.
12. Stand 3 h at room temperature, perform three (or more) phenol extractions (phenol/chloroform 1:1, as *Table 1*).
13. Obtain the DNA by ethanol precipitation (as *Table 1*), pellet by microcentrifugation for 10 min, dry, dissolve the DNA in 50 µl of 1 mM EDTA, 20 mM Tris−HCl, pH 7.5.

YEP−glucose = yeast extract−peptone−glucose medium: see Appendix I.
SP: see Appendix IV.

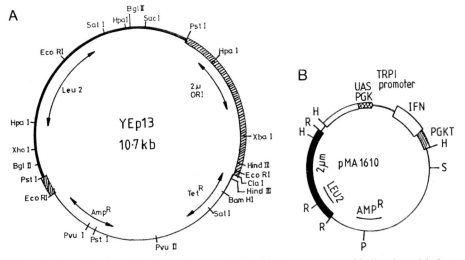

Figure 1. Examples of yeast cloning vectors. **A.** YEp13, DNA sequences are: thin line, bacterial, from pBR322; thick solid line, yeast, from pYE*leu10* (*LEU2* gene); thick shaded line, yeast from 2 µ plasmid. **B.** pMA1610, illustrating *PGK-TRPI* hybrid promoter, DNA sequences are: thin line, bacterial, from pBR322; thick solid line, yeast, 2 µ − *LEU2* fragment; thick open line, *TRPI* flanking sequence; thick shaded lines, *PGK* upstream activator sequence and terminator sequence; open box, human interferon α-2 coding sequence. Restriction sites are R, *Eco*RI; H, *Hind*III; P, *Pst*I; S, *Sal*I.

colonies, however, in which case faster, less laborious methods are required. These preparations ('rapid plasmid preparations') of small quantities of partially purified plasmid DNA are sufficient to confirm presence of the plasmid, estimate its approximate size and carry out limited restriction analysis. The procedures for rapid plasmid preparations from *E. coli* and from yeast are given in *Tables 4* and *5* respectively.

Table 3. Large-scale plasmid[a] preparation from *Escherichia coli*.

1. Grow plasmid-bearing *E. coli* strain overnight, with shaking, in 10 ml of L broth + 50 μg/ml ampicillin (or appropriate antibiotic) at 37°C.
2. Inoculate this culture into 50 ml[b] of the same broth, incubate (shaker) at 37°C to $2-4 \times 10^8$ cells/ml.
3. Add 200 μg/ml (final concentration) chloramphenicol, shake further $16-20$ h at 37°C.
4. Harvest by centrifugation (10 000 g for 10 min) in two 250 ml centrifuge bottles, resuspend in 10 ml of lysis buffer (5 mg/ml lysozyme) and transfer the total volume to a 100 ml[b] centrifuge tube.
5. Stand in ice for 30 min, add 20 ml of alkaline 10% SDS pH 9.0 and leave in ice for a further 5 min.
6. Add 15 ml of 3 M sodium acetate pH 4.8, leave on ice for 60 min with occasional swirling.
7. Centrifuge at 30 000 g for 20 min at 4°C, carefully pour off supernatant and divide into two equal volumes in 100 ml[b] centrifuge tubes.
8. Add to each tube two volumes cold (-20°C) absolute ethanol, stand several hours, or overnight, at -20°C.
9. Centrifuge at 30 000 g for 20 min at 4°C, drain the pellet and dry in vacuum desiccator.
10. Gently resuspend both pellets in total of 5 ml of TE buffer.
11. Store as crude DNA in 1.5 ml Eppendorf tubes or further purify by caesium chloride/ethidium bromide gradient centrifugation, as in steps $12-17$.
12. To 5 ml of DNA/TE buffer add $CaCl_2$ (~ 5 g) to reach a density of 1.71, add approximately 0.35 ml of 10 mg/ml ethidium bromide to give a final density of 1.59.
13. Add ~ 4.5 ml of mixture to each of two 'ultraclear' plastic centrifuge tubes, balance carefully and spin in an ultracentrifuge for 48 h (Beckman 50T: rotor, 45 000 r.p.m.) or longer at lower r.p.m.
14. View the tubes with ultraviolet transilluminator, remove the lower band by piercing side of tube with a 2 ml disposable syringe, 21-gauge needle, and transfer to a 1.5 ml Eppendorf tube (foil wrapped).
15. Remove ethidium bromide with isoamyl or isopropanol alcohol by thoroughly mixing with equal volumes, withdrawing the aqueous layer (Pasteur pipette) to a clean Eppendorf tube, repeating until all colour is removed.
16. Dilute the clean aqueous phase 5-fold with TE buffer, divide into 0.5 ml aliquots, add 1 ml cold (-20°C) absolute ethanol and stand for several hours or overnight at -20°C.
17. Centrifuge at 12 000 g (microcentrifuge) for 2 min, *partly* dry pellet, and resuspend in 0.5 ml of TE buffer.

[a]Amplifiable, e.g. YEp13.
[b]Vary volumes according to type of centrifuge available. Modified method of refs 16 and 17.
L,TE: see Appendix IV.

Table 4. Rapid plasmid preparation from *Escherichia coli*.

1. Grow, with shaking, cells of plasmid-bearing *E. coli* in 10 ml of L broth at 37°C to stationary phase ($\sim 10^9$ cells/ml).
2. Centrifuge (e.g. 4000 r.p.m. for $5-10$ min), resuspend in 2 ml of SET buffer and transfer 1 ml to a 1.5 ml Eppendorf tube.
3. Centrifuge in a microcentrifuge at 6500 r.p.m. for 3 min, then resuspend the pellet in 150 μl of SET.
4. Add 20 μl of RNase buffer, mix, add 350 μl of lytic mixture; vortex until the solution is clear.
5. Stand in ice for $10-30$ min, add 250 μl of sodium acetate buffer and leave in ice for a further 30 min.
6. Spin in a microcentrifuge at 13 000 r.p.m. for 3.5 min, transfer the supernatant to another Eppendorf tube; avoid any disturbance of the pellet.
7. Add an equal volume of isopropanol at -20°C, mix. The solution should go cloudy. Standing at -20°C for up to 2 h may be necessary.
8. Centrifuge at 13 000 r.p.m. for 5 min, drain off isopropanol and wash with 70% ethanol.
9. Vacuum-dry for 10 min and dissolve pellet in 50 μl TE buffer (if for agarose gels) or restriction buffer (if for restriction analysis).

L, SET, TE: see Appendix IV.
Modified method of ref. 18.

Table 5. Plasmid preparation from yeast.

1. Grow, with shaking, cells in 50−100 ml of appropriate selective medium at 30°C to ∼5 × 10^7 cells/ml. Harvest by centrifugation (e.g. 40−50 ml polypropylene tubes).
2. Wash in distilled water, resuspend in 20 ml of pre-treatment buffer and stand at room temperature for 10 min.
3. Centrifuge, wash twice and resuspend in SCE buffer, to a cell density ≤10^8 cells/ml.
4. Add Zymolyase 60 000 to a final concentration of 20 µg/ml, incubate at 37°C for 25 min and pellet the spheroplasts by gentle centrifugation (e.g. 1500 r.p.m. for 5 min).
5. Gently resuspend in 0.5 ml of ST buffer, add 9.5 ml of lysis buffer dropwise while magnetic-stirring (100 r.p.m.) for 90 sec.
6. Incubate at 37°C for 25 min, add 2 M Tris−HCl, pH 7.0 (with magnetic stirring) to bring pH to 8.5−8.9.
7. Add NaCl crystals to a final concentration of 3% (w/v), stand at room temperature for 30 min.
8. Add equal volume of NaCl-saturated phenol, stirring for 10 sec at 300 r.p.m. and a further 2 min at 100 r.p.m.
9. Centrifuge to separate the phases, remove the aqueous phase, add 2 vol of cold 95% ethanol.
10. Stand at −20°C for 2 h (or longer); separate the DNA by centrifugation (e.g. 11 000 *g* for 30 min).
11. Wash in 10 ml of cold 70% ethanol, drain, vacuum-dry, dissolve the DNA in 50 µl of TE buffer.

SCE, ST: see Appendix IV.
Modified method of ref. 19.

3. GENOMIC DNA LIBRARIES

Such libraries or gene banks consist of a large number of *E. coli* clones, each of which comprises a population of cells bearing a particular recombinant plasmid. The whole population of clones is selected following transformation of *E. coli* with *in vitro* recombinant plasmid molecules constructed in such a way that there is a high probability that each random fragment of the yeast chromosomal DNA is included as an insert in the vector molecule. There are numerous advantages to the *E. coli* host system, including relatively efficient transformation frequencies, easy amplification of plasmid molecules per bacterial cell, simple selection for cells bearing recombinant plasmids and lack of homology between most bacterial and yeast gene sequences. The latter greatly assists the screening of random yeast sequences by hybridization with a particular cloned sequence (probe). Genomic libraries of several strains of *Sacch. cerevisiae*, both laboratory and industrial, have been constructed but the gene banks, constructed by Nasmyth and Reed (20) and Nasmyth and Tatchell (21) of strain AB320 have been extensively used to clone many yeast genes. The genotype of this diploid (HO) laboratory strain is as follows: *MATa/MATαHO/HO ade2/ade2 lys2/lys2 trp5/trp5 leu2/leu2 can1/can1 met4/met4 urax/urax* (x = 1 and/or 3) and the most widely used bank is in plasmid YEp13 (21).

The construction of (genomic) gene libraries involves:

(i)   Extraction of genomic DNA.
(ii) Fragmentation of this DNA by a restriction enzyme to fragments of suitable size.
(iii) Frequently, selection of fragments within a suitable size range.
(iv) Ligation of this population of fragments into suitably cut (restricted) plasmid molecules.
(v)   Transformation of *E. coli* cells.
(vi) Selection of *E. coli* colonies composed of cells bearing recombinant plasmids, i.e. those with random inserts of genomic fragments.

111

In the procedure given in *Table 6, E. coli* cells transformed by YEp13 will be ampicillin-resistant and those carrying recombinant plasmids (genomic fragments inserted into the *tet* gene) will also be tetracycline-sensitive. This is an example of selection by 'insertional inactivation'. The same principle, with the same plasmid, has been used to construct genomic libraries of a flocculent, brewery-derived strain, ABXL-1D (22), and of *Schwanniomyces occidentalis* NCYC 953 (23).

The total number of clones required to give a high probability that the gene library contains a specific gene sequence can be calculated from the formula of Clark *et al.* (24):

$$N = \frac{\ln (1 - p)}{\ln (1 - \frac{i}{g})}$$

where N = number of recombinant clones, p = probability that the bank will contain a specific sequence, i = average insert size and g = total genome size. Thus, assuming a genome size of 9.9×10^9 daltons (25), for an average insert size of 9 kb (5.94×10^6 daltons), the number of clones required for probabilities 0.9, 0.95, 0.99 and 0.999 are approximately 4000, 5000, 8000 and 12 000, respectively. The number of recombinant transformants required in the bank as a function of given probabilities is shown in *Figure 2*. In the example of the gene bank of strain AB320 in YEp13 (21), the total pool of recombinant plasmid was approximately 5×10^4. The procedures for construction of a genomic library are listed in *Table 6*.

Prior to screening library clones for a particular gene clone, it is wise to perform a preliminary physical characterization of the bank. Plasmids are isolated from a random sample, normally 20 – 30, of recombinant *E. coli* colonies. The approximate size of the genomic inserts and limited restriction maps of the inserts are determined. These inserts should fall within the desired range and should be unique genomic fragments (according to their restriction sites).

Table 6. Construction of a genomic DNA library (e.g. in YEp13).

1. Extract high molecular weight genomic DNA from the donor strain (as *Table 1*)[a].

Partial Sau3 digestion (27)

2. Mix 33 μl of DNA (10 μg) with 15 μl of 10 × MS buffer and 102 μl of distilled water.
3. Dispense 30 μl into Eppendorf tube no. 1 and 15 μl into tubes nos 2 – 8 and retain remaining volume as control.
4. Stand the tubes in ice, add 5 μl of *Sau*3 to tube no. 1 and mix.
5. Transfer 15 μl of the reaction mixture to tube no. 2, mix; repeat 2-fold dilutions through to tube no. 8.
6. Incubate all the tubes for 1 h at 37°C.
7. Transfer the tubes to ice to stop reactions and add EDTA to a final concentration of 20 mM.
8. Transfer 15 μl from each tube to a set of fresh tubes; add 12 μl of 50% glycerol and 3 μl of 300 mM EDTA, pH 8.0.
9. Run samples on 0.6% agarose – ethidium bromide gel. Identify which gives fragments in the range 5 – 15 kb.
10. Use optimum small-scale result as conditions for scale-up of 350 μl containing 25 μg of DNA.
11. After digestion, remove 7.5 μl aliquots, add 7.5 μl of 50% glycerol and examine on 0.6% agarose gel.
12. Add 380 μl of phenol/chloroform to the remainder of samples, mix and spin for 1 min in microcentrifuge.
13. Remove the aqueous layer, add 100 μl of TE buffer, add 1 ml of absolute ethanol at −20°C and add 5 μl of 5 M NaCl.
14. Stand at −20°C for 2 h or longer, centrifuge and resuspend in 64 μl of TES buffer. Examine 15 μl sample on 0.6% agarose gel.

Sizing fragments by sucrose gradient centrifugation (8)

15. Following scale-up digestion of DNA, add the remaining 62.5 μl vol to 9 μl of 40% sucrose to give a final concentration of 5% sucrose.
16. Load onto a 10−40% sucrose gradient (10 steps with sucrose/TES).
17. Ultracentrifuge at 90 000 g for 12−13 h (e.g. Beckman SW 50.1 rotor).
18. Elute the gradient by syphoning ten aliquots of ~0.5 ml, add an equal volume of ethanol at −20°C to each and stand at −20°C overnight.
19. Spin in microcentrifuge at 13 000 r.p.m. for 2 min and resuspend in 75 μl of TE buffer.
20. Examine 5 μl of each sample on agarose gel.
21. Pool gradient fractions with fragments of 5−15 kb; dialyse against four changes of TE buffer at 4°C for 12−16 h.
22. Extract the dialysed DNA several times with an equal volume of isobutanol, reducing the volume to 5−8 ml.
23. Precipitate the DNA with ethanol at −20°C; dissolve in TE buffer to give a final concentration of 300−500 μg/ml.
24. Digest the plasmid DNA (e.g. YEp13) with *Bam*HI (*Bam*HI and *Sau*3 produce complementary single strand ends) (see *Table 14*).
25. Treat the digested plasmid with alkaline phosphatase (see *Table 15*).

Ligation of genomic fragments and plasmid DNA.

The following is based upon favourable results obtained with DNA concentrations of 10−25 μg/ml and molar ratios of insert to plasmid DNA between 2:1 and 5:1. Ligation reactions of total volume 20 μl contain 2 μl of ligation buffer ($\times$10), 2 μl of DTT ($\times$10), 2 μl of ATP ($\times$10), 0.1−1.0 units of T4 DNA ligase, 200−300 ng of DNA (molar ratio 2:1−5:1); remaining volume distilled water.

26. Mix the DNA and ligation buffer in appropriate volumes.
27. Heat at 56°C for 2 min and transfer to ice.
28. Mix a ligase cocktail of DTT, ATP and ligase, add appropriate volumes to the DNA/buffer tubes, mix well.
29. Stand in cold room (4−10°C) overnight.
30. Test 1−2 μl samples in transformations.
31. Heat at 65°C for 5 min, store at −20°C.
32. Select *E. coli* transformants with recombinant DNA (see *Table 7*).

---

[a]As an alternative to extensive RNase treatment to obtain RNA-free DNA, the following procedure has been shown to be effective and faster. Elute the nucleic acid in TEN buffer through a Biorad Biogel A15m 25 × 1 cm column with a flow rate of TEN buffer of 25 ml/h. Collect 1 ml fractions, assay 5 μl aliquots on surface of agarose plates containing 0.5 μg/ml of ethidium bromide. Identify DNA by UV irradiation of plate (RNA fluorescence is distinguished by diffuse appearance).

The storage of gene libraries in more than one form is recommended, such as nitrocellulose filters of colonies (8) at −70°C; total pool of cells and individual, or groups of, clones (e.g. in microtitre wells) in L broth + 15% glycerol at −20°C. In addition, total plasmid DNA may be stored at 4°C.

L, MS, TE, TEN, TES: see Appendix IV.

4. TRANSFORMATION

Transformation (as used in recombinant DNA technology) requires uptake, stabilization and replication of plasmid DNA molecules in host cells and expression in these cells of at least the marker gene(s) present on the plasmid. Plasmids, complete with particular marker genes, will be selected or constructed appropriately to ensure replication and expression but the first stage of plasmid uptake requires particular cell ('competence') and environmental conditions. The procedures for transformation of *E. coli* and yeast

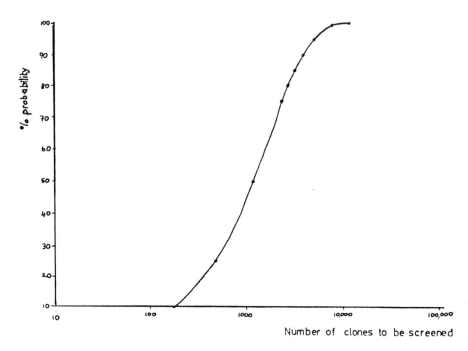

Number of clones to be screened

Figure 2. Probability (percentage) of detecting a particular gene clone per total number of clones screened for a gene library with average insert size of 9 kb (from ref. 26).

Table 7. Transformation of *Escherichia coli*.

1. Inoculate 50 ml of L broth, incubate with shaking at 37°C until A_{600} is 0.2−0.25 (late log phase).
2. Chill the culture on ice for 10 min.
3. Transfer 1.5 ml aliquots to sterile microcentrifuge tubes and pellet in the microcentrifuge (e.g. 6500 r.p.m. for 5 min).
4. Resuspend the cells in 0.75 ml of cold 50 mM Tris−HCl/10 mM $CaCl_2$, pH 8.0. Keep tubes at 0°C for 45 min; pellet cells by microcentrifuge.
5. *Gently* resuspend cells in 0.1 ml of 10 mM $CaCl_2$ solution at 0°C.
6. Add 10−20 ng plasmid DNA and keep the tubes at 0°C for 45 min.
7. Transfer the tubes to a 42°C waterbath for 2 min and return to 0°C for 3 min.
8. Add 1 ml of L broth (at 37°C) and incubate, with shaking, at 37°C for 1 h.
9. Gently spread 100 µl, 10 µl or other appropriate dilutions onto selective agar (e.g. L + amp), incubate at 37°C overnight.

Competent cells (step 5) may be stored at 0°C overnight or at −70°C for longer periods by addition of 15% glycerol to the $CaCl_2$ solution. When required, suspensions are thawed, spun and resuspended in cold $CaCl_2$ (step 5).

Volumes may be scaled-up for larger numbers of plates, if necessary. This procedure has given transformation efficiencies of ~ 10^7/µg DNA with *E. coli* strains HB101, C600 and C1400, using plasmid YEp13 (23).
L broth: see Appendix IV
L + amp: L agar + 100 µg/ml ampicillin.

are given in *Tables 7, 8* and *9* respectively. In the case of yeast the different methods of transforming spheroplasts (*Table 8*) and whole cells (*Table 9*) are both routinely used. The general experience is that the preferred method, in terms of frequency of

Table 8. Transformation of yeast (spheroplasts).

1. Grow the yeast culture with shaking in 100 ml of YEP−glucose at 30°C to ~2 × 10^7 cells/ml.
2. Harvest by centrifugation (5000 *g* for 5 min) and wash in distilled water.
3. Resuspend in 10 ml of SED, add 25 μl of β-mercaptoethanol and incubate 10 min at 30°C.
4. Centrifuge, wash in 10 ml of 1 M sorbitol and resuspend in 10 ml of SCE.
5. Add 0.1 ml of Glusulase (or Zymolyase 60 000, 0.5 mg/ml), incbuate at 30°C with gentle or occasional shaking. Monitor for spheroplast conversion as in *Table 1*.
6. Centrifuge at 2500 *g* for 3 min, wash twice in 10 ml of 1 M sorbitol and wash once in STC.
7. Resuspend in 0.5 ml of STC, divide into 0.1 ml aliquots in disposable plastic tubes and add 10 μg of DNA to 0.1 ml spheroplasts (tubes of spheroplast suspensions may be stored at 0−4°C at this point). Use one tube without addition of DNA as control.
8. Stand at room temperature for 10 min, add 1 ml of 10% PEG and gently mix by hand.
9. Stand at room temperature for 15 min, centrifuge (2000−3000 *g*) and resuspend in 150 μl of SOS.
10. Incubate at 30°C for 20 min (spheroplasts may now be plated, or held at 4°C, or given 30 min−1 h cold shock at 0°C; the latter effects on transformation frequencies are strain-dependent).
11. Add 0.1 ml of transformation mixture to 7−10 ml of regeneration agar in plastic tubes held at 45−50°C in water bath. Mix briefly and immediately pour over plates of selective media.
12. Incubate plates for 3−5 days.

YEP−glucose = yeast extract−peptone−glucose medium: see Appendix I.
SCE, SED, SOS, STC, regeneration agar: see Appendix IV.

Table 9. Transformation of yeast (lithium-treated cells).

1. Inoculate 50 ml of YEP−glucose broth with appropriate number of freshly-grown cells (depending on growth rate of strain), incubate, with shaking, overnight at 30°C.
2. Harvest the cells at 2−5 × 10^7/ml (late log phase) by centrifugation.
3. Wash the cells twice in 10 ml of TE buffer.
4. Resuspend the cells in 5 ml of LA solution, incubate at 30°C (shaker) for 1 h.
5. Dispense 0.3 ml aliquots into sterile microcentrifuge tubes and add 50 ng of plasmid DNA.
6. Add 0.7 ml of PEG solution, mix by inverting tubes several times and incubate at 30°C for 2 h.
7. Transfer the tubes to 42°C waterbath for 15 min.
8. Cool to room temperature, spread 100 μl, or dilutions, onto selective agar (e.g. *leu*⁻ plates for plasmid YEp13).
9. Incubate the plates for 3−5 days at 30°C.

This procedure has been observed to produce transformation frequencies of ~3−4 × 10^3/μg DNA with YEp13 and strains such as AH22 and DBY746 (23).
YEP−glucose = yeast extract−peptone−glucose medium: see Appendix I.
LA, TE: see Appendix IV.

transformants obtained, depends on the specific recipient strain to be transformed. Wide variations among strains in the efficiency of transformation by either method have been observed, for example see ref. 28.

5. CLONING YEAST GENES

The first stage in cloning a particular gene is isolation of a colony of the required phenotype from among the population of transformants produced by recombinant plasmids from a DNA library hopefully including the gene in question. The method of selection of this colony will depend upon the nature of its phenotype but complementation of a recessive mutation of the desired gene frequently offers the best approach. Isolation of the DNA insert sequence of the recombinant plasmid responsible

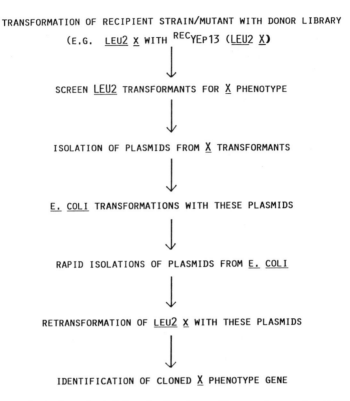

TRANSFORMATION OF RECIPIENT STRAIN/MUTANT WITH DONOR LIBRARY

(E.G. <u>LEU2</u> <u>X</u> WITH ^{REC}YEP13 (<u>LEU2</u> <u>X</u>)

↓

SCREEN <u>LEU2</u> TRANSFORMANTS FOR <u>X</u> PHENOTYPE

↓

ISOLATION OF PLASMIDS FROM <u>X</u> TRANSFORMANTS

↓

<u>E. COLI</u> TRANSFORMATIONS WITH THESE PLASMIDS

↓

RAPID ISOLATIONS OF PLASMIDS FROM <u>E. COLI</u>

↓

RETRANSFORMATION OF <u>LEU2</u> <u>X</u> WITH THESE PLASMIDS

↓

IDENTIFICATION OF CLONED <u>X</u> PHENOTYPE GENE

Figure 3. Generalized scheme for isolation of a cloned gene. The example uses plasmid YEp13 (*LEU2*) and cloned gene *X*; the recipient strain for transformation is, therefore, *leu2 x* (recessive mutant genes).

for complementation, however, does not by itself prove that the correct gene has been cloned. It will be necessary to confirm which gene has been cloned either by identification of its product (generally protein) or its correct location on the genetic map (29). A generalized scheme of obtaining a cloned gene is illustrated in Figure 3. The plasmid YEp13, with *LEU2* as its yeast marker gene, has been used as an example. That the colony is indeed a transformant is generally quickly tested by simultaneous loss of the marker and selected phenotypes under conditions of non-selective growth (at least for unstable plasmids, such as YEp13). Confirmation and isolation of the plasmid conferring the selected phenotype is obtained by (i) extraction of plasmid DNA from the yeast transformant culture (grown in selective medium to maximise plasmid-bearing cells), (ii) transformation of *E. coli* with this plasmid and its amplification in *E. coli*, and (iii) confirmation that this amplified plasmid transforms yeast to the required phenotye. The size of the insert may be estimated by removing it from the plasmid using the appropriate restriction enzyme (e.g. *Bam*HI in the example of the gene library discussed). Usually a restriction map of the insert is constructed and frequently the insert is sub-cloned, i.e. isolation of the smallest restriction fragment (re-inserted in the plasmid) which still confers the particular phenotype. Mapping of the cloned sequence can be achieved by either (i) its insertion into an integrating (YIp) vector, transformation and selection for its yeast marker gene and mapping of the new position of this gene by conventional methods, or (ii) by 'gene disruption' of the cloned gene in its original plasmid by insertion of a selectable marker gene into the cloned sequence, transformation for the selectable marker and its subsequent mapping (30). If the cloned gene is identical with the particular donor gene being sought, then the map positions of the former and latter should coincide.

The average frequency for many specific inserts from a gene bank such as described, for example (21), appears to be $0.5-1.0 \times 10^{-3}$ of total transformants, although much lower frequencies of $1-2 \times 10^{-4}$ have been reported. If no clone of the particular gene is obtained within a population of 10^4 transformants, there is a high probability that a recombinant plasmid containing this gene insert is not included in the library constructed. Occasionally, an alternative reason is that the gene dosage effect of the presence of the gene on a multi-copy plasmid (such as YEp13) is lethal to the cell. In a few cases, clones of such genes have been isolated from gene banks constructed in single-copy plasmids (such as CEN). In certain cases in *Saccharomyces*, but particularly where genomic DNA is derived from organisms other than *Saccharomyces* (as from other fungi or higher eukaryotes), genomic DNA inserts may not be expressed in recipient strains because the donor DNA contains one or more introns. In such cases, gene libraries are constructed using cDNA.

6. HYBRIDIZATION OF NUCLEIC ACIDS

Another way to screen transformant colonies for detection of a particular cloned sequence is to utilize the hybridization of complementary single-stranded DNA or RNA molecules, i.e. formation of DNA/DNA or DNA/RNA duplexes. Genetic homology, and therefore genetic relatedness, can also be tested by the formation of hybrid duplex molecules (single-stranded DNA or RNA from different sources).

The general procedure is to transfer DNA or RNA fragments separated by gel electrophoresis (or colonies in the case of colony hybridizations) to membrane filters prior to hybridization reactions. Transfer of DNA is termed a Southern blot (31) and that of RNA is referred to as a Northern blot. Only the procedure for Southern blotting is described here. The original use of nitrocellulose 'papers' has been extended to include membranes of other materials, such as diazobenzyloxymethyl-paper (see ref. 32) and nylon. Although more expensive, the latter has a number of advantages for ease of handling, high sensitivity and re-use of the same blot with a different hybridizing (probe) DNA. The procedure detailed in *Table 10* is for the nylon membrane 'Gene Screen Plus' (Du Pont, NEN Products) (33). It describes the method of 'capillary blotting' (step 3 denatures DNA to single strands). Methods of 'electro-blotting' and 'DOT blotting' may also be used, where appropriate (33). A diagram illustrating the capillary action arrangement is given in *Figure 4*.

Hybridization between blot and probe nucleic acids is detected by labelling the latter and detecting it only if it becomes fixed to the blot (by hybridization). Generally DNA or RNA is labelled to high specific activity with ^{32}P and labelled duplex molecules are detected by autoradiography. An alternative, which is a relatively recent development and much less commonly used, is biotin-labelling and detection by dye reaction (34,35). The latter has the advantage of eliminating the difficulties, with risks, of working with high levels of radioactivity, but it is not yet clear if it can adequately substitute for

Table 10. Preparation of Southern blots.

1. Differentiate between sides A and B of the membrane by its curl when *dry*. Side A is convex and side B is concave.
2. If larger DNA fragments (>20 kb) are present, 'acid nick' to facilitate transfer. Incubate the gel in 0.25 M HCl for 15 min at room temperature.
3. Incubate the gel in 0.4 M NaOH, 0.6 M NaCl for 30 min at room temperature (gentle agitation on slow-speed tray shaker) Pyrex dishes are convenient for gel immersion.
4. Incubate (gentle agitation at room temperature) the gel in 1.5 M NaCl, 0.5 M Tris−HCl, pH 7.5 for 30 min.
5. *Wear vinyl gloves*. Cut the membrane to exact size of gel, using a sharp scalpel or razor blade, keeping the membrane between linear sheets. Mark side B (pencil).
6. Place the membrane onto surface of deionized water in a tray, allow it to wet by capillary action (the membrane may be handled by flat-end forceps).
7. Place the membrane onto 10 × SSC solution; soak for 15 min.
8. Place a sponge (surface size larger than gel) into dish of 10 × SSC, with level below top of sponge.
9. Cut several pieces of filter paper (Whatman 3MM) to the size of the sponge, wet two pieces in 10 × SSC, place on sponge.
10. Place the gel on filter paper and gel spacers along each side of gel.
11. Carefully place side B of the membrane onto the gel, ensuring that no air bubbles are trapped between the gel and membrane.
12. Place five or six pieces of cut, dry filter paper on top of the membrane.
13. Place a 5−7 cm stack of absorbent paper towels (cut to approximate size of gel) on top of filter papers.
14. Lightly weight top of towels to press them down.
15. Allow capillary transfer from gel to membrane over 16−24 h, changing towels frequently (higher stack for overnight) and keeping 10 × SSC level topped-up.
16. Remove the towels and filter paper without disturbing membrane.
17. *Wearing gloves*, carefully remove the membrane from gel (flat forceps are useful); immerse in excess 0.4 M NaOH for 30−60 sec.
18. Transfer the membrane to excess 0.2 M Tris−HCl, pH 7.5, 2 × SSC.
19. Place the membrane (transferred DNA face up) on filter paper, allow to dry at room temperature.
20. When almost dry (curling), gently flatten by covering with another piece of filter paper and a light weight.

Dry membranes may be stored (in plastic bags or between filter papers for protection) at room temperature for many months.
SSC: see Appendix IV.

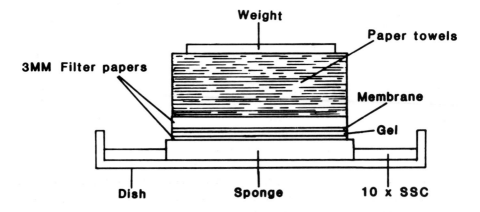

Figure 4. Diagram of Southern blot arrangement.

Table 11. Hybridization of blot DNA and probe[a] nucleic acid.

1. Pre-hybridize the membrane in 10 ml of PHB (see *Table 12*) in sealed plastic bag, with agitation, at 65°C for 15 min or longer.
2. Add 0.5 − 1.0 ml of denatured labelled probe DNA and denatured salmon sperm DNA (≥ 100 μg/ml). Re-seal bag, incubate with agitation at 65°C for 6 − 24 h. (Denature DNA in water by heating for 10 min at 90 − 100°C. Final concentration of probe DNA in bag ≤ 10 ng/ml, $1-4 \times 10^5$ d.p.m./ml.)
3. Remove the membrane, wash (a) twice with 100 ml of 2 × SSC at room temperature, with agitation, for 5 min, (b) twice with 200 ml of 2 × SSC, 1% SDS at 65°C, with agitation, for 30 min, (c) twice with 100 ml of 0.1 × SSC at room temperature, with agitation, for 30 min.
4. Place membrane (DNA face up) on filter paper, allow to dry at room temperature.
   N.B. Complete drying may result in *irreversible* binding of probe DNA, therefore do *NOT* dry completely if re-hybridization is intended.
5. Expose, develop autoradiograph.

[a]In this description, labelled by radioactive isotope (e.g. [32P]dNTP).
N.B. Labelled DNA has high specific activity. ALL PRECAUTIONS FOR HANDLING OF RADIOACTIVE MATERIALS MUST BE TAKEN.
SSC: see Appendix IV.

Table 12. Preparation of pre-hybridization buffer (PHB) (Table 11).

1. Prepare 10% SDS and 50% dextran sulphate solutions.
2. Add 1 ml of 10% SDS, 1 ml of 50% dextran sulphate, 7 ml of water (or 5 ml of formamide, 2 ml of water) to a centrifuge tube, mix by inversion.
3. Place the tube in water bath at 42°C (if with formamide) or 65°C (if without formamide) for 10 − 15 min.
4. Add 0.58 g NaCl, mix by inversion, return the tube to the waterbath for 10 − 15 min.
5. If buffering is desired, add Tris − HCl, pH 7.5 to a final concentration of 50 mM.
6. Pour the solution into a sealable bag containing the membrane, carefully remove any air bubbles, seal the bag.

radioactive isotope labelling with its proven record of success. Labelling is performed by 'nick translation', i.e. *in vitro* introduction of nicks into DNA and subsequent repair of the lesions with a mixture of labelled and unlabelled deoxynucleoside phosphates (dNTP). The labelling of nucleic acids, details of handling and disposal requirements are outside the scope of this chapter. Procedures are, however, readily available from several laboratory manuals (e.g. 7,8). The procedure for hybridization between Southern blot DNA (*Table 10*) and radioactive probe DNA is shown in *Tables 11* and *12* (33).

7. SEPARATION OF INTACT CHROMOSOMAL DNA MOLECULES

A recent development in molecular yeast genetics has been separation of very large DNA molecules (those of whole yeast chromosomes) by pulsed field gel electrophoresis (for a recent review, see ref. 36). This has allowed individual yeast chromosomes, or at least their DNA, to be viewed visually *in vitro*, by staining of their gel bands, and thereby constitutes a form of 'molecular cytology'. Because of the extensive linkage mapping and gene cloning in *Sacch. cerevisiae*, individual bands have been identified with specific chromosome (linkage groups) by hybridization with probes of mapped genes (37). There has therefore been a most satisfying merger of classical and molecular yeast genetics. Extraction of chromosomal DNA in intact molecules requires particular methods of preparation. The procedure is presented in *Table 13*.

Large DNA molecules, approximately 200 kb − >2 Mb, are separated from each

Table 13. Preparation of intact chromosomal DNA molecules.

---

1. Grow cells (with shaking) to late log phase ($1-5 \times 10^7$/ml) in $50-100$ ml of YEP−glucose at 30°C.
2. Wash twice in 50 mM EDTA, pH 7.5 at $0-5$°C.
3. Resuspend the pellets from $40-50$ ml of washed cells in $0.3-1.5$ ml (according to original cell density in range $1-5 \times 10^7$/ml) of 50 mM EDTA, pH 7.5, warm to 37°C.
4. Add 0.3 ml of cells to a 1.5 ml plastic cuvette[a], add 0.1 ml of solution I[b], then add 0.5 ml of low melting-point agarose (1%)[c] and mix rapidly by vigorous inversion.
5. Allow to solidify (room temperature or 4°C), place[d] the agarose block into a 5 cm diameter plastic Petri dish.
6. Cover the block with 4 ml of solution II[e] and incubate at 37°C overnight.
7. Remove solution II, wash in 0.5 M EDTA, pH 9.0, cover with 4 ml of solution III[f] and incubate at 50°C overnight[g].
8. Remove solution III, wash once with 0.5 M EDTA and cover with 0.5 M EDTA, pH 9.0, store at 4°C.

---

[a]Cut off the end of the cuvette, seal with a double layer of parafilm (cover the other end with parafilm when inverting).
[b]1 ml of SCE, pH 5.0; 50 μl of β-mercaptoethanol, 100 μg of Zymolyase 60 000.
[c]1% dissolved in 0.125 M EDTA, pH 7.5.
[d]Knock sharply to slide blocks from cuvette; or push gently with flat-ended object.
[e]0.45 M EDTA, pH 9.0; 10 mM Tris−HCl, pH 8.0; add 7.5% (v/v) β-mercaptoethanol.
[f]1% sodium-*N*-lauroyl sarcosinate, 1 mg/ml proteinase K dissolved in 0.45 M EDTA, pH 9.0, 10 mM Tris−HCl.
[g]Place a small Petri dish inside a larger dish with soaked tissue to maintain humidity.
YEP−glucose = yeast extract−peptone−glucose medium: see Appendix I.
SCE: see Appendix IV.

other by alterations in the direction of the electric field during agarose gel electrophoresis. For example, one system is by orthogonal-field-alternation (OFAGE) (38) and another by field-inversion (FIGE) (39). Most recently a sophisticated modification of OFAGE, referred to as Contour-clamped Homogeneous Electric Fields (CHEF) (40), has been developed. Details of the construction of an OFAGE system are described in reference 38 and a commercial model, named Pulsaphor (LKB) has now been marketed. The use of OFAGE (41,42) and FIGE (43) has allowed genetic characterization by karyotyping of both laboratory and industrial strains of *Saccharomyces* and of a variety of other yeasts. An example is illustrated in *Figure 5*.

8. BASIC METHODOLOGY

In this chapter, familiarity has been assumed with the most basic methodologies of molecular biology, such as gel electrophoresis/ethidium bromide staining and restriction enzyme digestion. These are described in the laboratory manuals already referred to, in particular detail in (7). A few examples and details, relevant to the preceding sections, are, however, added here. The conditions for digests with several common restriction enzymes are shown in *Table 14*. Typically, the DNA solution is mixed with distilled water in an Eppendorf tube to a volume of 18 μl. A 2 μl volume of the appropriate buffer (×10) is added and mixed by tapping the tube. One unit of enzyme is added and mixed by tapping. The tube is then incubated in a water bath at the required temperature.

Treatment for removal of phosphate groups on single-stranded DNA ends following restriction digestion with many endonucleases uses either bacterial alkaline phosphatase or calf intestine phosphatase. Procedures are given in *Table 15*.

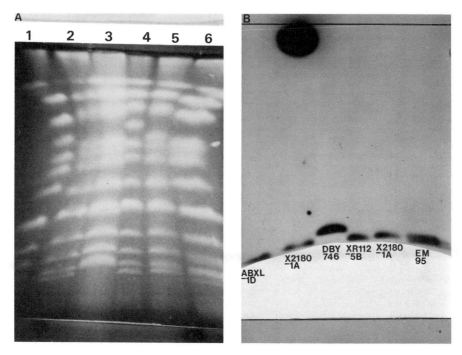

Figure 5. Example of an OFAGE gel (44). **A**. Gel stained with ethidium bromide. Band with the highest mobility is DNA of the smallest chromosome (I) of *Sacch. cerevisiae*. Strains are: **lane 1**, ABXL-1D; **lane 2**, X2180-1A; **lane 3**, DBY746; **lane 4**, XR112-5B; **lane 5**, X2180-1A; **lane 6**, EM93. The gel shows that strain DBY746 (a common transformation recipient strain) lacks the band identified with chromosome I of usual size. **B**. Autoradiograph of a Southern blot of gel hybridized with probe containing *PYK* gene (located on chromosome I). The band hybridizing with the probe in **lane 3** confirms the larger size of chromosome I in strain DBY 746. (The assistance of Dr A.Brown, Department of Genetics, University of Glasgow, in probing with *PYK* is gratefully acknowledged.)

Table 14. Conditions for several restriction digests.

Enzyme	Type, 10 × salt buffer	Incubation (1 h) temperature (°C)	Inactivation by 65°C for 15 min
*Bam*HI	Medium	37	−[a]
*Sau*3	Medium	37	−
*Eco*RI	High	37	+
*Hind*III	Medium	37−55	−
*Pst*I	Medium	21−37	+

[a]Heat-stable enzymes require restricted DNA to be extracted with phenol/chloroform.

Gels to be stored at 4°C should be wrapped in thin plastic such as Clingfilm. If gels are intended for hybridization procedures, then it is better to prepare Southern blots from them with as little storage time as possible.

9. ACKNOWLEDGEMENTS

I wish to thank Mr S.D.Finlayson, Dr J.M.Beaumont and Miss C.J.Daubney (Mrs Fleming) for their assistance with technical details.

121

Table 15. Procedure for phosphatase treatment.

1. Resuspend digested plasmid DNA in 10 mM Tris−HCl, pH 8.0, at concentration of ~250−300 ng/μl.
2. If BAP, add 1 unit, incubate tube at 65°C for 60 min.
3. If CIP, add 5 μl of 10 × CIP buffer, 29 μl of distilled water, 3 μl (0.5 unit) of CIP and incubate at 37°C for 30 min. Add a further 3 μl of CIP and incubate again for 30 min at 37°C. Add 40 μl of distilled water, 10 μl × STE buffer, 5 μl of 10% SDS and heat at 68°C for 15 min.
4. After either BAP or CIP treatment, extract twice with phenol/chloroform, precipitate with ethanol at −20°C and suspend the pellet in 5 μl of 10 mM Tris−HCl, pH 8.0 prior to ligation.

BAP = bacterial alkaline phosphatase.
CIP = calf intestinal phosphatase.
CIP buffer, STE buffer: see Appendix IV.

10. REFERENCES

1. Struhl,K., Cameron,J.R. and Davis,R.W. (1976) *Proc. Natl. Acad. Sci. USA,* **73**, 1471.
2. Ratzin,B. and Carbon,J. (1977) *Proc. Natl. Acad. Sci. USA,* **74**, 487.
3. Hinnen,A., Hicks,J.B. and Fink,G.R. (1978) *Proc. Natl. Acad. Sci. USA,* **75**, 1929.
4. Beggs,J.D. (1978) *Nature,* **275**, 104.
5. Rodriquez,R.L. and Tait,R. (1983) *Recombinant DNA Techniques.* Addison-Wesley, Reading, MA.
6. Wu,R., Grossman,L. and Moldave,K. (eds) (1983) *Methods in Enzymology.* Academic Press, New York, Vol. 101.
7. Dillon,J.R., Nasim,A. and Nestman,E.R. (eds) (1985) *Recombinant DNA Methodology.* J.Wiley, New York.
8. Maniatis,T., Fritsch,E.F. and Sambrook,J. (1982) *Molecular Cloning.* Cold Spring Harbor Laboratory, New York.
9. Cryer,D.R., Eccleshall,R. and Marmur,J. (1975) In *Methods in Cell Biology.* Prescott,D.M. (eds), Academic Press, New York, Vol. 12, p. 39.
10. Struhl,K., Stinchcomb,D.T., Scherer,S. and Davis,R.W. (1979) *Proc. Natl. Acad. Sci. USA,* **76**, 1035.
11. Botstein,D. and Davis,R.W. (1982) In *The Molecular Biology of the Yeast Saccharomyces, Metabolism and Gene Expression.* Strathern,J.M., Jones,E.W. and Broach,J.R. (eds), Cold Spring Harbor Laboratory, New York, p. 607.
12. Struhl,K. (1983) *Nature,* **305**, 391.
13. Gunge,N. (1983) *Annu. Rev. Microbiol.,* **37**, 253.
14. Broach,J.R., Strathern,J.H. and Hicks,J.B. (1979) *Gene,* **8**, 121.
15. Ogden.J.E., Stanway,C., Kim,S., Mellor,J., Kingsman,A.J. and Kingsman,S. (1987) *Mol. Cell. Biol.,* in press.
16. Clewell,D.B. and Helinski,D.R. (1969) *Proc. Natl. Acad. Sci. USA,* **62**, 1159.
17. Goebel,W. (1970) *Eur. J. Biochem.,* **15**, 311.
18. Kado,C.I. and Liu,S.T. (1981) *J. Bacteriol.,* **145**, 1365.
19. De Venish,R.J. and Newlon,C.S. (1982) *Gene,* **18**, 277.
20. Nasmyth,K.A. and Reed,S.I. (1980) *Proc. Natl. Acad. Sci. USA,* **77**, 2119.
21. Nasmyth,K.A. and Tatchell,B.D. (1980) *Cell,* **19**, 753.
22. Hodgson,J.A., Daubney,C.J., Berry,D.R. and Johnston,J.R., unpublished results.
23. Beaumont,J.M., Merrill,C., Finlayson,S.D., Johnston,J.R. and Berry,D.R., unpublished results.
24. Clarke,L., Hitzeman,R. and Carbon,J. (1979) In *Methods in Enzymology.* Wu,R. (ed.), Academic Press, New York, Vol. 68, p. 436.
25. Petes,T.D. (1980) *Annu. Rev. Biochem.,* **49**, 845.
26. Daubney,C.J. (1985) MSc Thesis, University of Strathclyde.
27. Maniatis,T., Hardison,R.C,. Lacy,E., Lauer,J., O'Connell,C., Quon,D., Sim,D.K. and Efstratiadis,A. (1978) *Cell,* **15**, 687.
28. Johnston,J.R., Hilger,F. and Mortimer,R.K. (1981) *Gene,* **16**, 325.
29. Mortimer,R.K. and Schild,D. (1985) *Microb. Rev.,* **49**, 181.
30. Rothstein,R.J. (1983) In *Methods in Enzymology.* Wu,R., Grossman,L., Moldave,K. (eds), Academic Press, New York, Vol. 101, p. 202.
31. Southern,E.M. (1975) *J. Mol. Biol.,* **98**, 503.
32. Frei,E., Levy,A., Gowland,P. and Noll,M. (1983) In *Methods in Enzymology.* Wu,R., Grossman,L. and Moldave,K. (eds), Academic Press, New York, Vol. 100, p. 309.

33. Du Pont (NEN Products) (1985) Catalog No. NEF-976.
34. Leary,J.J., Brigati,D.J. and Ward,D.C. (1983) *Proc. Natl. Acad. Sci. USA,* **80**, 4045.
35. Bethesda Research Laboratories (1985) Catalogue No. 8329 SA.
36. Anand,R. (1986) *Trends Genet.,* **2**, 278.
37. Carle,G.F. and Olson,M.V. (1985) *Proc. Natl. Acad. Sci. USA,* **82**, 3756.
38. Carle,G.F. and Olson,M.V. (1984) *Nucleic Acids Res.,* **12**, 5647.
39. Carle,G.F., Frank,M. and Olson,M.V. (1986) *Science,* **232**, 65.
40. Chu,G., Vollrath,D. and Davis,R.W. (1986) *Science,* **234**, 806.
41. Johnston,J.R. and Mortimer,R.K. (1986) *Int. J. Syst. Bacteriol.,* **36**, 569.
42. De Jonge,P., De Jonge,F.C.M., Meijers,R., Steensma,H.Y. and Scheffers,W.A. (1986) *Yeast,* **2**, 193.
43. Johnston,J.R., Contopoulou,R.C. and Mortimer,R.K. (1987), *Yeast,* **3**, 207.
44. De Zoysa,P. and Johnston,J.R., unpublished results.

CHAPTER 6

Isolation of yeast nuclei and chromatin for studies of transcription-related processes

D.LOHR

1. INTRODUCTION

The purpose of this chapter is to describe isolation protocols which have been used to prepare yeast nuclei and chromatin for studies of chromosomal structure (both the bulk genome and specific genes) and to carry out generalized and gene-specific transcription. Isolated nuclei could also be used for such purposes as subcellular localization studies, as a source for transcription factors, gene regulatory factors, nuclear enzymes, or for isolating gene-specific chromatin or as source material for isolating components of chromosomal architecture (scaffold, matrix, centromeres, etc.).

Traditionally, yeast nuclei have been considered to be difficult to isolate. Certainly in comparison with isolation of nuclei from the commonly used animal cell systems, yeast nuclear isolation presents unique problems: yeasts have a cell wall; yeast nuclei constitute a much smaller fraction of the cellular material and thus substantial purification is required to get rid of non-nuclear contaminants; there can also be proteolysis problems. However, these problems are soluble to a greater or lesser degree. The unique experimental possibilities presented by the sophisticated genetics (traditional and molecular) and varied life style (log, stationary, sporulating cells) more than justify the extra effort required to obtain yeast nuclei.

The discussion in this chapter will be limited to protocols with which I have personal experience. They have been used to isolate nuclei in a variety of circumstances: from a very large number of different haploid or diploid yeast strains, with doubling times from 75 min to 3 h; from cells grown in glucose, galactose, glycerol, ethanol or acetate carbon sources; from cells grown to early or mid-log phase, from (early) stationary phase and from sporulating cells. This wide applicability constitutes an important recommendation for these protocols.

In considering the protocols, keep in mind your purpose for isolating nuclei or chromatin. Although the procedures described below produce material which has proved to be satisfactory for our purposes (chromosome structural analysis and nuclear transcription), substantial protocol changes might be required for success in other applications. For example, in subcellular localization studies, one would be very concerned with purity and therefore might try to enhance the purity, at the expense of yield, by combining protocols, for example those shown in Sections 3.3.2 and 3.3.4. For studies involving small hydrophilic molecules, it would be useful to try non-aqueous nuclear isolation procedures to minimize the leakage of these molecules. When beginning a project involving isolation of yeast nuclei or chromatin, explore any and all protocol modifications which might seem beneficial considering your particular applications. In that way, new isolation protocols, which might be superior to those available now, can be developed.

2. GENERAL CONSIDERATIONS

The yield of nuclei, their behaviour in the protocols described below and their micro-scopic purity (presence of cells, cell ghosts, vesicular blebs) can vary depending on strain, carbon source and life state. For example, in some circumstances it may not be possible to prepare nuclei without microscopic contamination. Therefore, I recom-mend that a new worker begin with one of a number of common, well behaved laboratory strains (Y55, A364a, S288C), grown to early log phase ($2-3 \times 10^7$ cells/ml) in yeast extract−peptone (YEP)−D, and learn to recognize nuclei and become familiar with the techniques by using these strains. After such an introduction, one can more con-fidently adapt the protocols, if needed, to be successful in the circumstances of interest. Some of the possible adaptations which might be needed are discussed at the appropriate places in the protocols.

If fluorescence microscopy is available, use 4′,6-diamidino-2-phenylindole (DAPI) staining to identify nuclei and to monitor the purification steps. This is especially useful for people just beginning work in the area. Phase contrast microscopy can also be used to monitor nuclear isolations and identify nuclei (*Figure 1*). In phase contrast, nuclei are rather uniformly grey, usually slightly rough-edged and sometimes show darker

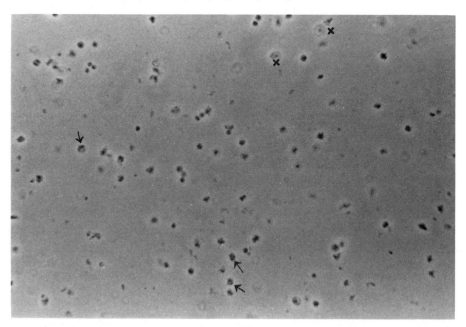

Figure 1. Isolated yeast nuclei in phase contrast. A phase contrast micrograph of the product of the differ-ential centrifugation (long preparation) described in Section 3.3.3 for the strain Y-55 is shown. Most of the spherical objects in the field are nuclei. Note that they are various sizes. The irregularly shaped material is probably cellular debris. A nucleus with cytoplasmic tags is pinpointed (→). A large piece of cell 'shell' is also marked ('x'). Magnification is ×400 in the microscope.

regions and/or cytoplasmic (vesicular) tags. The easiest way to learn to identify nuclei in phase contrast is by comparing a phase contrast field and a field stained with a DNA-specific stain, such as Giemsa, both made from the same preparation. Since the protocols below (Sections 3.3.2−3.3.4) produce nuclear preparations which contain no detectable mitochondrial DNA (Lohr, unpublished observations), the only DNA-containing organelles will be nuclei (and whole cells).

In the protocols described below, centrifugations are done at 4°C in a Sorvall SS-34 rotor, unless otherwise stated. Cells, cell suspensions and nuclei are normally kept on ice except when being treated at specified temperature-controlled steps. All solutions are made with double-distilled water. Solutions used in the protocols, except those explicitly identified, are not sterilized but are kept frozen at −20°C and removed to the 4°C compartment of the refrigerator the night before a nuclear preparation.

3. ISOLATION PROTOCOLS

3.1 Cell harvest

(i)   Grow the strain of interest in the medium of interest to the desired density.
(ii)   Harvest the cells as rapidly as possible by any convenient method, for example centrifugation, filtration, etc.
(iii)   Determine the wet cell weight.

In deciding how many cells to grow and therefore how much medium to use, keep in mind that very small amounts of cells are difficult to work with because of the many manipulations involved in these protocols. Furthermore, many of the sources of loss encountered during the procedures depend only slightly on the volume of cells being processed. For example, the solutions used are very viscous. Each time cell suspensions or cell lysates in these solutions are put into tubes and flasks, there is some loss due to wetting. In sum, these losses become quite significant for very small preparations (<1 g wet wt cells). Thus, it is better to do a larger preparation and discard, or freeze, nuclei not required.

For estimation purposes, 3.0 litres of synthetic complete medium containing haploid cells at 1×10^7 cells/ml give $1-2$ g wet weight of cells. For another strain, 1.5 litres of YEP-D harvested at 4×10^7 cells/ml give $3-5$ g wet weight. One gram wet weight of haploid cells will yield $300-400$ μg of DNA as purified nuclei prepared by the short protocol (Section 3.3.2) described below.

For log phase cells, harvest at $3-5 \times 10^7$ cells/ml. There may be some slowing down of *in vivo* transcription at the higher density range and so if transcription studies are of particular interest, harvest at $2-3 \times 10^7$ cells/ml. This is assuming that the cells will grow to 30×10^7 cells/ml or more in the medium used. If this is not the case, adjust the harvest cell density accordingly. For example, the cells which I harvest at 1×10^7 only grow to $8-10 \times 10^7$ cells/ml in that particular medium.

We determine cell density with a Petroff−Hauser cell counter. Any type of cell counter should do, so long as it is useful for cells in the size range of yeast. We prefer to determine cell density by direct counting of cells, for it allows one to check for contamination in the growth medium (this rarely happens) and serves as a good visual monitor of cell condition and morphology at the beginning of the preparation. This is a useful reference during the subsequent spheroplast and nuclear isolation procedures.

127

For step (ii) harvest cells by any available method convenient for the volume used. For example, a Sorvall GS-3 rotor can spin down 3.0 litres of medium per run. Spin for 5 min at 3300 g. Resuspend the cells in water, one-fifth to one-third of the original volume and re-centrifuge. Remember that cells at this point remain sensitive to their environment. Therefore, in isolating cells grown under special conditions, for example from an α-factor arrest or at the restrictive temperature (temperature-sensitive cells), perform these and subsequent operations involving intact cells or spheroplasts under conditions which maintain the appropriate environment, that is in the presence of α-factor or at 36°C in the above examples.

For step (iii) it is necessary to develop a fairly consistent way to determine wet cell weight since solution volumes used throughout the preparation are based on it. We simply spin the water-washed cells in pre-weighed centrifuge bottles. Invert for 30 sec to drain and then with laboratory quality absorbent tissue, wipe out excess water, particularly that which has accumulated near the rim of the bottle. Weigh the bottle and cell pellet.

3.2 Spheroplasting

Probably the single most crucial step in a yeast nuclear isolation is the spheroplasting step. We have never obtained a good nuclear preparation from poorly spheroplasted cells. In isolating nuclei from yeast cells in various life states other than log phase, such as sporulating cells or stationary phase cells, it is generally the spheroplasting step which requires the most adjustment. Thus, for successful nuclear isolations, devote significant attention to this step.

3.2.1 *Pre-treatment*

It is often possible, with rapidly growing strains harvested at early to mid-log phase ($1-4 \times 10^7$ cells/ml), to resuspend the washed cells directly in S buffer, the spheroplasting solution described below (Section 4.2), and get good spheroplasting ($>90\%$) in an acceptable time (30 min or less). This rapid approach may be particularly useful in transcription studies since lengthy isolation times have a negative effect on nuclear transcription rates (cf. 1). For slowly growing log phase strains or for stationary or sporulating cells, such direct treatment will not usually produce satisfactory spheroplast formation. It is necessary to add a pre-treatment step. However, even in log phase cells in which it might not be absolutely required, we usually use the pre-treatment step because it is easy to carry out and assures reproducibly rapid spheroplasting. At a later stage in the protocol, it is too late to go back and carry out pre-treatment.

For pre-treatment of 1 g of wet weight cells, carry out the following steps.

(i) Resuspend the cell pellet vigorously, rapidly and as completely as possible in 1.4 ml 0.1 M Tris−HCl/0.1 M EDTA, pH 8.0.

(ii) Transfer the resuspended cells with a plastic pipette, measuring the volume, to a 50 ml polycarbonate centrifuge tube (log phase) or appropriately sized flask (stationary/sporulating).

(iii) Add water to a final vol of 3.5 ml.

(iv) Add 1/200 volume (final concentration 0.5% v/v) β-mercaptoethanol.

(v)   Incubate for 5 min on ice (log phase), stirring occasionally, or for 45 min at 30°C (stationary), with shaking in a water bath.

(vi)   Centrifuge at 12 000 g for 5 min. Wash the cell pellet in each tube in 20 ml of S buffer (1.0 M sorbitol, 25 mM KH_2PO_4, pH 6.5).

(vii) Re-centrifuge at 12 000 g for 5 min.

For step (i) be sure that the cells are well suspended; that is no clumps should remain. Since the suspension is probably quite turbid, hold the tube up to a strong light to check for clumps.

For the short times of pre-treatment required for log phase cells, one can simply stir the cells in the centrifuge tube in step (ii). For the longer times required for stationary cells, it is more convenient to let them shake in a water bath.

The presence of β-mercaptoethanol in step (iv) may inhibit transcription in nuclei isolated from these cells (G.Ide, personal communication). Since rapidly growing cells are usually used for such studies, it might be best to skip the pre-treatment. If there is poor spheroplasting without it, carry out the pre-treatment but minus β-mercaptoethanol, although this is not as effective. The presence of β-mercaptoethanol at some point in the procedure appears to be required for successful spheroplasting with the enzyme Oxalyticase (see below).

If log phase cells are incubated at 30°C, spheroplasting in the subsequent procedure may be so rapid that the cells lyse. However, higher temperatures and longer pre-treatments are essential for successful spheroplasting of stationary or sporulating cells. If necessary, pre-treatment time can be increased to 1 h for these cells. Thus, 5 min on ice should be considered the minimum and 1 h at 30°C the (practical) maximum for step (v). Some strains or carbon sources (or combinations of these), even for log phase cells, will require some intermediate combination. This should be empirically determined, either increasing the time of pre-treatment at 4°C or increasing the temperature to 30°C without changing the time.

After each spin, in steps (vi) and (vii) let the tube drain upside down for 1 min and remove liquid near the lip of the tube with a tissue.

3.2.2 *Spheroplast formation*

There are several enzymes which can be used for making spheroplasts. Glusulase (cf. 2) is an extract prepared from snail gut juice and was used in the earliest nuclear isolation protocols. Scott and Schekman (3) described an activity called Lyticase, which is secreted by *Arthrobacter luteus* (syn. *Oerskovia xanthineolytica*). Because it can be isolated from supernatant medium, it should be 'cleaner', that is contain less nuclease and protease contamination than Glusulase. We have used a related commercially available enzyme preparation, Oxalyticase (Enzogenetics, USA), and an enzyme produced by *A. luteus*, Zymolyase (Miles). Using the published procedures for Lyticase, one can undoubtedly prepare enzyme which is quite suitable for making spheroplasts. I recommend the commercial enzymes solely for people who cannot or do not want to prepare their own lytic enzyme. Note that the enzyme used can affect the properties of the nuclei isolated from these spheroplasts in a number of ways, as discussed below.

For spheroplast formation (per g of original wet weight cells) follow the procedure given below.

(i)   Resuspend the cells vigorously and completely in 4 ml of S buffer (see Section 3.2.1).

(ii)   Add Ca^{2+} to 0.4 mM (Zymolyase, Oxalyticase) or EDTA to 1 mM (Glusulase). Dissolve the enzyme in a small amount of S buffer (Zymolyase, Oxalyticase) and transfer it to the cell suspension. Gently swirl. Rinse out the tube which contained the enzyme solution with 0.5 ml of cell suspension.

(iii)   Spheroplast at 30°C with shaking which is vigorous enough to keep the cells suspended but gentle enough to avoid turbulence in the flask.

(iv)   After 10 min, dip two Pasteur pipettes into the cell suspension and allow capillary action to fill a small column in each. Dispense one into a disposable glass tube containing $6-7$ drops of water and the other into a tube containing $6-7$ drops of S buffer. If spheroplast formation is working properly, the suspension in the tube with water should become visibly clearer to the eye. Examine a drop from each in the microscope.

(v)   Sample again at 20 min. When spheroplasting is better than 95% or, preferably, no cells remain in the water field in the microscope, harvest by centrifugation at 3600 *g* for 10 min.

(vi)   (Optional) Feed the cells to restore them to a metabolically active state.

(vii)   Wash the cells one (Zymolyase, Oxalyticase) or more times (Glusulase) with S buffer, using half of the spheroplast volume. Re-centrifuge.

It is essential that cells be well suspended in step (i) and that no clumps remain. Again, hold the flask up to a strong light to check for clumps since the solution will be turbid.

In making up the enzyme solution for step (ii) avoid over-vigorous shaking which could denature the enzyme, and do not dissolve the enzyme until you are ready to add it (Zymolyase or Oxalyticase). The required levels of enzyme may be strain dependent and should be pre-determined. Use the lowest amount which will give the desired spheroplasting efficiency. For most of the strains we use, 1 ml of de-salted Glusulase (2), 1.5 mg/ml of Zymolyase-5000, 25 units/ml of Zymolyase-100 000 or 1100 units/ml Oxalyticase completely converts pre-treated log phase cells to spheroplasts in $20-25$ min. Stationary phase and sporulating cells may require up to twice as much enzyme (and longer incubations, see below).

In our experience, spheroplasts can undergo lysis if shaking in step (iii) is too vigorous. On the other hand, one needs to keep the cells resuspended. We solve this problem by putting the spheroplast solution in an Erlenmeyer flask large enough so that the volume of spheroplast solution just fills the bottom of the flask up to its widest point. This allows a gentle movement back and forth to keep the cells suspended. There are obviously many other ways to solve this problem

Since our goal is to spheroplast the cells $20-25$ mins (log phase cells), we carry out the early check of step (iv) at 10 min. If the reaction is going too slowly, that is there is no visible clearing in the water tube, we usually add $25-50\%$ more enzyme and continue the incubation.

By allowing the Pasteur pipettes to fill to the same level and dispensing these equal aliquots into volumes of water or sorbitol which are equal to each other, one can quantitatively estimate the extent of spheroplasting in the microscope; five cells in the water field or 100 in the sorbitol represents 95 % spheroplasting (step v). However, it is not necessary to count the cells accurately; just estimate.

Under the microscope, spheroplasts in the sorbitol field will generally be highly refractile, so long as they have been treated gently. Physical stress, such as over-vigorous shaking, can cause them to lose this high refractivity. Then they will appear much darker but will still be approximately the same size and shape as the intact (and highly refractile) spheroplasts. If there are large numbers of these darker objects, be more gentle in the above procedures.

Incomplete spheroplasting can produce objects in the water field which resemble the objects previously described, dark (not highly retractile), but still approximately the size of spheroplasts. This probably results from the enzyme breaking down enough wall to punch 'holes' in the cell and allow leakage but not enough for lysis and complete release of cellular contents. These incompletely spheroplasted cells will not release nuclei in the subsequent steps. If such objects appear, increase the pre-treatment conditions (more time or higher temperature) in the subsequent preparations. If less than 30 min of treatment have elapsed, increase the enzyme concentration and allow more time for spheroplasting. However, in our experience this seldom solves the problem satisfactorily and it is usually best to simply discard the preparation at this point. Incomplete spheroplasting can also result from cells not remaining suspended during spheroplasting. If you are not sure that cells are remaining resuspended, gently swirl the flask by hand every 5 min.

If the enzyme is working properly, the water field in the microscope should show no spheroplast-like objects after $20-25$ min. There should be lots of small dot-like debris in the field and sometimes one even can see nuclei. Whether nuclei remain intact depends on exactly how much water and how large a spheroplast sample one uses.

For sporulating or stationary cells, it may be necessary to spheroplast for periods up to 1 h. In this case, the first check should probably be at 20 min or so since the reaction goes more slowly. Depending on the strain and how far toward stationary phase one has grown the cells, complete spheroplasting may not be achievable. In this case, one must simply do the preparation in the presence of some intact cells.

In centrifuging spheroplasts, avoid centrifugation at higher speeds than stated. The supernatant after centrifugation should be clear. If spheroplast lysis has occurred (over-vigorous shaking?), the supernatant will be cloudy. If this occurs, be more careful in handling the cells the next time you make spheroplasts from this strain. Always avoid leaving spheroplasts shaking at 30°C any longer than is absolutely required to achieve the desired level of spheroplasting. However, some strains do have a tendency to lyse no matter how carefully they are handled and in these some cloudiness may be unavoidable.

The optional ('feeding') in step (vi) may be particularly important in transcription studies. For example, only 5 min feeding of spheroplasts can produce a 2.5-fold increase in the general transcription rate in nuclei isolated from these spheroplasts (4). The benefits of this treatment continue to increase up to at least 25 min of incubation (8-fold over unfed spheroplasts). Such an enhancement in nuclear transcription is not surprising because cells have been starved to this point in the procedure and they could respond to this starvation by shutting off transcription. This treatment does not, so far, appear to make a difference in chromatin *structural* features detected on the *GAL1-10* genes (Lohr, unpublished observations), but this analysis has not been completed.

A number of media have been used to restore the cells to a metabolically active state. We recommend 0.3% casamino acids +2% carbon source in 20% sorbitol, 20 mM $Na_2HPO_4-KH_2PO_4$, pH 6.2 or YM-5 medium (2% glucose, 1% succinic acid, 0.6% NaOH, 0.67% Yeast−Nitrogen Base, 0.2% Bactopeptone, 0.1% yeast extract, 40 μg/ml adenine, 40 μg/ml uracil) containing 1.2 M sorbitol (1). The nitrogen and carbon sources are probably the essential ingredients. These solutions must be filter sterilized for storage. Cells can either be harvested, and then resuspended in the restoration media or restoration media can be added to the spheroplasting solution, and the spheroplasts harvested and washed as described (step vii). Since resuspension steps involve time and usually some spheroplast lysis, we simply add a X10 casamino acids + carbon source combination directly to the spheroplasting solution when spheroplasting is almost complete, for example at 20−22 min in a 25 min treatment. If present earlier, spheroplasting seems to be inhibited.

There is one note of caution. We have observed repeated failures (no nuclei) when Zymolyase is used to prepare spheroplasts from log phase cells grown in galactose (and spheroplasts incubated in galactose as the carbon source). The cause of this is unknown.

Since spheroplasts are sensitive to mechanical stress, they should be gently resuspended during the wash in step (vii). Less than complete resuspension, that is some clumps remaining, is probably sufficient. Again, monitor the extent of spheroplast lysis by observing the clarity of the supernatant after centrifugation. After each centrifugation, the tube should be inverted and allowed to drain for 30 sec. Wipe out as much of the liquid along the walls as possible with tissue.

Oxalyticase is clean enough with respect to nuclease and protease activities that, even without washing the spheroplasts, there is little or no nuclease or protease degradation detected in nuclei isolated from these spheroplasts. On the other hand, nuclei from glusulase-prepared spheroplasts have enough DNase to produce visible degradation of the nuclear DNA in 15−30 min incubation at 37°C (Lohr, unpublished observations) and nuclei from Zymolyase-prepared cells may have suffered proteolysis (see below). In our experience, washing of spheroplasts does not completely remove either problem.

However, it is difficult to make a single general recommendation of which spheroplasting enzyme to use because none of the enzymes are without drawbacks and all of them have been used, apparently successfully, in nuclear isolation protocols. We and others (F.Ziemer, unpublished observations) have noticed that spheroplasting with Zymolyase-100 000 produces clumping of nuclei in many strains. It is likely that Zymolyase is the cause of this for there is no nuclear clumping in the same strains spheroplasted with Oxalyticase or Glusulase. Thus, if clumped nuclei would be unacceptable for the intended use, use the latter enzymes. Oxalyticase produces 'frayed' looking and often irregularly shaped nuclei. We do not know the cause of this but it could result from the way nuclei are extruded from Oxalyticase-prepared spheroplasts. However, because of its low protease (see below) and nuclease levels, Oxalyticase is the preferable enzyme for chromosome structure analyses. Glusulase has potential nuclease, and perhaps protease, contamination problems.

3.3. **Isolation of nuclei**

There seems to be general agreement that the basic isolation approach initially developed by May (5) and modified by Wintersberger *et al.* (6) provides the most consistently successful methodology for nuclear isolation (cf. 7). The most essential feature of the approach is the discovery that 18% Ficoll lyses spheroplasts but does not lyse nuclei. In the years since the Duffus review article (7), there have been few new published protocols and the major one (8) still depends on the efficacy of the spheroplast lysis in 18% Ficoll from method (b) of Wintersberger *et al.* (6).

3.3.1 *Spheroplast lysis*

(i)   Spread the cells evenly around the bottom and along the lower 1/3 of the sides of the centrifuge tube(s).

(ii)   *Slowly* add phenylmethylsulphonyl fluoride stock solution (PMSF) 100 mM in isopropanol, to a final concentration of 1 mM, to a rapidly stirring solution of 18% Ficoll, 20 mM KH_2PO_4 (pH 6.5), 0.5 mM Ca^{2+}. Add a few ml of this solution to the cells in the tube and vigorously resuspend the cells. Continue until clumps are no longer visible. Add Ficoll to the final volume required and stir to homogeneity.

(iii) Homogenize with a few strokes of a loose-fitting tissue homogenizer (e.g. Potter—Elvehjem), if desired.

During step (i) you can and should be rough with the spheroplasts since the object now is to break them open. Try to work as fast as possible! Keep cells on ice as much as possible. Spreading the cells around before adding the solution seems to avoid much of the clumping that can occur whenever new solutions are added to cell pellets.

PMSF is a protease inhibitor. It is also a cumulative neurotoxin so avoid ingestion or skin contact with it. It has a very short effective lifetime in aqueous solutions so add it to the solution immediately before use. The exact quantity of 18% Ficoll solution to be used depends on which protocol will be followed after this step: 3 volumes for Percoll gradients (Section 3.3.4); 5—7 volumes for the differential centrifugation protocols (Sections 3.3.2 and 3.3.3). Volumes are based on the original wet cell weight (1 ml Ficoll = 3—7 × g wet wt). In the differential centrifugation protocols (see below), centrifuge tubes should be no less than 40% and no more than 60% filled. So use the precise volume of Ficoll, within the range of 5—7 volumes, which will fill to 40—60% as many tubes as necessary, depending on the preparation size.

Again, during the resuspension work rapidly and stir vigorously to resuspend. A very useful motion is to hold the tube at 30—45° above horizontal and rapidly circle around the inside wall of the tube, using a fairly thick (1 cm diameter) glass rod. This seems to catch clumps on the side and disperse them.

It is possible that some of the tight-fitting commercial homogenizers could cause nuclear lysis. Therefore for step (iii), we use a Teflon pestle which fits loosely into the polycarbonate tubes used for centrifugations (cf. *Figure* 2A). Thus, we can homogenize in the centrifuge tubes and avoid extra transfer steps. This homogenization step aids release of nuclei from spheroplasts and helps prevent cytoplasmic aggregates from trapping nuclei. It is an essential step in isolating nuclei from stationary

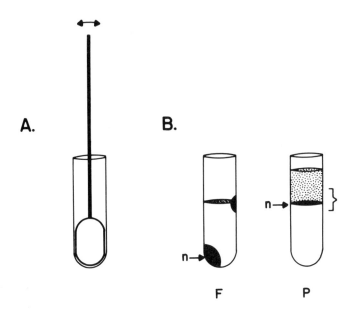

Figure 2. Some intermediate stages in the nuclear isolation. **(A)** The pestle used for homogenizations of the Ficoll lysate is shown. The curved arrows at the top are to indicate that there is some looseness in the fit of the pestle in the polycarbonate centrifuge tube. **(B)** The appearance of the solutions after the fast Ficoll spin of Section 3.3.2 **(F)** or the Percoll gradient centrifugation **(P)** are shown. In each case, the location of nuclei are denoted by **n**. The bracket to the right of the Percoll tube signifies that there is some latitude in the exact location of the nuclear band(s). The stippled area is meant to show that this part of the gradient is usually cloudy.

cells (see below). Note that because 18% Ficoll is hypotonic, the nuclei may appear swollen under the microscope.

3.3.2 *Differential centrifugation (short preparation)*

(i)   Spin tubes (40−60% full) at 2500 *g* for 5 min.
(ii)   Pour about 3/4 of the supernatant (containing nuclei) into another centrifuge tube.
(iii)   Gently swirl the remaining supernate around the bottom of the tube to resuspend any loosely packed part of the pellet.
(iv)   Combine all supernatants and centrifuge at 30 000 *g* for 20 min (tubes 40−60% filled).
(v)   Pour off the supernatant. Wipe out the tube with a tissue, being especially careful to remove all of the solid material which floated to the top of the solution (cf. *Figure 2B*, 'F') and not allow any of this material to contact the nuclear pellet at the bottom of the tube (*Figure 2B*, 'n').
(vi)   Wash the pellet once by pouring in the next solution to be used and rolling it carefully around the tube over the pellet without disturbing the pellet. Pour out this solution and wipe out the tube with a tissue.

The purpose of the slow spin in step (i) is to remove whole cells and other large cellular debris from the nuclear suspension. Some whole cell contamination will be removed by this step. However, separation of whole yeast cells from nuclei by differential centrifugation is not very efficient because the sedimentation properties of the two are similar. The best approach is to achieve highly efficient spheroplasting so that there will be as little whole cell contamination to remove as possible.

In some cases, there can be significant clumping of nuclei (cf. production of spheroplasts with Zymolyase), causing them to pellet in the low-speed spin. When they do this, they usually form a separate loosely packed pellet, which is often whiter than the cell and cellular debris part of the pellet. Thus, one can carefully swirl away most of the nuclear part without disturbing the darker cellular pellet. Some loss of nuclei and/or carry-over of cellular debris is unavoidable. Note that one can encounter strains which show clumping of nuclei even when using the enzymes (Oxalyticase, Glusulase) which minimize this. Be alert for clumping because it changes the sedimentation behaviour of nuclei quite drastically.

Step (iv) will pellet the nuclei (and whole cells and cell walls) and float out membrane vesicles, vacuoles, mitochondria, etc. (6). This floated material will be present as a 'skin' on top of the solution and as a compact pellet on the side of the tube opposite the nuclear pellet (*Figure 2B*, 'F').

Since this top pellet contains the vacuoles, be sure to remove it completely from the tube containing the final nuclear pellet before proceeding to subsequent steps. This can be done with a tissue. If necessary, wet the tissue to help remove all the viscous 18% Ficoll solution from the tube. Work as fast as possible up to this point. Once free of the cellular lysate, one of the major sources of protease and nuclease degradation in nuclear isolations will have been removed.

In stationary phase preparations, some nuclei are attached to vesicular material and consequently float out during this spin. To avoid losing these nuclei, carefully remove the top pellet after the first Ficoll spin and re-homogenize it as before with a few up and down strokes of a pestle, using fresh Ficoll for the liquid phase. Re-centrifuge at 30 000 *g* for 15 min. Combine the bottom pellet from this spin with the bottom pellet from the previous spin and use as the nuclear pellet.

Step (vi) is mainly to wash out the tube. Nuclei can be resuspended and re-centrifuged as a wash step, if desired.

3.3.3 *Differential centrifugation (long preparation)*

This corresponds closely to method (b) of ref. 6. Steps (i)−(v) are exactly the same as for the short preparation above (Section 3.3.2).

(vi)   Spread the crude nuclear pellet carefully around the bottom and up the lower part of the side wall of the centrifuge tube(s), using a small glass rod.

(vii) Slowly add PMSF to a final concentration of 1 mM, with stirring, to 8−10 vol (8−10 × g original cell wet wt) of HM (1 M sorbitol, 20% glycerol, 7% Ficoll, 0.5 mM Ca^{2+}, 20 mM KH_2PO_4, pH 6.5). Add a small amount of this HM to the dispersed crude nuclear pellet.

(viii) Gently resuspend the nuclei until all clumps are gone. Bring to final volume with HM.

(ix)   Centrifuge at 5900 *g* for 5 min. Pour off the supernatant, swirling to take along any loosely packed pellet.

(x)   Centrifuge at 5100 *g* for 5 min. Again pour off the supernatant with swirling.

(xi)   Centrifuge at 4500 *g* for 5 min. Pipette or pour off the supernatant leaving the pellet behind.

(xii)   Spin the supernatant from step (xi) at 22 000 *g* for 20−25 min.

(xiii) Pour off the supernatant. Wash the pellet by pouring in the next solution to be used and rolling it carefully around the tube and over the pellet without disturbing the pellet. Pour out this solution and wipe out the tube with a tissue.

As before, step (vi) seems to avoid some of the clumping which accompanies solution changes. However, during the dispersion process in this case, you are putting mechanical stress directly on nuclei and at this point you do not want to break (lyse) them open. Thus, use a small glass rod (1.5 mm diameter) and work slowly! Note that this is different from step (i) of Section 3.3.1, since in that case the mechanical stress was being applied to spheroplasts which were to be broken anyway. Start the nuclear resuspension with a small volume of HM. It is easier to catch up with the clumps.

Again PMSF should be added slowly to a vigorously stirred solution of HM just before it is to be used. Remember PMSF is a neurotoxin; avoid ingestion or skin contact.

Step (viii) is time-consuming. If there are several tubes, work on each one for a short time and then move to another. As the partially resuspended nuclei sit in the tube, they become easier to resuspend completely. Or, using a glass rod and a small amount of HM, transfer all the pellets into one tube to resuspend and then split them up again when diluting to the final volume. Choose the precise volume within the range of 8 − 10 volumes which will fill up centrifuge tubes no less than 40% and no more than 60% full for the centrifugations in steps (ix)−(xii).

The purpose of the slow spins in steps (ix)−(xi) is to remove whole cells and other large debris left over from the 18% Ficoll step. Separation of cells and nuclei by differential centrifugation is more effective in HM than it is in the 18% Ficoll although it is still not highly efficient. With each slow spin there is a decrease in cellular contamination (and a loss of nuclei). After the last spin, remove the supernatant without swirling. Examine the supernatant in the microscope to see if it is satisfactory. If so, proceed to step (xii). If not, do another slow spin. If nuclei are clumped, the usefulness of these slow spins will be severely compromised.

Carefully invert the tube for 30 sec to drain and wipe out the excess liquid. Step (xiii) is again mainly a tube rinse. The nuclear pellet will usually remain packed upon inversion and washing, if handled carefully. The nuclear pellet can also be washed by resuspension and re-centrifugation. However, every resuspension of nuclei is time-consuming and also results in some lysis. Unless it is absolutely necessary, it should be avoided.

3.3.4 *Percoll gradients*

This protocol begins with the cell lysate from step (iii) of Section 3.3.1 and is taken from ref. (8).

(i)   Dilute the 18% Ficoll lysate with an equal volume of 0% Percoll (1.0 M sorbitol, 0.5 mM Ca^{2+}, pH 6.5).

(ii)   Layer 2 − 10 ml of the diluted lysate on pre-spun 32.5% (haploid cells) or 35% (diploid cells) Percoll gradients made by mixing appropriate amounts of 0% Percoll and 100% Percoll (the above solutes dissolved in Percoll).

(iii) Centrifuge at 9500 *g* in a Sorvall HB-4 rotor for 30−40 min.

(iv) Pull out the nuclear band with a wide-mouth plastic pipette. Dilute with 2 vol of 0% Percoll.

(v) Centrifuge at 6000 *g* for 5 min in a Sorvall SS-34 rotor.

(vi) Wash the pellet once by pouring in the next solution to be used and roll it carefully around the tube and over the pellet without disturbing the pellet. Pour out this solution and wipe out the excess liquid.

The dilution in step (i) reduces the density of the 18% Ficoll lysate so that it does not sink into the Percoll when layered on the gradient.

Prior to being loaded with the 18% Ficoll lysate, the Percoll gradients are generated (pre-spun) in step (ii) by centrifugation of 34 ml of the appropriate Percoll solution at 27 000 *g* for 50 min in a Sorvall SS-34 rotor. Layer the Ficoll lysate carefully on these gradients to avoid excessive mixing at the top.

I find that separations are improved by increasing the time of centrifugation in step (iii) to 30−40 min. Ide and Saunders (8) originally recommended 20−25 min. A diagram of how the gradient might appear after centrifugation is shown in *Figure 2B* ('P'). Note that there may be multiple bands and they may be located at various distances into the tube.

Nuclei will band somewhere near the centre of the tube (*Figure 2B* 'P'). Remove them carefully (step iv) to avoid disturbing the gradient, using a wide-bore pipette to avoid shear forces. Nuclei stick to plastic less well than to glass.

Step (v) removes some additional cellular contamination (8), and step (vii) is again a wash step for the tube.

3.3.5 *Comparison of the three principal methods for nuclear isolation*

I have compared the three protocols described in Sections 3.3.2−3.3.4 using the well behaved (in nuclear preparations) diploid strain Y-55. Other strains should show similar behaviour but may vary in details such as DNA yield, purity, etc. The yield of nuclei is calculated as the ratio of μg of DNA present in the isolated nuclear preparation to μg of nuclear DNA present in the cells used to begin the preparation × 100.

The short preparation described in Section 3.3.2 gives a high yield (70−80%) of morphologically sound nuclei. Nuclease digestion gives DNA patterns which are like those from the original nuclear preparations used to characterize yeast chromatin (the long preparation described in Section 3.3.3). However, in this short preparation, there is usually significant contamination of nuclei with whole cells and large cellular debris. For nuclease digestion, and probably for nuclear transcription also, these contaminants do not appear to cause problems. Although purer nuclear preparations can be obtained by using the other procedures, the speed and yield of the short preparation are significant advantages. We have used this preparation exclusively for the last 3 years in our studies of yeast single gene chromatin.

The long preparation (Section 3.3.3) produces morphologically intact nuclei in 40−50% yield, with minimal whole cell contamination. The lower yield reflects losses at resuspension and slow spin steps. These nuclei show a fairly typical yeast DNA:RNA: protein ratio of 1:4:30, carry out active transcription elongation and initiation and exhibit the structural features associated with yeast chromatin, as determined by nuclease digestion.

Nuclei isolated from Percoll gradients are morphologically intact, are free from cellular contamination and from contamination by soluble cytoplasmic enzymes like ADH (8) and are obtained in 30−40% yield. The nuclei carry out active transcription initiation and elongation and show typical chromatin structure, as evidenced by nuclease digestion profiles (8). This preparation is much faster than the long differential centrifugation preparation described above but not as fast as the short one. However, the Percoll nuclear preparation contains little or no cellular contamination.

There are some negative aspects associated with the Percoll procedure. The capacity of the preparation is limited practically to 6−7 g wet weight of cells; the differential centrifugation preparations are not limited. Also, intact nuclei are present in other parts of the gradient besides the main nuclear band. Since Percoll gradients separate material on the basis of density, this suggests that there are different density classes of nuclei. Such classes could arise from variations in nuclear density through the cell cycle, for example. Thus, one may be limiting the isolation to a sub-set of nuclei. We have also noticed a higher failure rate, that is preparations which fail to produce a band or a well separated band, for the Percoll procedure than for the differential centrifugation preparations. This observation has been confirmed by one of the developers of the procedure (G.Ide, personal communication). Evidently the Percoll procedure is more sensitive to the inevitable variations associated with isolating yeast nuclei, such as cell type, growth conditions, spheroplasting efficiency, etc. However, when successful, the Percoll procedure is very attractive, owing to the combination of speed and high purity of nuclei produced.

3.3.6 *Other approaches*

A popular isolation procedure for multicellular eukaryotic nuclei involves cell lysis in the detergent Nonidet P-40 (NP-40). This procedure produces very clean nuclei from animal cells; ribosomes and other material appear to be removed from the outer nuclear surface. I find that 0.5% NP-40 in an osmotically stabilizing medium like HM causes yeast spheroplasts to lyse, producing nuclei which are largely free of the cytoplasmic tags usually present on isolated yeast nuclei. Nuclei from spheroplasts lysed in this way avoid being swollen during the isolation, as they are in the hypotonic 18% Ficoll lysis. The nuclei can be collected by differential centrifugation as described in Section 3.3.3 steps (ix)−(xiii). I have had success with this approach in isolating nuclei only from Glusulase- or Zymolyase-treated cells. I have not extensively analysed the nuclei produced by this protocol because of the potentially harmful effects of detergents on gene organization and nuclear transcription. However, the ability to produce nuclei free of cytoplasmic tags and bound ribosomes might prove overwhelmingly useful in other applications, in spite of these reservations.

3.3.7 *Histone proteolysis*

Proteolysis is an important concern in isolations of yeast nuclei. There are chromatin-associated proteases and cytoplasmic proteases, mainly vacuolar. Strategies are devised mainly for proteases from the latter source. One should routinely analyse the histones from each nuclear preparation because occasionally, for unknown reasons, extensive proteolysis can occur.

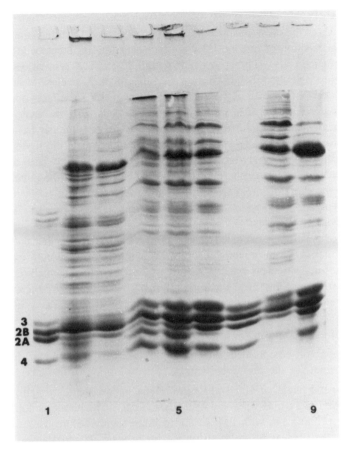

Figure 3. Yeast histones from isolated nuclei. SDS−polyacrylamide (Laemmli) gels of histones isolated from various preparations of yeast nuclei are shown. This gel compares the effect of various spheroplasting enzymes and the effect of the NP-40 nuclear wash [once in buffer A (9) in 1 M sorbitol, once in buffer A alone). Histones are identified in the standard calf thymus histone **track 1**. Electrophoresis is from top to bottom. **Track 2**, yeast (Zymolyase-100 000, no Np-40); **track 3**, yeast (Oxalyticase, no NP-40); **track 4**, yeast (Zymolyase-100 000, NP-40 wash); **track 5**, yeast (Zymolyase-5000, NP-40 wash); **track 6**, yeast (Oxalyticase, NP-40 wash); **track 7**, chicken erythrocyte histone standard; **track 8**, yeast (Glusulase, NP-40 wash); **track 9**, yeast (Oxalyticase, NP-40 wash; different nuclear preparation from **track 3**).

   The length of time spent in nuclear isolation has some effect. Thus, the short differential preparation and the Percoll preparation will often show less proteolysis. At 0°C, the effect is small (Lohr, unpublished observations). Incubation of isolated nuclei at 30°C will produce visible degradation of histones, on the time scale of a few hours (Lohr, unpublished observations). Thus, the time scale of a nuclease digestion (minutes, except with restriction enzymes) or a nuclear transcription is short enough to avoid significant degradation. These observations reinforce the necessity to work as fast as possible and to keep nuclei on ice as much as possible during the isolation. The use of *pep* mutants (protease deficient) can help minimize cytoplasmic protease contributions. I find that protease inhibitors like PMSF, $NaHSO_3$ and aprotinin are surprisingly ineffective in preventing histone proteolysis (Lohr, unpublished observations).

The choice of spheroplasting enzyme again has an effect. Nuclei from Oxalyticase-treated cells show the most intact histones (*Figure 3*), at least for this strain. The band at the normal position of histone H2A is often very low in intensity in nuclei from Zymolyase-treated cells. This suggests the possibility of H2A proteolysis. There is a strong band just below on the gel (*Figure 3*). Nuclei from Glusulase-treated cells can show some decrease in histones 2A and 4. Again, these are effects specific to the spheroplasting enzyme because we used aliquots of cells from the same cell harvest for the experiments in *Figure 3*: these aliquots were treated similarly before and after spheroplast formation, and they were processed simultaneously throughout. The only difference is the enzyme used to make spheroplasts. Washing Zymolyase- or Glusulase-treated spheroplasts before spheroplast lysis does not improve the pattern. For this experiment, the short differential centrifugation preparation was used. The histones from this preparation or the Percoll gradient preparation look the same (Lohr, unpublished observations).

Histones are usually acid extracted directly from nuclei. With yeast nuclei this produces a complex protein gel pattern with many proteins in the histone region (*Figure 3*). Davie *et al.* (9) described a nuclear wash procedure using 0.5% NP-40 which removes many of the interfering proteins. One wash in buffer A (9) plus 1 M sorbitol and one wash in buffer A alone produces quite interpretable patterns (*Figure 3*).

3.4 Isolation of chromatin

This discussion will be confined to isolation of chromatin from nuclei. For most applications, nuclei are the preferable source.

(i)   Resuspend nuclei at 100−400 μg/ml DNA in low ionic strength buffer (e.g. 10 mM Tris, 1 mM EDTA, 10 mM NaCl, pH 7.5) without osmotic stabilizers (e.g. sorbitol) and with PMSF, added immediately before use.
(ii)   Allow the nuclear lysate to sit on ice for at least 10 min.
(iii) Centrifuge at 10 000 g, 10 min.
(iv)   Wash the chromatin pellet in the above buffer. Re-centrifuge.

Nuclei are osmotically sensitive and a low salt lysis is the most common way to liberate their chromatin.

For minichromosome isolations, a homogenization step using a loose-fitting pestle as described in Section 3.3.1 might aid in releasing minichromosomes from entanglements with genomic chromatin.

The released chromatin is insoluble and can thus be pelleted. Step (iii) achieves only a modest increase in purity, as judged by the DNA:RNA:protein ratio (6). However, the chromatin is now in a much more accessible state for various treatments, for example restriction enzyme digestion, salt elution studies, etc.

Chromatin may be subjected to shearing forces as it is released from nuclei during low salt lysis (10). To avoid this, one can carry out a limited nuclease digestion, usually with staphylococcal nuclease (described in Section 4.1.1). After the digestion, spin down the nuclei and then lyse them by the low salt treatment. The smaller size of nuclease-fragmented chromatin evidently lessens the shear during the release of chromatin from nuclei. Nuclease-digested chromatin is usually further purified and/or fractionated on a sucrose gradient. Since chromatin is generally solubilized by nuclease treatment, it can no longer be spun out of solution by low-speed centrifugation. Note that small chromatin fragments may be released from nuclei by nuclease treatment and will thus be present in the supernatant after centrifugation, even before the low salt lysis. Also, nuclease can remain bound to purified chromatin and may become activated again at a subsequent step, by the presence of Ca^{2+}. This is an important consideration for transcription studies since staphylococcal nuclease also has RNase activity.

For a number of kinds of studies, including transcription, it would be very useful to be able to isolate yeast plasmids as minichromosomes, with their attendant histone and non-histone components. However, such isolation has proved very difficult to achieve in high yield and purity. A method for isolating 2 μm minichromosomes using metrizamide gradient banding of chromatin from isolated nuclei has recently been described (11) and may prove useful.

4. PROCESSES RELATED TO TRANSCRIPTION

In this section, I will briefly discuss two possible applications for the material isolated by the preceding protocols.

4.1 Structural studies of active chromatin

Many studies of the chromosome structure of genes rely on nuclease or chemical cleavage agents to fragment chromosomal DNA. Structural inferences are gained from the fragmentation pattern. This approach can also provide starting material for the isolation of gene-specific active chromatin (12). We will illustrate the approach with staphylococcal nuclease (MNase) as the cleavage agent.

4.1.1 *Nuclease digestion of chromosomal DNA*

(i)   Resuspend nuclei at 300 μg/ml DNA in 1 M sorbitol, 0.05 mM Ca^{2+}, 5−10 mM Pipes (piperazine-*N,N'*-bis 2-ethane sulphonic acid) pH 6.5, containing 0.5 mM PMSF, added just prior to use.

(ii)   Add the cleavage agent of choice (e.g. MNase) and allow the reaction to proceed for the appropriate pre-determined time(s).

(iii) Stop the reaction (e.g. by chelating Ca^{2+} with Na_2 EDTA).

Prior to adding the digestion solution, spread the nuclear pellet around the bottom of the tube. This helps avoid clumping when the solution is added. During the resuspension, nuclei are subjected to mechanical stress. Use a thin glass rod to resuspend them and work slowly. The approximate amount of DNA present in nuclei can be estimated from the efficiency of the particular preparation used (see above) and the original cell wet weight: 1 g wet weight of haploid cells yields roughly 200 μg of DNA into nuclei isolated by the long preparation (Section 3.3.3) or 300−400 μg by the short preparation (Section 3.3.2). These are rough estimates because they depend on how much water is present in the whole cell pellet. However, so long as the original wet cell weight is determined consistently from experiment to experiment, it will become possible to know the DNA concentration precisely enough. Thus, we estimate the DNA concentration and use different concentrations of nuclease or different times of digestion to obtain the digestion extent we want. Because of the large amount of RNA in yeast nuclei, an A_{260} reading of nuclei is not useful for estimating DNA concentration. Most other DNA determinations are lengthy and it is important to begin the digestion as soon as possible.

MNase needs only Ca^{2+} to function. For other cleavage agents, further additions need to be made: for DNase I, Mg^{2+} to 0.5 mM; for methidiumpropyl EDTA·FeII (MPE), dithiothreitol to 2 mM; for restriction enzymes, Mg^{2+} to 2 mM, Tris−HCl buffer (pH 7.5) 6 mM and NaCl and β-mercaptoethanol if required. Be sure to mix the nuclear suspension while adding these substances to avoid locally high concentrations which might cause nuclear lysis. Note that a lower concentration of Mg^{2+} is used for nuclear restriction digestions: 10 mM Mg^{2+}, used in naked DNA restriction enzyme digestions, is likely to condense chromatin and thus could actually inhibit digestion. Note that a procedure to isolate residual nuclei (i.e. scaffold or matrix) has recently been described (13): a Percoll-type prototocol, but Mg^{2+} is present at 10 mM. This may be important in the isolation of such material.

Most applications for gene-specific studies require fairly limited digestion: MNase, 30 units/ml for $1-10$ min; DNase I, $1-5$ units/ml for 30 sec−3 min; MPE·Fe(II), $5-10$ μM for $10-30$ min. Note that MPE·Fe(II) digestion can be done in the presence of 1 mM EDTA to chelate divalent cations and thus avoid the possibility of concurrent endogenous nuclease cleavage. During MPE digestions, nuclei should be resuspended in 1 M sorbitol, 0.05 mM Ca^{2+}, $5-10$ mM Pipes, pH 7.2 because MPE works much more slowly at pH less than 7.

The nuclease digestions can be stopped by adding EDTA in excess (e.g. to 2 mM for isolation of intact chromatin; to 10 mM for samples from which DNA is to be extracted. MPE·Fe(II) cleavage is stopped by addition of the iron chelator bathophenanthroline in excess.

4.2 Nuclear transcription

Most studies of eukaryotic transcription have used *in vitro* approaches, attempting to reconstruct the transcription machinery from its basic components. For the most part, these approaches have been unsuccessful in yeast, except for polymerase III systems (14). However, one can also use nuclei or chromatin to study transcription. In yeast, all three classes of RNA polymerase can elongate RNA in these materials (15). Polymerase I (16,17) and polymerase III (1,16,18) can also initiate transcription in isolated nuclei. Primary transcripts of tRNA can also be processed in nuclei (1,18). No initiation of a polymerase II transcribed gene has been demonstrated. Some limited success using exogenous polymerases has been achieved (19). Chromatin prepared by the methods described above, even when nuclease-fragmented, can also be used in transcription studies (15).

We will only discuss transcription studies using nuclei isolated by the methods described above and relying on endogenous RNA polymerase present in the nuclei.

(i)   Resuspend nuclei at $200-400$ μg/ml DNA in 1 M sorbitol, 0.05 mM Ca^{2+}, pH 6.5.

(ii)   Place 5 μl of the $10\times$ transcription stock salt solution [TSS, 500 mM Tris, pH 7.9 (23°C), 10 mM $MnCl_2$, 100 mM $MgCl_2$, 50 mM PEP, 10 μg/ml pyruvate kinase, 1.0 M KCl; store frozen] in a microcentrifuge tube containing 50 μM [^{3}H]UTP or [α-^{32}P]UTP (50 Ci/mmol).

(iii)   Add 5 μl of an experimental variable solution [e.g. α-amanitin, $(NH_4)_2SO_4$, etc.], if desired. If nothing is to be added, add 5 μl of water.

(iv)   Add 35 μl of nuclei. Equilibrate for 5 min on ice.

(v)   Begin the reaction by adding 5 μl of a 10$\times$ nucleotide stock. Incubate at 25°C for the desired time.

The 10$\times$ nucleotide stock is 2.4 mM ATP, 2.4 mM CTP, 2.4 mM GTP dissolved in water. For transcription initiation studies [γ-S]ATP or [γ-S]GTP replaces ATP or GTP. For concentration determination of nucleotides use extinction coefficients (CRC Handbook of Biochemistry). Store frozen and re-make the stock at least every 6 months. If a labelled nucleotide other than UTP is to be used, cold UTP should *replace* that (the labelled) nucleotide in the stock solution.

Nuclei do not have to be prepared fresh for each transcription. Divide the resuspended nuclei from a preparation into convenient aliquots and freeze. Aliquots can then be thawed as desired. There may be some nuclear lysis resulting from the freeze − thaw cycle but it seems to have little effect on the transcription (20). Yeast nuclei incorporate 20 − 40 pmol of labelled UTP/μg of DNA into RNA in a 15 min incubation (20). From this and the specific activity of the radiolabel used, one can estimate the amount of nuclei needed in any particular experiment.

The transcription mix described in step (ii) gives quite active transcription elongation (15) and initiation (18) in nuclei. However, these conditions were not optimized to give maximal transcription, particularly for individual genes. Optimal conditions (divalent cation concentration, etc.) for nuclear transcription have been determined recently (21) using nuclei isolated by the (long) differential centrifugation protocol (Section 3.3.3). Returning spheroplasts to a metabolically active state (feeding) prior to their lysis enhances the general transcription capabilities of nuclei (4), this step will probably also increase transcription from individual genes. Although not definitively demonstrated, it is probably also wise to avoid the use of detergents such as Triton X-100 or NP-40 during the nuclear isolation. Several substances were found to have no stimulatory effect on the nuclear transcription (1): a phosphoenolpyruvate − pyruvate kinase energy regenerating system:bovine serum albumin at 1%; S-adenosyl- L-methionine at 0.1 mM or 1.1 mM. There may also be significant variations between preparations in the levels of transcription (1; G.Ide, personal communication).

There is a large amount of endogenous RNA in yeast nuclei. By radiolabelling RNA newly synthesized in isolated nuclei, one can distinguish it from the RNA made *in vivo* and thus study transcription elongation. To study transcription initiation in nuclei, the newly initiated RNA must be separated from that RNA which was merely elongated in nuclei. Use of γ-sulphydryl nucleotides, combined with Hg − agarose affinity chromatography, provides the best method for such an analysis (18).

A persistent problem in nuclear transcription studies is the short duration of synthesis, both initiation and elongation. Optimizing transcription conditions, especially for initiation, could help. It is also possible that transcription ceases because transcription-associated factors become limiting. An obvious approach is to try to supply the missing factors with nuclear or cellular extracts. In an organism like yeast, with its high degree

Table 1. *In vitro* initiation in nuclei from various yeast strains[a,b].

Strain	Characteristic	Label incorporated into newly initiated (bound) RNA (%)	No. of determinations
Y-55	Wild-type	1.6	2
A364a	Wild-type[c]	2.7	7
108-3c	Wild-type[c]	1.0	3
Pep-IV	Deficient in RNase and protease	5.6	4
RNase 3⁻	~50% deficient in RNase	8.7	5
GRF 110	Deficient in rRNA processing activity	10.0	1

[a]Reprinted with permission from ref. 15.
[b]Nuclei were isolated by the Percoll gradient procedure described in Section 3.3.4.
[c]With respect to synthesis of RNA.

of genetic sophistication and capacity for experimental manipulation, it should be quite advantageous to study nuclear transcription. For example, low signal is a problem in initiation studies. By using RNase⁻ mutants in a study of 5S and 35S transcription initiation, the level of initiation detected was boosted by 3- to 5-fold (*Table 1*). Up to 10% of the label is incorporated into nuclear initiated RNA.

5. MEDIA

(i) *YEP-D.* 10 g/l yeast extract, 20 g/l Bactopeptone, 20 g/l dextrose. Dissolve in water. For large quantities, materials can be dissolved at twice the final concentration, dispensed in flasks and diluted to the final volume. Preferably, the dextrose should be autoclaved separately and added aseptically, just before inoculation of the medium.

(ii) *Synthetic complete medium.* 10 g of succinic acid, 6 g of NaOH. Dissolve completely in 700 − 800 ml of water, then add 6.7 g of Yeast − Nitrogen Base (without amino acids) and 0.72 g of 'dropout mix' (see below). Bring the volume to 960 ml and sterilize. Add aseptically 40 ml of separately sterilized 50% dextrose solution, immediately before inoculation.

(iii) *Dropout mix.* 0.80 g each of adenine, uracil, arginine, histidine, methionine and tryptophan; 1.20 g each of lysine and tyrosine; 2.00 g of phenylalanine; 2.40 g of leucine, 8.00 g of threonine. Mix well. This is called dropout mix since individual ingredients are omitted in studies using auxotrophic yeast strains.

6. ACKNOWLEDGEMENTS

I would like to acknowledge the support of PHS Grant number Ca00911, a Research Career Development Award from the National Cancer Institute, the technical assistance of John Lopez in some of this work, the patient typing of revisions by Carol Unruh and Anna Leon, and Dr E.Skibo for critical reading of the chapter.

7. REFERENCES

1. Bennetzen,J. (1980) Ph.D. Dissertation, University of Washington, Seattle, Washington.
2. Cabib,E. (1971) In *Methods in Enzymology,* Jackoby,W.B. (ed.), Academic Press, New York, Vol. 22, p. 120.

3. Scott,J. and Schekman,R. (1980) *J. Bacteriol.*, **142**, 414.
4. Beltz,W. (1977) Ph.D. Dissertation, Florida State University, Tallahassee, Florida.
5. May,R. (1971) *Z. Allg. Mikrobiol.*, **11**, 131.
6. Wintersberger,U., Smith,P. and Letnansky,K. (1973) *Eur. J. Biochem.*, **33**, 123.
7. Duffus,J.H. (1975) *Methods Cell Biol.*, **12**, 77.
8. Ide,G. and Saunders,C. (1981) *Curr. Genet.*, **4**, 85.
9. Davie,J., Saunders,C., Walsh,J. and Weber,S. (1981) *Nucleic Acids Res.*, **9**, 3205.
10. Noll,M. and Kornberg,R. (1977) *J. Mol. Biol.*, **109**, 393.
11. Shalitin,C. and Vishlizky,A. (1985) *Curr. Genet.*, **9**, 107.
12. Tekamp.P., Garcea,R. and Rutter,W. (1980) *J. Biol. Chem.*, **255**, 9501.
13. Potashkin,J. and Huberman,J. (1986) *Exp. Cell Res.*, **165**, 29.
14. Klekamp.M. and Weil,A. (1982) *J. Biol. Chem.*, **257**, 8432.
15. Lohr,D. and Ide,G. (1979) *Nucleic Acids Res.*, **6**, 1909.
16. Lohr,D. and Ide,G. (1983) *J. Biol. Chem.*, **258**, 4668.
17. Kempers-Veenstra,A., Musters,W., Dekker,A., Klootwijk,J. and Planta,R. (1985) *Curr. Genet.*, **10**, 253.
18. Ide,G. (1981) *Biochemistry*, **20**, 2633.
19. Tekamp,P., Valenzuela,P., Maynard,T., Bell,G. and Rutter,W. (1979) *J. Biol. Chem.*, **254**, 955.
20. Ide,G. (1982) Ph.D. Dissertation, Oregon State University, Corvallis, Oregon.
21. Jerome,J. and Jaehring,J. (1986) *Mol. Cell. Biol.*, **6**, 1633.

CHAPTER 7

Isolation of yeast mRNA and *in vitro* translation in a yeast cell-free system

BARRY FEINBERG and CALVIN S.McLAUGHLIN

1. INTRODUCTION

The development of the yeast mRNA-dependent cell-free translation system, which is described in this chapter, serves a 2-fold purpose. First, it allows the combination of genetics, both classical and molecular, with biochemistry to answer many questions which remain on the mechanism and control of eukaryotic protein synthesis. Secondly, it provides an inexpensive and easily prepared system, for high activity translation of most eukaryotic mRNAs. With attention to the few details which follow, an active *in vitro* translation system and mRNA can be prepared from *Saccharomyces cerevisiae*.

2. ISOLATION OF MESSENGER RNA FROM YEAST

The keys to obtaining good preparations of mRNA are to minimize ribonuclease activity during the initial stages of extraction and to avoid the accidental introduction of small amounts of ribonuclease from your hands, glassware and solutions. Ribonucleases are very active enzymes and difficult to destroy. Trace amounts are sufficient to degrade mRNA; therefore, good laboratory technique is essential.

Sterile, disposable plasticware is essentially free of RNase and can be used in all RNA work without any pre-treatment. Glassware, however, is often a source of contamination and should be baked at $180-250°C$ for at least 4 h. It is advisable to set aside glassware and batches of plasticware for use only in experiments involving RNA.

A major source of potential RNase contamination is the hands of the investigator. Gloves should be worn at all stages during the preparation of materials and solutions used for RNA isolation and analysis as well as during manipulations involving RNA.

All solutions should be prepared using glass-distilled autoclaved water and dry chemicals used that are reserved for work with RNA and handled only with baked spatulas. The solutions should be treated with 0.1% diethylpyrocarbonate (DEPC) for at least 12 h, when possible. These treated solutions should then be autoclaved for 15 min to remove all traces of the DEPC, as any remaining DEPC will inactivate RNA by carboxymethylation. DEPC cannot be used to treat solutions containing Tris: it reacts with all primary amines and decomposes to ethanol and CO_2.

2.1 Isolation of mRNA from polysomes

Translation of mRNA in the yeast cell-free protein synthesis system is not inhibited by the presence of rRNA. Therefore, total polysomal RNA does not have to be fractionated into poly(A)$^+$ mRNA using oligo(dT)−cellulose columns if translation is the only interest.

2.1.1 *Polysome isolation*

The presence of the cell wall in yeast necessitates the use of rather harsh conditions (French press, glass beads, etc.) to rupture intact cells. This causes the rupture of vacuoles and the release of intracellular RNases. We prefer to isolate polysomes from spheroplasts which have the cell wall removed enzymatically. Much gentler conditions of lysis, such as the Dounce homogenizer, can then be used. Since polysomes are pelleted in the final centrifugation step in the preparation of an active protein synthesis lysate, these are a ready source of polysomal RNA. We store the drained polysomal pellets remaining after our S-100′ (lysate) preparations for protein synthesis (Section 3.1.4) at −70°C until needed for RNA extraction.

2.1.2 *Isolation of polysomal mRNA*

Total polysomal RNA is easily extracted from pools of frozen polysomal pellets (Section 3.1.4) by phenol extraction (1) (*Table 1*). The phenol should be as pure as possible. Commercial reagent-grade phenol must be re-distilled before use, with a mantle temperature of 160°C and the condenser fed with warm water to keep the phenol from solidifying. Phenol can cause severe burns and strict safety precautions are necessary, including the use of double gloves, safety goggles and a fume hood for the distillation. In addition, if solid phenol crystals are used, the distillation must be stopped before

Table 1. Extraction of RNA from polysomes.

1.	Resuspend polysome pellets from the S-100′ preparation in LET buffer to 50−60 A_{260}units/ml. Use a baked glass rod to ensure complete resuspension. LET buffer; 0.1 M LiCl; 1 mM EDTA; 0.1 M Tris−HCl (pH 7.5).
2.	Make 0.1% in SDS by adding 1/10 vol of 1%.
3.	Add an equal volume of phenol:chloroform:isoamyl alcohol (50:50:1 by vol) which has been pre-equilibrated with LET buffer.
4.	Shake for 10 min at room temperature.
5.	Centrifuge the solution at 10 000 g for 10 min at 20°C in baked Corex tubes or bottles.
6.	Collect the upper (aqueous) phase. Do not contaminate with interface material.
7.	Re-extract the bottom (organic) phase and interface materials with an equal volume of fresh LET buffer by shaking again.
8.	Centrifuge as in step 5.
9.	Remove the aqueous phase, combine with the first aqueous phase (step 6), and re-extract by shaking with an equal volume of fresh chloroform:isoamyl alcohol, 50:1.
10.	Centrifuge as in step 5.
11.	Remove the aqueous phase. Add 1/10 vol of 3 M sodium acetate (pH 5.2) followed by 2.5 vol of −20°C, 95% ethanol. Allow the RNA to precipitate overnight at −20°C.
12.	Centrifuge at 10 000 g for 10 min at 0°C.
13.	Remove the supernatant and wash the pellet twice in −20°C, 95% ethanol with 0.1 M sodium acetate by resuspension and centrifugation as in step 12.
14.	Decant the final wash and dry the RNA either in a stream of clean nitrogen or by lyophilization.
15.	Resuspend in a minimal volume of sterile water and read the A_{260} and A_{280}. Adjust the concentration to 200−250 A_{260} units/ml and store at −20°C.

the residue is too small and dark coloured, for the preservative which is used can form explosive products in the distillation vessel. Also, the distillation flask should be cooled before opening and exposing the residue to air. Because of these dangers it is easier to purchase nucleic acid grade, re-distilled phenol which is now commercially available from several vendors including Bethesda Research Laboratories. Re-distilled phenol should be saturated with water or buffer before use. Mix it with the aqueous solution, and let it stand overnight at room temperature. Store the water-saturated phenol at $-20°C$, but first blow N_2 onto the top of the jar and seal it tightly to keep out O_2.

Traces of phenol present in the RNA will inhibit translation. This can be minimized by performing the final organic extraction with only chloroform/isoamyl alcohol, precipitating the RNA several times and rinsing the RNA pellet with ethanol (*Table 1*).

Polysomal RNA, dissolved in water, can be safely stored at $-20°C$ for prolonged periods.

2.1.3 *Poly(A)$^+$ mRNA isolation by oligo(dT)−cellulose chromatography*

This is the most widely used method for the purification of mRNA which has a 3' poly(A)$^+$ tract (2), which includes almost all eukaryotic mRNA species. Up to 10 mg of polysomal RNA can be loaded per ml of oligo(dT)−cellulose. A second passage through the column is necessary if most of the contaminating rRNA is to be removed (*Table 2*).

Table 2. Oligo(dT)−cellulose affinity purification of poly(A)$^+$ messenger RNA.

1.	Equilibrate the oligo(dT)−cellulose (Collaborative Research) in sterile loading buffer. Prepare loading buffer from DEPC-treated and autoclaved stock solutions of NaCl and EDTA, an autoclaved (not DEPC-treated) stock of RNase-free Tris−HCl, and a 20% stock of SDS that has been heated at 65°C for 1 h. Loading buffer: 20 mM Tris−HCl (pH 7.6), 0.5 M NaCl, 1 mM EDTA, 0.1% SDS.
2.	Pour a 1−2 ml column of this oligo(dT)−cellulose in a sterile plastic syringe or baked Pasteur pipette loosely plugged with sterile glass wool or in a disposable plastic 5 ml column (Bio-Rad Laboratories).
3.	Wash the column with three column volumes each of: (a) sterile water, (b) 0.1 M NaOH, 5 mM EDTA and (c) sterile water. Continue washing with water until the pH of the effluent is <8.
4.	Wash with five column volumes of sterile loading buffer to re-equilibrate.
5.	Dissolve the RNA in sterile water at $50−100$ A_{260} units/ml. Heat at 65°C for 3−5 min. Add an equal volume of 2 × loading buffer, cool to room temperature and load this RNA on the column. Collect the unbound RNA which elutes, heat to 65°C again, cool and re-apply to the column.
6.	Wash the column with loading buffer. Collect 1 ml samples and check the absorbance at 260 nm. Continue to wash until there is little if any absorbance (~ 10 column volumes).
7.	Elute the poly(A)$^+$ RNA with sterile low salt elution buffer. Elution buffer: 10 mM Tris−HCl (pH 7.5). 1 mM EDTA, 0.05% SDS. Check the absorbance of the fractions.
8.	Adjust the NaCl concentration of the eluted RNA back again to 0.5 M and repeat steps 5−7.
9.	Pool the fractions containing mRNA and add 1/10 vol 3 M sodium acetate (pH 5.2) and 2.5 vol of $-20°C$, 95% ethanol. Allow the RNA to precipitate overnight at $-20°C$.
10.	Centrifuge at 10 000 g for 10 min at 0°C. Rinse the pellet with 70% ethanol, and dissolve in a minimal amount of sterile water. Store at $-70°C$.
11.	Regenerate the column by washing the oligo(dT)−cellulose with 0.1 M NaOH, 5 mM EDTA, then water, then loading buffer as in steps 3 and 4 above.

2.2 **Isolation of total cellular mRNA by chaotropic disruption—guanidinium thiocyanate**

This method is based on the ability of high concentrations of guanidinium salts to dissociate ribonucleoprotein complexes and denature proteins (3). This chaotropic disruption gives active RNA preparations from cells high in RNase, such as pancreas, and yields intact RNA preparations from yeast, with very few, if any, nicked molecules. This method is more difficult than RNA extraction from polysomes, but is necessary for procedures such as Northern blots and S1 nuclease mapping where completely intact mRNA is required.

2.2.1 *Cell lysis and chaotropic disruption*

The intact yeast cells are lysed by vortexing with glass beads in the presence of high concentrations of guanidinium thiocyanate (*Table 3*). A strong vortexer must be used and the procedure should be performed in the fume hood. The guanidinium and β-mercaptoethanol inhibit the RNases completely and dissociate all ribonucleoprotein complexes. This method yields total cellular nucleic acids, including tRNA, small nuclear RNAs, heterogeneous nuclear RNAs (mRNA precursors), and rRNA, as well as mRNA.

The guanidinium thiocyanate buffer, minus the sarkosyl and β-mercaptoethanol, should be stirred overnight at room temperature when it is prepared, and then warmed the next day to $60-70°C$ to assist in the dissolving of the salt. If there is still insoluble material, it can be removed by centrifugation at 3000 g for 10 min at 20°C. Then add the sarkosyl and β-mercaptoethanol, filter through a disposable Nalgene filter and store at 4°C.

2.2.2 *Separation of RNA in caesium chloride gradients*

This procedure takes advantage of the fact that the buoyant density of RNA in CsCl is greater than that of the other cellular macromolecules (4). During centrifugation, the DNA and protein float upward in the CsCl solution and the RNA forms a pellet at the bottom of the tube. After the centrifugation be sure to dry the walls of the ultracentrifuge tube with a sterile applicator after carefully removing the supernatant (*Table 3*). The purity of an RNA preparation can be checked by measuring the ratio of the absorbance of a dilution of the sample at 260 nm and 280 nm; 1 mg of RNA is equivalent to about 23 A_{260} units and pure RNA will have an $A_{260}:A_{280}$ ratio of 2.0.

3. TRANSLATION OF mRNA IN A CELL-FREE YEAST LYSATE

This section describes the preparation of a cell-free translation system from yeast cells and its use in the translation of exogenous natural and synthetic mRNA templates (5,6). The use of a cell-free system allows the investigator to identify mRNAs by the physical properties of the radiolabelled proteins which are produced and by the reaction of the products with specific antibodies. Cloned genes can be identified by the use of hybridization to select specific transcripts which can then be identified in a cell-free system (7). Genes can also be identified by their ability to hybridize specifically to the templates that they encode for, and thereby inactivate the mRNA activity (hybridization arrest) which is then assayed in a cell-free system (8).

Table 3. Guanidinium thiocyanate extraction of RNA.

1. Grow 4 litres of yeast overnight in YM-1 (*Table 4*) to an absorbance of 1 A_{660} unit/ml.
2. Harvest the cells by centrifugation at 3500 *g* for 5 min at 4°C.
3. Wash the cells in sterile, distilled water and re-centrifuge in four aliquots.
4. Resuspend each aliquot of cells in 10 ml of sterile water and centrifuge each in a separate, baked 30 ml Corex tube.
5. Add 0.5 − 1 ml of guanidinium thiocyanate buffer to each tube and then add baked, acid-washed, 0.45 mm glass beads just up to the bottom of the miniscus in each tube. Vortex each at maximum speed, in a fume hood, for four 30-sec bursts with 30-sec rests on ice.
    Guanidinium thiocyanate buffer (see Section 2.2.1): 6 M guanidinium thiocyanate (Fluka Chemical Co.), 5 mM sodium citrate pH 7.0 (prepare a 0.5 M stock at pH 7.0), 0.1 M β-mercaptoethanol (stock reagent is 14 M), 0.5% sarkosyl (sodium lauryl sarkosinate).
6. Add 1.5 ml of additional guanidinium thiocyanate buffer to each tube. Swirl gently and remove with a baked Pasteur pipette.
7. Wash the glass beads in each tube twice more with 1.5 ml of guanidinium thiocyanate and pool the three washes, keeping those from each tube separate.
8. Add an additional 10 ml of guanidinium thiocyanate to each of the four pools yielding a total volume of about 15 ml in each.
9. Centrifuge the four separate extracts at 10 000 *g* for 10 min at 20°C.
10. Remove the supernatants and dissolve 1 g of solid CsCl/2.5 ml in each of the four.
11. Carefully layer each guanidinium thiocyanate extract with the dissolved CsCl over a 3.7 ml cushion of 5.7 M CsCl in 0.1 M EDTA (pH 7.5), 5 mM sodium citrate (pH 7.0) in a polyallomer SW41 (Beckman Instruments) centrifuge tube. Centrifuge for 18 h at 28 000 r.p.m. at 20°C.
12. Carefully remove the supernatant and dry the walls of the tube with a sterile applicator (see Section 2.2.2).
13. Dissolve the pellet of RNA in: 10 mM Tris − HCl (pH 7.4), 5 mM EDTA, 1% SDS.
    The pellet may be difficult to dissolve and may require brief heating to 60°C.
14. Pool the dissolved RNA pellets and extract the RNA once with a 4:1 mixture of chloroform and 1-butanol. Transfer the aqueous phase to a fresh, sterile tube, and re-extract the organic phase with an equal volume of: 10 mM Tris − HCl (pH 7.4), 5 mM EDTA, 1% SDS.
15. Combine the two aqueous phases, and add 1/10 vol of 3 M sodium acetate (pH 5.2) and 2.5 vol of −20°C, 95% ethanol. Allow the RNA to precipitate overnight at −20°C.
16. Centrifuge at 10 000 *g* for 10 min at 0°C and decant the supernatant.
17. Dissolve the pellet in a minimal volume of sterile water. Re-precipitate with ethanol as in step 15.
18. Centrifuge and decant as in step 16.
19. Dry the RNA pellet in a stream of clean nitrogen or lyophilize to dryness.
20. Dissolve the RNA in a minimal volume of sterile water and store at −70°C.

This cell-free system carries out all the reactions involved in polypeptide chain − initiation, elongation, and termination — faithfully and reproducibly, and assays for the partial reactions allow the detailed study of the mechanism and control of protein synthesis and of the action of antibiotics which inhibit the process (9 − 11). Yeast is ideally suited to these studies because of the ease with which the powerful tools of classical and molecular genetics can be employed. The availability of a pool of mutants (12,13), and the possibility of identifying many more, has made it possible to clone several genes involved in this very complex process of protein synthesis. This will allow a level of understanding of the relationship of structure, function and control not before possible.

Table 4. Components of media and buffers used in the preparation of yeast lysate.

Medium or buffer	Components	Storage
YEP−glucose	2% glucose	Room temperature
	1% Bactopeptone	
	1% yeast extract	
YM1	2% glucose	Room temperature
	1% Bactopeptone	
	0.5% yeast extract	
	1% succinic acid	
	0.6% NaOH	
	0.67% Yeast−Nitrogen Base without amino acids	
YM5	2% glucose	Room temperature
	0.1% yeast extract	
	0.2% Bactopeptone	
	1% succinic acid	
	0.6% NaOH	
	0.67% Yeast−Nitrogen Base without amino acids and add 1/5 final volume of sterile 2 M $MgSO_4$ before use to give 0.4 M $MgSO_4$	
Buffer A	20 mM Hepes-KOH (pH 7.4)	4°C
	100 mM potassium acetate	
	2 mM magnesium acetate	
	2 mM dithiothreitol (DTT, added after autoclaving)	
Buffer B	As for buffer A except made up in 20% (v/v) glycerol	4°C

Media are prepared as a 2× concentrate without glucose, and sterilized by autoclaving. Glucose is prepared as a 40% stock solution, autoclaved separately. The glucose and sterile (autoclaved) water are added just before use, as is the sterile $MgSO_4$ stock solution to YM5. All buffers are sterilized by autoclaving, except that DTT is sterile-filtered and added to sterile buffer just before use. Note that although standard autoclaving conditions are normally 15 min at 120°C, we have found longer times to be necessary: 90 min for the peptone concentrate and 30 min for the glucose.

3.1 **Preparation of an active lysate**

A number of strains have been used in the preparation of an active lysate, but strain choice is an important consideration. In general we have found that strains that have a doubling time in rich media of less than 3 h at 25°C yield the most active lysates. In addition, the most active lysates are prepared from cells which have been spheroplasted; therefore, a strain which can easily be spheroplasted is preferable. We have obtained our most reproducible results using the prototrophic diploid strain SKQ2n (α/a, *ade1/+, +/ade2, +/his1*; isolated by Dr S.-K.Quah, University of Alberta). We have also used a number of haploid strains such as A364A and several temperature-sensitive mutants derived from it (14), 465/4d (15), LL-20 and CHY9B (unpublished), all of which are multiply auxotrophic.

3.1.1 *Growth of yeast cells*

(i)   Grow a 20−100 ml starter culture to stationary phase in a rich medium such as YEP−glucose or YM1 (see *Table 4*). This culture can be stored at 4°C, with little loss of viability for at least 4−6 weeks.

(ii)   Inoculate 5 × 1 litre or 3 × 1.5 litre batches of the same medium in 2 litre flasks with starter culture. We normally use a dilution factor of at least 1:2000 for a wild-type strain such as SKQ2n. Incubate the culture overnight at $20-25°C$ with vigorous agitation until a cell density of around $1-3 \times 10^7/ml$ ($A_{660} \sim 1.0$) is reached — normally after $12-16$ h. This should yield $2-2.5$ g wet weight of cells per litre.

3.1.2 *Spheroplast production*

The removal of the yeast cell wall to produce membrane-bound spheroplasts is a well established procedure (16) and gives lysates with the highest activity. Several different enzymes are commercially available which can be used to digest the cell wall to produce intact spheroplasts, provided that an osmotic stabilizer (1 M sorbitol) is present to prevent lysis. Although Glusulase (β-glucuronidase/arylsulphatase, Boehringer Mannheim) is more expensive than Zymolyase 60 000, we produce lysates with higher activity using the former. If Zymolyase 60 000 is used, then 0.5 mM phenylmethylsulphonyl fluoride (PMSF) must be included in the lysis and column buffers to reduce protease activity. The degree of spheroplasting must be carefully monitored to ensure that the digestion does not proceed too far. We routinely monitor spheroplasting during the reaction by diluting 20 μl of the reaction into 1 ml of water and 20 μl into 1 ml of 1 M sorbitol and briefly vortexing. Stop the cell wall digestion when the absorbance of the water dilution, at 600 nm, is reduced by at least 50% but not more than 75% of that of the 1 M sorbitol dilution. It is important that the yeast be in the early part of the log phase of growth for effective digestion with Glusulase. In addition, as cells move into the stationary phase of growth their protein synthetic activity declines (17). The activity of both enzymes can be increased by pre-treating the cells with β-mercaptoethanol. Due to strain variations, each new strain must be tested for the optimum method of spheroplasting. There are also batch to batch variations in the activity of the sphero-plasting enzymes, and therefore it is necessary to standardize each new batch of enzyme. The enzymes are stable at $-70°C$ and so it is advisable to purchase a fairly large quantity of an active batch, and store it for up to several years.

(i)   Wash the harvested cells once with 250 ml (50 ml/litre of original culture) cold ($2°C$) sterile water and centrifuge at 3000 g for 5 min.

(ii)   Resuspend the cells in 100 ml (20 ml/litre of original culture) of freshly prepared 10 mM β-mercaptoethanol, 2 mM EDTA, pH 7.0 and incubate at $20°C$ for 30 min. This step is not necessary with most strains of yeast that are harvested when they have grown to an A_{660} of 1.0 or less. If spheroplasting (step iii) takes longer than a 1-h incubation, this step should be included.

(iii)   Centrifuge yeast as above and resuspend in 200 ml of sterile 1 M sorbitol ($40-50$ ml/litre of original culture). Add $0.4-0.8$ ml of Glusulase/litre of original culture and gently swirl every few minutes. Monitor spheroplasting as described above. Enough Glusulase should be added so that spheroplasting is completed ($50-75\%$ reduction in A_{660}) in 60 min, although with some strains we have allowed the reaction to proceed for as long as 2 h with no deleterious effects.

(iv)   Harvest spheroplasts by centrifugation (3000 g for 10 min) and gently wash with 200 ml (40 ml/litre of original culture) of 1.2 M sorbitol. The spheroplasts must be treated very gently to avoid lysis. We find that they are easiest to resuspend after centrifugation if they are gently stirred with a sterile 20-cm lab-spoon spatula and carefully swirled in the centrifuge bottle. A wide-bore 10 ml pipette may also be used to pipette the spheroplasts *gently* until they are suspended.

(v) Resuspend the cells in YM-5 media containing 0.4 M $MgSO_4$ ($80-120$ ml of YM-5/litre of original culture). Incubate this suspension for $60-90$ min at $20°C$ with occasional swirling or gentle agitation on an orbital shaker ($30-50$ r.p.m.). This step is necessary to restore the protein synthetic capacity of the cells which has declined during spheroplast production.

3.1.3 *Cell homogenization*

It is imperative that only a small volume of lysis buffer be used if one is to obtain a concentrated lysate. The use of a protease inhibitor, such as 0.5 mM PMSF, is necessary when Zymolyase 60 000 is used to generate spheroplasts. We have also found that all-glass Dounce homogenizers work best, but they must be tight fitting. There is variability in the tightness even from the same manufacturer, and the homogenizer should be hand-picked to be tight fitting. We have purchased tight-fitting homogenizers from Wheaton, Bellco Glass and Kontes, but have also received homogenizers from all three manufacturers which were not tight enough. All glassware is baked overnight at $180°C$ before use.

(i) Harvest the spheroplasts by centrifugation (3000 *g* for 5 min). The pelleted spheroplasts can be quickly frozen and stored at $-70°C$ for at least 1 week before homogenization.

(ii) Resuspend the cell pellet in $1.5-2$ ml of buffer A (lysis buffer) per litre of original culture (see *Table 4*). If the cells are frozen add the frozen pellet directly to the baked, tight fitting, pre-chilled Dounce homogenizer. If the cells have not been frozen add the pellet directly to the Dounce homogenizer using a sterile, 20-cm lab-spoon spatula. Rinse the centrifuge bottle with a portion of the required lysis buffer and transfer the remaining spheroplasts and buffer to the homogenizer with a sterile Pasteur pipette. Maintain the homogenizer on ice while breaking the cells. This usually requires $15-30$ strokes of the homogenizer but occasionally more. Monitor the degree of breakage microscopically until few intact spheroplasts remain.

3.1.4 *Centrifugation of lysate*

The cell lysate is centrifuged in two steps. The first, a short 30 000 *g* spin, is to remove intact cells and cell debris including mitochondria and nuclei. A lipid layer floats to the top of this post-mitochondrial fraction and there is flocculent material just above the pellet. These should both be avoided. Use a sterile Pasteur pipette to withdraw the supernatant fraction (S-30) from between the two layers.

This S-30 fraction contains a small amount of 40S and 60S ribosomal subunits together with a high proportion of 80S monosomes and polysomes. It also contains tRNA, aminoacyl-tRNA synthetases, initiation, elongation and termination factors; however, it is unable to initiate translation on endogenous or exogenous mRNA, only elongate and terminate. However, if the S-30 is centrifuged a second time at 100 000 *g* for 30 min most of the polysomes are removed and the resultant S-100' initiates, elongates and terminates both the little remaining endogenous as well as exogenously added mRNA. This S-100' still contains the 40S and 60S ribosomal subunits as well as most of the 80S monosomes. The S-100' must be carefully removed from the centrifuge tube, using a baked Pasteur pipette, to avoid both the lipid layer at the top and the fluffy flocculent material at the bottom which overlays the polysome pellet. The resulting S-100' should be a pale yellow colour and clear. The surface of the polysome pellet should be briefly rinsed with buffer A and stored at −70°C for future isolation of mRNA (*Table 1*).

(i) Transfer the homogenate from the Dounce homogenizer to a sterile Nalgene centrifuge tube. Centrifuge at 27 000−30 000 g for 15 min at 4°C.

(ii) The supernatant (S-30) should be carefully aspirated avoiding the lipid layer at the top and the flocculent material at the bottom; 7−8 ml should be obtained from a 5 litre culture. Transfer the S-30 to a 10 ml 'Oakridge' type, clear, polycarbonate centrifuge tube.

(iii) Centrifuge the S-30 again at 100 000 g (40 000 r.p.m. in a Beckman 50Ti rotor) for 30 min after speed is reached.

(iv) The clear supernatant (S-100') is carefully aspirated with a Pasteur pipette to avoid the lipid layer which is on the top and the 'fluffy' material which is on top of the clear polysome pellet at the bottom. At least 3−4 ml of S-100' should be obtained from a 5 litre culture. Transfer this to a sterile, RNase-free tube. We routinely use sterile, disposable plastic tubes which are always RNase-free.

3.1.5 *Removal of endogenous amino acids*

The polysome-free S-100' contains a large number of low molecular weight components such as amino acids, polyamines, nucleotides, etc. These are removed by molecular sieve chromatography to reduce the concentration of endogenous amino acids. This allows for an increase in the specific activity of radiolabelled amino acids used to measure translation. The S-100' is loaded onto a Sephadex G-25 (medium) column (Pharmacia) previously equilibrated with buffer B (buffer A containing 20% glycerol). The glycerol helps to stabilize proteins, especially at the low temperatures used during storage.

It is essential that the column is not overloaded. Do not apply more than 3 ml to a 30 × 1.5 cm column and not more than 10 ml to a 40 × 2.4 cm column. The column should be checked for its ability to separate completely the high and low molecular weight components. This can be done with Blue Dextran 2000 (Pharmacia) mixed with vitamin B-12 or phenol red at basic pH. The mixture of the high molecular weight Blue Dextran with the low molecular weight red solutions yields a purple solution which should be completely separated into a blue solution, some clear buffer, and a red low molecular weight component after passage through the column. Be sure to prepare the test mixture in the same volume as that of lysate to be applied to the column.

(i) Prepare a column of sterile (autoclaved) Sephadex G-25 (medium) at 4°C. The column may be poured at room temperature and then thoroughly chilled, but a cold column should never be warmed, for air bubbles will form. Use a column of sufficient size to chromatograph the volume of lysate you have obtained. Five litres of original yeast culture should yield about 3 ml of lysate to be chromatographed.

(ii) Equilibrate the column with cold buffer B (*Table 4*) prior to application of the

S-100′. Three column volumes of buffer should be sufficient.

(iii) Halt the flow of buffer B from the reservoir and allow the column to drain until the layer of buffer B is just at the top of the Sephadex. Carefully apply the S-100′ with a baked Pasteur pipette, trying not to disturb the bed of G-25. When the S-100′ has just run into the column, apply 2−3 ml of buffer B with a sterile Pasteur pipette. Allow this to run in, and then add 4−5 ml of additional buffer using the Pasteur pipette. Secure the top of the column and allow the reservoir to flow. The pale yellow S-100′ can be seen proceeding down the column. When the band of yellow nears the bottom begin collecting fractions, 1 ml for the smaller column and 2 ml for the larger. The fractions containing the S-100′ are slightly opaque and faintly yellow, but the A_{260} of each fraction should be measured by preparing a 1:100 dilution.

(iv) Collect and pool the peak fraction, those fractions preceding the peak having an absorbance greater than 20 A_{260} units/ml, and *no more than one fraction after the peak fraction* (even if there are trailing fractions with an absorbance of greater than 20 A_{260} units).

(v) Measure the A_{260} of the pooled fractions, and aliquot 200 μl fractions in 0.5 or 1.5 ml microcentrifuge tubes. Quick-freeze these aliquots in liquid nitrogen or a dry ice−acetone bath and store at −70°C.

It is imperative that no more than one fraction following the peak fraction be collected from the G-25 column. We have found an inhibitor of translation in fractions on the trailing edge of absorbance in many of our preparations.

3.1.6 *Storage of lysate*

We have found that yeast cell-free lysates (S-100′) can be stored for several years at −70°C without loss of activity. This lysate can also be thawed and re-frozen several times with little loss of activity, but it is best to avoid many repeated cycles of thawing and freezing.

3.2 Translation of homologous and heterologous mRNA

A large number of homologous and heterologous mRNAs have been successfully translated in this system. One- and two-dimensional gel electrophoresis of the products synthesized *in vitro* with yeast polysomal RNA, when compared with the products synthesized *in vivo*, demonstrate that essentially all the mRNAs coding for products below 95 000 daltons are translated faithfully, both quantitatively and qualitatively. Heterologous RNA has been faithfully translated from many sources including rabbit reticulocytes (globin mRNA), Chinese hamster ovary cells, mouse myeloma RNA, brome mosaic virus RNA, tobacco mosaic virus RNA and turnip yellow mosaic virus RNA.

3.2.1 *Removal of endogenous mRNA*

Most of the polysomes, and along with them most of the mRNA, are removed during the second centrifugation step in the preparation of the lysate. This results in much lower levels of endogenous template activity than the reticulocyte system, but can be reduced even further by using micrococcal nuclease treatment (18). This nuclease is

Table 5. Incubation conditions for translation of mRNA in yeast lysate.

The translation reaction is usually performed in a final vol of 50 μl, but any volume can be used if the ratio of the components is maintained. If the products of translation are to be observed on polyacrylamide gels, then the reaction is best performed in plastic microcentrifuge tubes. If the entire sample is to be counted to assay for translation, then the reaction is best performed in 10 or 12 $\times$ 75 mm glass tubes.

1. Add the stock solutions of the following components, on ice:

Component	Stock concentration	μl per assay
Salt mix (see below)	20$\times$ (see below)	2.5
Cold amino acids	1.6 mM each, minus label (see *Table 7*)	1.25
Creatine phosphate	0.5 M	2.5
Creatine phosphokinase	8 mg/ml	1.25
ATP/GTP	20 mM/4 mM	1.25
CaCl$_2$	12.5 mM	1.0
Micrococcal nuclease	0.5−1.0 mg/ml	1.25
S-100′ (lysate)/buffer B[a] (see *Table 4*)	S-100′ (lysate) plus buffer B should be 1/2 the final reaction volume	25

2. The above components are incubated at 20°C for 10−15 min to destroy the endogenous mRNA.
3. Then add the following:

EGTA	12.5 mM[b]	3.0

4. Mix each tube gently to inactivate the nuclease. Then add the following:

Labelled amino acid i.e. [^{2}H]leucine	10−50 Ci/mM, 1 mCi/ml	5.0
or		
i.e. [^{35}S]methionine	800−1200 Ci/mM	3.0
mRNA—total polysomal RNA	10−20 mg/ml	5.0
or		
poly(A)$^+$ RNA[c]	200−1000 μg/ml	5.0
Water	to bring final volume to	50.0 μl

5. Incubate at 20°C for 60−90 min. If the entire translation assay is to be counted, stop the reaction with 0.5−1.0 ml of 10% TCA containing either 2 g/l casamino acids or 10 mM of the unlabelled amino acid which is radiolabelled in the reaction. See *Table 9*. If the translation products are to be analysed by SDS−PAGE, *do not add TCA* but follow the procedure in *Table 10*.

  Salt mix (20$\times$) composition:

  0.4 M Hepes-KOH (pH 7.4)

  3.2 M potassium acetate

  0.04 M magnesium acetate

  0.04 M DTT

All buffers are autoclaved after preparation and stored at 4°C. DTT is added after autoclaving and just before use. The DTT can be added to the 20$\times$ salt mix when the stock is prepared, using autoclaved salt and buffer stock solutions. This salt mix, with DTT, must be stored at −20°C and should be prepared fresh every 6 months. Except for the S-100′, poly(A)$^+$ RNA, and [^{35}S]methionine which should all be stored at −70°C, all the other translation reagents should be stored at −20°C[d]. The biochemical reagents are dispensed with baked spatulas, and solutions containing them are never autoclaved but prepared with sterile water. The translation reagents are stored in screw-top, sterile, polystyrene or polyethylene tubes and handled only with gloves to minimize RNase contamination. We purchase all our biochemicals from Boehringer Mannheim.

[a]Buffer B for the translation reaction can be prepared as a 2$\times$ or even 3$\times$ stock if extra volume is needed for other components.
[b]EGTA stock solution can be prepared at 37.5 mM and 1 μl used , instead of 3 μl of 12.5 mM, if extra volume is needed for other components.
[c]10−20 μg of deacylated *E. coli* or yeast tRNA, which has been prepared by stripping commercial tRNA (Boehringer Mannheim Biochemicals), should be added to the translation reaction when poly(A)$^+$ RNA is used as template. This will help to inhibit the low levels of endogenous RNase present in the lysate (19). See *Table 8*.
[d]The [^{35}S]methionine and creatine phosphokinase should not be re-frozen after thawing. The micrococcal nuclease, poly(A)$^+$ RNA [oligo(dT)−cellulose purified], and S-100′ can be frozen and thawed four or five times. All the other reagents, including the total polysomal RNA, can be frozen and thawed repeatedly.

Table 6. Concentrations of added components in yeast *in vitro* protein synthesis.

Component	Final concentration
S-100′ (yeast lysate)	3−8 mg protein/ml
19 cold amino acids	40 μM
Hepes-KOH (pH 7.4)	30 mM
Potassium acetate	210 mM
Magnesium acetate	3.0 mM
ATP	0.5 mM
GTP	0.1 mM
Creatine phosphate	25 mM
Creatine phosphokinase	200 μg/ml
Glycerol	10%
DTT	3 mM
$CaCl_2$	0.25 mM
EGTA	0.75 mM
Micrococcal nuclease	12.5−25 μg/ml
[^{3}H]amino acid	100 μCi/ml
or [^{35}S]methionine	600 μCi/ml
Total polysomal RNA	0.7−2 mg/ml
or poly(A)$^+$ RNA	10−100 μg/ml

Ca^{2+}-dependent and can thus be inactivated by the addition of the Ca^{2+}-chelating agent EGTA (ethylene glycol bis [β-aminoethyl ether]-*N,N,N′,N′*-tetraacetic acid) that has been previously neutralized with 2 M NH_4OH and filter sterilized. The conditions which are used for the nuclease digestion eliminate most of the endogenous messenger activity but do not damage the activity of the ribosomes or the tRNA in the S-100′.

The translation reaction is carried out with a two-step incubation. The first step is the nuclease digestion of mRNA in the S-100′ supplemented with salts, amino acids and an energy-generating system. The nuclease is inactivated with EGTA, the exogenous mRNA and isotopic amino acid are added, and amino acid incorporation is assayed.

The micrococcal nuclease, isolated from *Staphylococcus aureus*, is added so that its concentration during digestion is 17−34 μg/ml. There is considerable variation between batches of nuclease (as purchased from Boehringer Mannheim), so each batch must be assayed to determine the exact amount to add and the duration of incubation (usually 10−15 min at 20°C). The concentration of Ca^{2+} in the digestion is approximately 0.3 mM and the final concentration of the EGTA used to stop the nuclease is approximately 1 mM.

3.2.2 *Translation conditions for exogenous natural mRNA*

The yeast lysate must be supplemented before the nuclease digestion with K^+ and Mg^{2+} ions, unlabelled amino acids other than the radiolabelled one which will be used to measure translation, ATP and GTP and a nucleoside triphosphate-regenerating system. We routinely use a final K^+ concentration of 210 mM and a final Mg^{2+} concentration of 3 mM; however, these ionic conditions should be optimized for the mRNA which you are translating. Use the acetate salts of these ions because the Cl^- salts inhibit translation (6). The protocol for the assays is given in *Table 5* and the final concentration of the components is given in *Table 6*.

Table 7. Amino acid stock solutions for translation.

L-Amino acid	Mol. wt	mg/20 ml stock	Concentration (mM)
Alanine	89.1	89	50 mM
Arginine	174.1	174	50
Asparagine	132.,1	132	50
Aspartic acid	133.1	133	50
Cysteine	121.2	Add solid directly to amino acid mix	
Glutamic acid	147.1	73.5	25
Glutamine	146.1	146	50
Glycine	75.1	75	50
Histidine	155.2	155	50
Isoleucine	131.2	131	50
Leucine	131.2	131	50
Lysine	146.2	146	50
Methionine	149.2	149	50
Phenylalanine	165.2	82.6	25
Proline	115.1	115	50
Serine	105.1	105	50
Threonine	119.1	119	50
Tryptophan	204.1	204	50
Tyrosine	181.2	Add solid directly to amino acid mix	
Valine	117.1	117	50

A 1.6 mM (40×) mixture of the 19 non-isotopic amino acids required for translation (deleting the isotopic amino acid used for the assay) is prepared by adding:
  320 μl each of the 50 mM stock solutions
  640 μl each of the 25 mM stock solutions
  2.9 mg of solid tyrosine
  1.9 mg of solid cysteine
Add sterile water up to 10.0 ml in a 15 ml graduated, screw-top, polystyrene tube. Vortex until dissolved. Divide into 1 ml aliquots in sterile tubes and store at −20°C.

Table 8. Deacylation of tRNA.

1. Dissolve tRNA to 0.6−1.0 mg/ml in 1.0 M Tris−HCl (pH 9.0 at 23°C), 0.01 M MgCl$_2$.
2. Incubate at 37°C for 1 h.
3. Cool on ice and lower pH to 5−5.5 with HCl so that tRNA will precipitate in next step.
4. Add 1/10 vol 2 M potassium acetate (pH 5.0), followed by 2 vol of 95% ethanol. Allow tRNA to precipitate overnight at −20°C.
5. Centrifuge tRNA at 10 000 g for 20 min at −20°C.
6. Resuspend the pellet of tRNA in sterile water and dialyse for 2 h in sterile water. Boil dialysis tubing for 10 min in 1 mM EDTA and 7 mM mercaptoethanol before use. Handle the tubing with gloves only. It can be stored in the refrigerator in sterile water for several weeks.
7. Store the tRNA at −20°C in sterile tubes.

We routinely use radiolabelled methionine or leucine in our assays. A mixture of the 19 non-isotopic amino acids is prepared as a 40-fold concentrate (1.6 mM). Stocks of the individual amino acids, except for cysteine and tyrosine, are used to make the mixture. The stock solutions are 50 mM for each amino acid except glutamic acid and phenylalanine, which are 25 mM. Add the cysteine and tyrosine separately, in solid form, after the mixture is prepared (*Table 7*).

Table 9. Assay incorporation of isotopic amino acid into protein.

1.	Set up 50 μl translation assays, in glass test tubes as described in *Table 5*.
2.	Incubate the reactions at 20°C for 60−90 min.
3.	Terminate the reactions by the addition of 0.5−1.0 ml 10% TCA with amino acids as described in *Table 5*. Leave on ice 10 min.
4.	Transfer the reaction tubes to a water bath at 85−100°C for 15 min.
5.	Briefly vortex each tube and fill with 5% TCA containing 1 g/l casamino acids. Filter each through a Whatman 934 AH glassfibre filter, and wash each filter three times by filling the reaction tube three successive times with casamino acid supplemented 5% TCA.
6.	Dry the filters under a heat lamp or place in scintillation vials directly and dry in a warm oven (50−60°C). Fill the vials with scintillation fluid and count.

Table 10. Preparation of [^{35}S]methionine-labelled translation products for SDS−PAGE.

1.	Set up 25 μl translation assays in clean, new 1.5 ml plastic microcentrifuge tubes, with [^{35}S]methionine as the label, and using half the volumes indicated in *Table 5*.
2.	Incubate at 20°C for 60 min.
3.	Remove 5 μl of the assay and determine the amount of incorporation of [^{35}S]methionine into hot TCA-insoluble material as described in *Table 9*.
4.	Add pancreatic RNase A to the remaining 20 μl to a final concentration of 100 μg/ml and incubate at 37°C for 15 min.
5.	Add 1.0 ml of ice-cold 90% (v/v) acetone to each reaction tube and leave at −20°C for 1−2 h.
6.	Centrifuge the precipitated protein in a microcentrifuge for 10 min. Thoroughly air dry the pellet and resuspend in an appropriate SDS−PAGE sample buffer.

Add the labelled amino acid after the nuclease activity is inhibited by EGTA, followed by the addition of the RNA you wish to assay. We routinely add total polysomal RNA to the translation system as there is no inhibition of protein synthesis by the rRNA present. Yeast polysomal RNA is usually prepared as described earlier (see Section 2.1 above) to a concentration of 200−250 A_{260} units/ml, and added to a final concentration of 600−1200 μg/ml. When poly(A)$^+$ RNA [oligo(dT)−cellulose purified] is used, the addition of deacylated (stripped) yeast or *Escherichia coli* tRNA to 200−400 μg/ml is required for maximal activity (*Table 8*).

After the addition of the mRNA the assay is incubated at 20°C for 60−90 min. It is important that the temperature is not allowed to rise above 22°C at any stage in the isolation and use of the lysate, because a nuclease that efficiently degrades mRNA becomes activated (19).

The assay is terminated with 10% trichloroacetic acid (TCA), as described in *Table 5*, and the samples are treated, filtered and counted as indicated in *Table 9*. However, if you wish to examine the products which have been synthesized by one- or two-dimensional electrophoresis, then add TCA to only a small aliquot of the total reaction to measure synthesis and treat the sample as described in *Table 10*.

Under these conditions, translation of exogenous mRNA continues for at least 1 h. There is minimal endogenous protein synthesis confirming the stringent dependence on added template. Typical results of a 50 μl assay using [^{3}H]leucine at 50 Ci/mmol are 2−5 × 10^3 c.p.m. without added mRNA and 4 × 10^5−1.5 × 10^6 c.p.m. with polysomal RNA. There is usually a 100- to 500-fold stimulation with exogenous template. A control tube without mRNA should always be run to check on the efficiency of the micrococcal nuclease reaction.

4. REFERENCES

1. Galis,B.M., McDonnel,J.P., Hopper,J.E. and Young,E.T. (1975) *Biochemistry*, **14**, 1038.
2. Aviv,H. and Leder,P. (1972) *Proc. Natl. Acad. Sci. USA*, **69**, 1408.
3. Chirgwin,J.M., Przybyla,A.E., MacDonald,R.J. and Rutter,W.J. (1979) *Biochemistry*, **18**, 5294.
4. Glisin,V., Crkvenjakov,R. and Byus,C. (1974) *Biochemistry*, **13**, 2633.
5. Gasior,E., Herrera,F., Sadnik,I., McLaughlin,C.S. and Moldave,K. (1979) *J. Biol. Chem.*, **254**, 3965.
6. Tuite,M.F., Plesset,J., Moldave,K. and McLaughlin,C.S. (1980) *J. Biol. Chem.*, **255**, 8761.
7. Prives,C.L., Aviv,H., Paterson,B.M., Roberts,B.E., Rozenblatt,S., Revel,M. and Winocour,E. (1971) *Proc. Natl. Acad. Sci. USA*, **71**, 302.
8. Paterson,B.M., Roberts,B.E. and Kuff,E.L. (1977) *Proc. Natl. Acad. Sci. USA*, **74**, 4370.
9. Gasior,E., Herrera,F., McLaughlin,C.S. and Moldave,K. (1979) *J. Biol. Chem.*, **254**, 3970.
10. Feinberg,B., McLaughlin,C.S. and Moldave,K. (1982) *J. Biol. Chem.*, **257**, 10846.
11. Herrera,F., Moreno,N. and Martinez,J.A. (1984) *Eur. J. Biochem.*, **145**, 339.
12. Hartwell,L.H. and McLaughlin,C.S. (1967) *Proc. Natl. Acad. Sci. USA*, **59**, 422.
13. Hartwell,L.H. and McLaughlin,C.S. (1968) *J. Bacteriol*, **96**, 1664.
14. Herrera,F., Martinez,J.A., Moreno,N., Sadnik,I., McLaughlin,C.S., Feinberg,B. and Moldave,K. (1984) *J. Biol. Chem.*, **259**, 14347.
15. Tuite,M.F., Cox,B.S. and McLaughlin,C.S. (1981) *J. Biol. Chem.*, **256**, 7298.
16. Hutchison,H.T. and Hartwell,L.H. (1967) *J. Bacteriol.*, **94**, 1697.
17. Boucherie,H. (1985) *J. Bacteriol.*, **161**, 385.
18. Pelham,H.R.B. and Jackson,R.J. (1976) *Eur. J. Biochem.*, **67**, 247.
19. Herrera,F., Gasior,E., McLaughlin,C.S. and Moldave,K. (1979) *Biochem. Biophys. Res. Commun.*, **88**, 1263.

CHAPTER 8

Isolation and analysis of cell walls

BRIAN J.CATLEY

1. INTRODUCTION

To understand the methods used for the isolation and fractionation of yeast cell walls and to appreciate that these protocols may modify the product, it is necessary to know something of the walls' components and architecture.

The proportion of dry cell weight that is found outside the cytoplasmic membrane is large, typically in the region of 25%. It is referred to loosely as the yeast cell wall. This extracellular mass, a proportion of which may contain polymers contributing little to the supportive structure but necessary for cell protection and control of nutrition, comprises mostly polysaccharides and glycoproteins with a high proportion of carbohydrate. Typical components are the β-1,3- and β-1,6-D-glucans, α-1,3- and α-1,4-D-glucans, cellulose, mannoproteins and chitin (1). Combinations of these polymers have been used to classify yeasts and fungi (2). There is evidence that these populations of polymers may be covalently linked to one another, for example chitin to protein and β-D-glucan (3) and β-D-glucan to mannoprotein (4). Mannoproteins may be linked by disulphide bonds (1). A multitude of hydrogen and hydrophobic bonds contribute to the integrity of the complex.

The isolation of cell walls, damaged as little as possible, requires that glycosidic and peptide bonds must be protected from conditions that promote their hydrolysis. The arithmetic is persuasive: hydrolysis of 1% of glycosidic bonds of a polysaccharide comprising 1000 residues yields products of an average M_r, 10% that of the original molecule. Acidic conditions and exposure to carbohydrases, themselves often components of the extracellular mass, must therefore be avoided if intact walls are required. Proteases from within the cell, released by its rupture, can equally damage peptide moieties of glycoprotein components. Particularly acid-labile bonds such as sialyl or furanosyl are not normally found in yeast cell walls.

Brief exposure of mannosyl and glucosyl bonds to mild acid will produce some hydrolysis. The extent of cleavage that is tolerable is related to the purpose for which the cell wall is isolated. If it is for the raising of antibodies or as a substrate in a nutrient medium for microbial growth it will matter less than if it is used for a study of cell wall architecture. A measure of susceptibility of a major constituent in yeast cell walls, the β-D-glucans, is that approximately 20% of glucosidic bonds are hydrolysed in 0.3 M acid held at 100°C for 30 min. The extent depends both on the sugar residue and the linkage, for example chitin is more stable. The stability of glucosidic bonds in alkali is very much greater, and concentrations of NaOH, greater than 1 M, at 100°C are routinely used to solubilize cell wall polysaccharides. Oxidation under these rigorous conditions is minimized by purging the solvent with, and maintaining an atmosphere of, nitrogen. Polymer bonds other than glycosidic are also labile at extremes of pH. Thus chitin, a polymer of *N*-acetyl glucosamine, is de-acetylated in strong alkali to produce the polymer chitosan. Carbohydrate moieties covalently bound to peptides through seryl or threonyl residues, for example the smaller mannosyl side chains of mannoprotein, are released by exposure to weak base. Thus storage at 20°C for 18 h at pH 12 is sufficient for cleavage to take place through an elimination reaction.

Moreover, it is not only extractants that can modify the extracted components. The initial centrifugation, pelleting cells from the growth medium, is often sufficient to remove the highly hydrated capsular polysaccharide that is loosely bound to the underlying cell wall. These gels are not part of the conventional view of yeast cell walls but should be considered if a total picture of the extracellular macromolecular array is required, for example in a study of cellular antigenicity or adhesiveness.

2. PREPARATION OF CELL WALLS

2.1 Cell physiology and time of harvesting

Before commencing the preparation of cell walls it is worthwhile considering the necessity of using a homogeneous population of defined cells and the importance of taking cells from a particular point in the cell cycle or phase of the growth cycle. Again this depends on the reasons for preparing the cell walls, but, whatever they are, it is important to realize that the composition and structure of the wall can change from strain to strain and from one phase of a cell or growth cycle to another. The mechanisms of fission, budding or germ-tube formation require changes in the cell wall and the deposition of scars of distinctive composition. Selection of cells from a particular point in the growth cycle is described below. The preparation of cells for synchronous growth is described in Chapter 3.

Measurement of absorbance, more correctly turbidity, using an arbitrarily selected wavelength, usually lying between 540 and 600 nm, provides a conveniently fast measure of cell growth. Preliminary studies using other measures of cell growth, such as dry weight or cell number, will have established a relationship between phases of the growth cycle and absorbance and the range over which these parameters bear a linear relationship with absorbance. Absorbances are typically measured in 1 cm path length cells at 540 nm. Cell numbers can be estimated using a haemocytometer. Dry cell weights can be measured by filtering an aliquot of cell suspension through a previously dried and weighed membrane filter, for example 2.5 cm diameter, 0.8 μm pore size, followed by washing several times with water and drying at 60°C for 18 h, preferably over P_2O_5. Severe caramelization of the nutrient medium during sterilization can produce absorbances that are not related to cell density and it is worthwhile autoclaving the carbohydrate component of the medium separately.

Harvesting, whether by centrifugation or filtration, may change the cellular environment. Alterations which take place to the cell wall upon removal of nutrient, for example by washing in water, buffer or saline are not matched by the rapidity of intracellular changes. Nevertheless, exposure of whole cells to these washes for more than 30 min might be expected to initiate changes in the cell wall. Indeed, since the subsequent preparation of cell walls requires extensive washing it is probably not necessary to clean the cells exhaustively at this stage though at least one wash is recommended to remove extracellular hydrolases. If washing or filtration is for the replacement of one isotope by another, for example in a pulse-chase investigation, then the supernatant fluid from an identical but unlabelled culture can be used. In this way the composition of the cellular milieu remains essentially unchanged.

Centrifugation is carried out at room temperature unless it is suspected that the difference between growth and ambient temperature is sufficient to initiate changes in cell wall structure as can happen in some dimorphic fungi. Typically harvesting is at $2000-3000$ g for 10 min. Washing is assisted by resuspending the pellet in medium followed by a few passages through a pipette to disperse small clumps.

Filtration is a faster method of harvesting cells but becomes impracticable for large volumes. The circular filter, purchased or cut from sheets, is clamped onto a fritted disc or stainless steel mesh support. The latter is preferred, being easier to clean. Useful porosities of cellulosic filter, such as those supplied by Millipore or Oxoid, are $0.4-0.8$ μm diameter. The glassfibre filters, for example GF/C supplied by Whatman, have less well defined porosity but usually a faster flow-rate. Care must be taken to avoid piercing the filters or dislodging the glass fibres during washing of cells by resuspension.

It may be prudent at this stage to consider storing the supernatant and washings. Reference has already been made in Section 1 to the looseness of extracellular capsules. They may well have been removed during centrifugation. They can be isolated in a fashion similar to that for the soluble fractions of the cell wall described in Section 3. Storage should be below 0°C to avoid endogenous enzymic modification.

Harvested and washed cells are resuspended in a medium (Section 2.2.1) designed to minimize the effect of carbohydrases and proteases released in the subsequent disintegration of the cell. Storage, if necessary at this point, should be below 0°C but it is recommended that cell breakage is attempted as soon as practicable if the fine structure of the wall is to be maintained.

2.2 Cell breakage procedures

2.2.1 *General considerations*

There are a number of methods for disrupting the yeast cell, the majority of which use hardware designed for the purpose. All methods will cause some disruption but, presented with cells, the breakage of which has not previously been described, the investigator cannot easily predict which method will be most effective. Selection is usually by trial and error. Methods to be preferred are those where no addition to the cell suspension is necessary.

There are two main categories of disruption method. Firstly, cold or frozen cells are forced through a small hole to produce shear forces or ice-phase changes that rupture the cell. Examples of this method are the French, Eaton and X-press. Sonication is sometimes effective. Alternatively cell suspensions are violently agitated with abrasive particles or beads that produce shearing forces within the liquid. The disadvantages of these methods are the removal of beads, whole and disintegrated, and the generation of heat.

It is advisable to add protease inhibitors to the cell suspension prior to disruption

Table 1. Assay of β-1,3-D-glucan hydrolase.

1.	Prepare an aqueous solution of a β-1,3-D-glucan, for example soluble laminarin or curdlan, at a concentration of 4 mg/ml. Heating at 100°C may be necessary. Some preparations of glucan contain glucose and low mol. wt sugars that produce unacceptably high blank readings. If this is the case dialyse the solution against 20 vol of water at room temperature with one change.
2.	Incubate 0.5 ml of broken cell suspension with 1.0 ml of glucan solution and 0.5 ml of water at 37°C in a shaking water bath. The pH is that of the breakage medium buffer.
3.	At suitable intervals, e.g. 5, 20 and 60 min, withdraw 0.5−0.6 ml, spin the sample in a microcentrifuge to pellet cell debris and assay 0.4 ml of supernatant solution for reducing power. The assay used, e.g. Somogyi-Nelson (5), should be capable of producing a reasonable absorbance (e.g. 0.5) with 100 μg of glucose.
4.	Alternatively the broken cell suspension is centrifuged at 3000 *g* for 10 min and the supernatant assayed for hydrolase activity as in steps 2−3. In this case the microcentrifuging of the samples is unnecessary.

Table 2. Assay of β-1,6-D-glucan hydrolase.

Pustulan, a β-1,6-D-glucan is often supplied as the naturally O-acetylated polysaccharide. De-acetylation is effected by mild acid hydrolysis.

1.	Finely grind 2 g of crude pustulan and transfer it to a 200 ml round-bottomed flask.
2.	Add 50 ml of 0.15 M H_2SO_4 and place in a boiling water bath for 30 min. It is advisable to mount a reflux condenser on the flask.
3.	Decant the contents of the flask and centrifuge down (2000 *g* for 5 min) any particulate matter. Discard the pellet.
4.	Neutralize the supernatant solution with 2 M NaOH.
5.	Dialyse the solution against 1 l of water with one change.
6.	Concentrate the dialysate to 15−20 ml by rotary evaporation.
7.	With stirring add 3 vol of ethanol over a period of 2−3 min and allow the white precipitate to settle overnight at 4°C.
8.	Decant or aspirate the supernatant liquid and wash the precipitate by centrifugation or on a sintered glass funnel with 20−30 ml of 80% ethanol (twice) and finally 20−30 ml of ethanol.
9.	Remove the residual ethanol by storage in an evacuated desiccator over $CaCl_2$.
10.	Prepare an aqueous solution of purified pustulan at a concentration of 4 mg/ml.
11.	Assay β-1,6-glucan hydrolase activity as described in *Table 1*, steps 2 and 3 or 2 and 4.

Table 3. Assay of protease.

A convenient 'catch-all' assay for proteases is to use as substrate some insoluble protein to which a dye has been covalently bound. Hydrolysis by the protease(s), each with its own specificity, will solubilize dye-tagged peptide fragments the absorbance of which is an arbitrary measure of proteolysis. It also has the advantage of being a visible process. One such dye-tagged protein is azocoll.

1.	Place 1.3 ml of 100 mM buffer in a capped 2.0−2.5 ml microcentrifuge tube. The pH of the buffer is determined by the nature of the investigation but might be that of the breakage medium.
2.	Add 3 mg of azocoll and 100 μl of cell lysate, and close the lid tightly. Wrap the top of the tube with slightly stretched parafilm to ensure a water-tight seal.
3.	Incubate at 37°C with sufficient shaking to ensure that the azocoll is kept in suspension. This is best done by placing the tubes in a beaker clamped in a reciprocating shaker moving at such a speed as to keep the tubes on the move.
4.	At the end of 1 h remove the tubes, centrifuge them at full speed in the microcentrifuge for 2−3 min, then measure the absorbance of the supernatant solution at 520 nm.

in order to eliminate or minimize hydrolysis of cell wall glycoproteins. Maintenance of low temperatures and careful selection of buffers also combat cell wall hydrolysis. Examples of cell breakage will be described to illustrate the methods available. Pressure devices such as the French, Eaton or X-press are substantial volumes of metal, and time must be allowed for the whole of their mass to be brought to 0°C or lower. It is easier to surround the vessels of top-driven homogenizers with coolant than those of bottom-driven models. Again, time must be allowed for the cooling of containers, drive shafts, beads and receptacles before disruption takes place. Suppliers of cell breakage appliances described below are listed in the Chapter Appendix.

Partial hydrolysis of cell wall glycans by released hydrolases presents a problem if it is essential to maintain the wall constituents in their original macromolecular state. Since the majority of the carbohydrases have acidic optimum activity, the breakage medium is often adjusted to pH 8−9 to minimize their effects. It may be worthwhile studying the pH dependence and susceptibility to inhibition of these hydrolases. General assays for carbohydrases and proteases are described in *Tables 1−3*. Although positive identification of hydrolases will not provide an assessment of their action on cell wall structure, it will allow the investigator to devise conditions that minimize their activity. The reason for preferring the broken cell suspension rather than the supernatant derived from it is that occasionally hydrolases bind to the cell wall and cell debris.

Protease activity can often be removed by the addition of specific inhibitors. Proteases active at acidic pH may be inhibited by including pepstatin at a final concentration of 10 μg/ml in the cell-breakage medium. Stock solutions are prepared at a concentration of 0.5 mg/ml in methanol and should be made up on the day of use. Other proteases may be inhibited by phenylmethylsulphonyl fluoride (PMSF). Care should be taken in handling this solid as it is a powerful inhibitor of serine proteases. The handling of stock solutions, for example 20 mM in ethanol, should be done with gloves and behind a screen in the fume cupboard. It is best prepared just prior to use. The final concentration in the cell-breakage medium should be about 1 mM. Spills should be treated with dilute alkali followed by a thorough washing with water.

A word of caution; some breakage procedures are not normally conducted in sealed vessels, for example blending with beads, ultrasonic disintegration etc. If the organism is listed as pathogenic, appropriate safety measures must be taken.

2.2.2 *Cell breakage by Braun homogenizer (MSK)*

(i)   Ensure that the container and lid are clean, fit well, and that there is an adequate supply of compressed CO_2 for cooling the container during cell breakage.

(ii) Cool glass beads (0.45−0.50 mm diameter), buffer and container to 0°C in melting ice.

(iii) Mix the pelleted, drained wet cells with buffer and beads in the ratio 1:2:1 wet-wt/v/v. If the wet weight of the cells is not known then calibrate the centrifuge tube in ml and assume that 1 ml is approximately equal to 1 g. Typical buffers might be 50 mM phosphate, pH 8.5 or acetate, pH 5.0 containing appropriate inhibitors (see Section 2.2.1). For efficient breaking a total volume of 30 ml is recommended for a 75 ml container. A typical homogenization might thus comprise 7−8 ml of packed cells, 16 ml of buffer and 7−8 ml of glass beads (10 ml of beads weighs ∼18 g). If sufficient breakage is not achieved, cell and bead weight/ml should be increased up to 2-fold.

(iv) Insert the container into the Braun oscillator and shake at about 4000 cycles/min. To minimize heating the contents a series of 30−40 sec shakes should be used, the container being cooled with CO_2 snow generated on release of gas from the cylinder. It is this aspect of the method that is least satisfactory as it is difficult to avoid local freezing of the contents. A series of trial runs is recommended to find the best conditions. A total of 1.5−2 min shaking is usually sufficient to achieve better than 90% breakage.

(v) Remove the beads and glass debris by filtration through nylon mesh (45 μm square) or a coarse sintered glass filter and wash the retained material with two 5 ml vol of cooled buffer. Discard the retained material. Alternatively, having stored the container in an upright position for 2−3 min, aspirate or decant the cloudy supernatant liquid from the beads into a cooled receiver. Wash the beads with two 5 ml vol of cold buffer.

(vi) Centrifuge the filtrate or aspirated suspension and washings at 3000 g for 10 min at 4°C and discard the cloudy supernatant liquid.

2.2.3 *Cell breakage by micronizing mill*

Micronizing mills, developed for producing powders from coarse granular materials, can be used for breaking yeast cells. A typical bench mill, powered by a 1/3 h.p. motor, accommodates a polypropylene vessel of about 125 ml capacity, the major part of which is occupied by 48 cylindrical grinding elements (1.2 × 1.2 cm) composed of agate, hot-pressed polycrystalline alumina, hot-pressed boron carbide, monocrystalline sapphire or self-bonded silicon carbide. Satisfactory breakage of yeast cells has been obtained with the first two. A disadvantage is that heat is generated. The grinding must therefore be periodically stopped and the breakage vessel removed to be cooled in melting ice, a procedure which is easily performed.

(i) Cool the breakage vessel and grinding cylinders in melting ice; then dry.

(ii) Form the small agate or alumina cylinders into a cylindrical stack comprising 2.7 layers containing eight cylinders in each layer. This should be done on a small board. Slide the breakage vessel over the array, covering it completely. Invert the whole and remove the board. The array of grinding elements should now be neatly stacked in the vessel.

(iii) Pour in a slurry of cold buffer (Section 2.2.2.iii) and pelleted cells (1:1 v/v), the maximum volume being about 5 ml.

(iv) Screw the cap firmly onto the vessel, place it in the vibro-mill and switch on. Five 30−45 sec periods of vibration with intermittent cooling are usually sufficient to achieve a good breakage.

(v) Remove the cap and decant the slurry of cell debris into a cooled receiver. Wash the cylinder stack with four 5 ml vol of cold buffer, adding the washings to the suspension of debris.

(vi) Centrifuge the debris and washings at 3000 g for 10 min at 4°C and discard the cloudy supernatant liquid.

2.2.4 *Cell breakage by beads-in-a-tube*

Breakage of cells using the somewhat cumbersome apparatus of the Braun homogenizer can sometimes be replaced on the small scale by use of a test tube, glass beads, and bench-top vortex mixer. It is worthwhile carrying out preliminary experiments to see if this method is feasible, especially if a large number of small samples have to be processed.

(i)   Cool to 0°C a thick-walled glass tube ($\sim 1.5-2.0$ cm diameter and 20 ml vol), a quantity of glass beads (0.45−0.50 mm diameter) and buffer.

(ii)   Mix moist pelleted yeast cells, cooled buffer and beads in the ratio 1:2:1 wet-wt/v/v (Section 2.2.2.iii) to a final volume of about 10 ml. Store for 5 min in melting ice with occasional brief mixing.

(iii)   Agitate the tube vigorously, using a bench-top vortex mixer, for periods of 30−40 sec and a total agitation time of 4−5 min. Between each agitation cool the tube in melting ice for about 1 min.

(iv)   Store the tube in an upright position allowing the glass beads to settle and using a Pasteur pipette aspirate the supernatant broken-cell suspension into a cooled receiver. Alternatively, use a nylon mesh filter (Section 2.2.2.v). Wash the beads by stirring with three to four 6−7 ml vol of cold buffer and add to the original supernatant.

(v)   Centrifuge the broken cell preparation at 3000 *g* for 10 min at 4°C and discard the cloudy supernatant liquid.

2.2.5 *Cell breakage by blending with beads*

The grinding or shearing action of glass beads produced by oscillation or vortexing can also be achieved using an impeller rapidly rotating in a sealed contoured vessel. Examples of this are the use of the Waring blender, usually driven through the base of the container, or the Sorvall Omnimixer fitted with a top-driven impeller shaft. These homogenizers use cutting blades, the edges of which are blunted by beads during blending. It is improbable that these edges contribute much to the distintegration of the cell and old sets of blades are acceptable. Homogenizers such as the Bead-Beater have custom-designed Teflon impeller discs in place of the blades.

(i)   Cool to 0°C the blending vessel, a quantity of glass beads (0.45−0.55 mm diameter) and impeller shaft and blades if these are separate, for example in top drive models. Packing the vessel with melting ice is the easiest method.

(ii)   Mix moist pelleted yeast cells, cooled buffer and beads in the ratio of 1:1:1 wet wt/v/v (Section 2.2.2.iii). It is important to avoid a slurry that is not properly mixed or maintained in circulation during blending because it is too thick. If plenty of cells are to hand for trial runs, then a drop of dye added to the surface just prior to blending will indicate whether efficient dispersal is occurring.

(iii)   Using 30−60 sec periods of blending, with interim cooling in melting ice, homogenize for a total blending time of 2−3 min. Very short bursts of blending action can be used during cooling to ensure thorough mixing.

(iv)   Remove the homogenate from the blender into a cooled receiver, separate the glass beads from the broken cells and wash the preparation as described in Section 2.2.2(v).

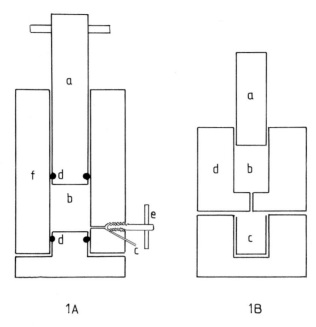

Figure 1. (A) French press; a, piston; b, well; c, delivery port; d, O-ring; e, screw-valve; f, pressure vessel. **(B)** Eaton press; a, piston; b, well; c, receiver; d, pressure vessel.

2.2.6 *Cell breakage by hydraulic press*

The passage of cell suspensions, subjected to considerable hydraulic pressure, through a small hole is often an efficient and preferred means of disrupting yeast cells. It has the twin advantages of not requiring the addition, and therefore removal, of abrasive or shear-inducing elements, and that temperatures can be kept near to or below 0°C throughout the procedure. A hydraulic press capable of exerting a pressure of at least 30 000 p.s.i. (2175 kg/cm^2) is required. Two methods are described: use of the X-press where cells remain frozen for most of the time, and the French press, where cells are maintained in liquid suspension close to 0°C (*Figure 1*). Since there are variations in operation of presses and pressure vessels, it is essential that the maker's literature is read before they are used.

In preparation for these procedures, allow sufficient time, 30 min or more depending on the devices to hand, for cooling the large mass of metal comprising the pressure vessel and piston.

The procedure of cell breakage using frozen cells and the X- or Eaton press is described.

(i)   Clean the interior surface of the pressure vessel and the piston with lint-free cloth. The piston should fit smoothly into the well of the pressure vessel, sliding down with slight hand-pressure. Considerable hand-pressure or a free-falling piston means dirty or mismatched surfaces and application of hydraulic pressure should not be attempted.

(ii)   Make a note of the distance travelled by the piston before it reaches the bottom of the well. There will usually be a mark on the piston to show this for it is important to know the travel distance when pressure is applied.

(iii)   Chill both pressure vessel and piston to about $-25°C$. A mixture of ethanol and water (3:2, v/v) is a suitable coolant. Pure ethanol may be used if lower temperatures are required for media with high solute concentrations. Dry ice can be used if a refrigerated bath is not available. For most sizes of vessel, for example 10×15 cm, $30-40$ min at this temperature is usually sufficient. Pre-cooling overnight at $4°C$ can save time. It is advisable to keep the piston and pressure vessel separate during refrigeration as problems are occasionally experienced in removing the piston from the cooled cylinder. The receiver, if this is separable from the pressure vessel, need only be cooled in melting ice.

(iv)   Using tongs and thermally insulated gloves, withdraw the pressure vessel and piston from the bath, invert to drain the coolant from the well and remove final traces with a dry lint-free cloth.

(v)   Working as quickly as possible, position the vessel on the base plate of the hydraulic press with the receiver, if separate, in position. Dispense the cooled buffered cell suspension into the well of the pressure vessel. The suspension freezes rapidly on contact with the metal surface. Build up an even bed of frozen paste, avoiding splashing the sides. Do not fill to the top as the piston must be inserted at least 2 cm. With that proviso, the cell suspension can be any volume, a typical capacity being about 30 ml. It should be thick but yet sufficiently fluid to be easily dispensed through a pipette with a tip of internal diameter of about 2 mm. The criteria for suspension buffers are described in Section 2.2.2(iii).

(vi)   Insert the piston, making sure that those which have no head-plate, appearing symmetrical, are in the correct position for the travel marker to be seen. Ensure that the pressure vessel and inserted piston are perpendicular, viewed from two positions at right angles to one another, to the platens of the press.

(vii)   Over the period of 1 min apply pressure, increasing it evenly to the region of 20 000 p.s.i. (~ 1450 kg/cm^2). At about this load the contents in the pressure vessel will be forced through the orifice producing a very audible cracking noise accompanied by a drop in pressure. Pressures of 30 000 p.s.i. (~ 2175 kg/cm^2) should be approached with caution and the press manufacturer's guide lines consulted. If 'cracking' does not occur, lower the pressure, allow the cells to warm up a little and re-apply the pressure. After the first 'crack' the remaining cells are pressed through in a series of similar steps until the piston has reached the bottom of the well. Do not go further! Release the pressure.

The broken-cell suspension, extruded through the orifice, is now a frozen mass in the receiver. Some presses, such as the X-press, allow the pressing of the first lysate back through the orifice. A number of passages may be required for complete breakage. For the simpler Eaton press, lysates must be defrosted and re-loaded.

(viii)   Carefully defrost the lysate with occasional stirring of the ice crystals to avoid local temperature rises. It is at this stage, where cell debris and cytosol are still in close contact, that hydrolases may begin to act on the cell wall.

(ix)   Centrifuge the melted lysate as soon as possible at 3000 *g* for 10 min; discard the cloudy supernatant.

Similar procedures using hydraulic pressure to lyse liquid cell suspensions maintained at about 0°C, such as those for the French press, are now described below.

(i)   Clean the pressure vessel and piston as described above. In addition, ensure that the screw release valve (*Figure 1*) and delivery port are clean and working smoothly. Since the lysate is ejected at some speed attach a length of small-bore plastic tubing to the delivery port. As the pressure vessel will be mounted in the hydraulic press ensure that there is sufficient length of tubing to reach to the bottom of the receiver in this position. Replace the seals and O-rings that are blemished or perished.

(ii)   If it is not already marked, make a note of the distance travelled by the piston before it reaches the well bottom.

(iii)   Cool the pressure vessel, piston and receiver in melting ice allowing sufficient time for the whole of the mass to reach 0°C, usually about 40−60 min. Storage overnight at 0−4°C saves time.

(iv)   Consult the manufacturer's manual for the precise methods of loading the cell suspension and inserting the piston. Ensure that the screw valve is closed. Composition and consistency of the cell suspension is described in step (v) above. The maximum volume may be about 40 ml, though this depends on the apparatus used.

(v)   Place the filled and assembled piston and vessel in the press and over 1−2 min increase the pressure to about 15 000 p.s.i. ($\sim$1100 kg/cm^2). Slowly open the screw valve until the pressure starts to fall and cell lysate is ejected into the receiver.

(vi)   Maintain the pressure at 15 000 p.s.i. with gentle pumping until the piston has travelled to the bottom of the well. A drop of pressure to below 10 000 p.s.i. ($\sim$750 kg/cm^2) will produce little breakage.

2.2.7 *Cell breakage by ultrasonic disintegration*

In many ways ultrasonic disintegration is the cleanest and easiest method of cell rupture, though it is not always successful. If a disintegrator is to hand it is certainly worth examining its effect. A medium-sized installation of about 150 W capacity with output frequency of 20 kHz an disintegrate suspensions of up to 100 ml. Although it is housed in an acoustic cabinet, it is worth considering the use of ear shields during operation. Heat is generated during transduction and the glass containing vessel must be surrounded with melting ice. The transducer probe should be centred in the vessel and the tip positioned no more than 1 cm below the surface. Trial disintegrations will indicate temperature rise. This will determine the duration of sonication before cooling is necessary. The temperature should not rise above 5°C.

2.3 **Monitoring cell breakage**

Ideally an assay of cell disintegration should be fast enough to allow monitoring during breakage. Rapid assessments can be made by microscopic observation. Using a magnification of ×400 with phase-contrast illumination, the loss of the characteristic halo around the cell is an indication of rupture. Empty cells are dark whereas whole cells are bright with some contents distinguishable. Alternatively the Gram stain can

Table 4. Assay of glucose-6-phosphate dehydrogenase.

1.	Prepare the following solutions: glucose-6-phosphate, 35 mM; (10 mg sodium salt/ml) NADP, 13 mM (10 mg sodium salt/ml); MgCl$_2$.6H$_2$O, 0.1 M (2 g/100 ml); triethanolamine buffer 0.1 M, pH 7.6.
2.	Mix ethanolamine buffer (2.55 ml), MgCl$_2$ solution (0.2 ml) glucose-6-phosphate solution (0.1 ml) and NADP solution (0.1 ml). Bring to 25°C in a quartz or plastic cuvette of 1 cm path length. Note that not all plastic cuvettes are transparent to 340 nm radiation.
3.	During the course of steps 1 and 2 centrifuge 0.5 ml of cell lysate at full speed in a bench microcentrifuge for 3−4 min to obtain supernatant liquid for enzyme assay.
4.	Add 50 μl of supernatant liquid to the cuvette, mix well and record the decrease of absorbance at 340 nm using a spectrophotometer coupled to a moving chart recorder. The initial slope is a measure of enzyme activity and, since most assays are comparative, for example used for assaying the attainment of the maximum released activity or comparison of one in-house preparation with another, is an adequate measure of dehydrogenase activity.

be used. Examination after fuchsin staining reveals broken cells as pink whereas remaining whole yeast cells are blue.

A method that can be used if time and cell availability permit, but which is not suitable for rapid monitoring *in situ*, is comparison of viable count of treated and untreated cells, counting dilutions to 10^{-8} .

Another useful assay of cell disintegration is the measurement of released glucose-6-phosphate dehydrogenase (*Table 4*). It is quicker than a viability count but, again, is not a direct measure of cell wall release, and indeed not a direct measure of cell disruption. Nevertheless it is a fast assay and can produce results within 10 min. The activity of enzyme released rises to a plateau at which point it is assumed cell disintegration is complete.

2.4 Cell wall isolation

The separation of cell walls from the contents of the cytoplasm, cytoplasmic and organelle membranes, etc. is by centrifugation. Conditions routinely listed after each breakage procedure are 3000 g for 10 min. As with most centrifugation data this represents a practical compromise and variations should be tried to find conditions best suited to the preparation.

2.4.1 *Washing of cell walls*

The extent of washing and criteria of purity will depend on the reason for preparing cell walls. Some components are more loosely bound into the structure than others. Physical bonds play a major role in maintaining the architecture. Even the mildest washing procedures may therefore remove not only unwanted cellular components but also parts of the wall. If a complete survey of the wall is required it is advisable to store all aqueous washings, concentrate them by rotary evaporation, add 3 volumes of ethanol with stirring, and store overnight at 4°C. Solubilized polysaccharides should precipitate and may be purified in a similar manner to those extracted by more vigorous means, (see above). Bearing in mind that washing may remove wall components, *Table 5* describes some possible procedures. All centrifugal washings can be carried out at room temperature, but more advisably at 4°C, using forces of 3000 g for 10 min. The ratio of packed walls to wash is typically about 1:10.

Table 5. Washing cell walls.

1.	Remove contaminating membrane lipids by agitation (e.g. occasional vortexing) of the preparation with ethanol for about 10 min. Centrifuge the cell walls, discard the supernatant, and repeat the ethanol wash.
2.	In a similar manner wash the walls three times with a chloroform:methanol mixture, 1:1 (v/v), three times with ethanol:ether, 1:1 (v/v) and finally once with water.

The following is an optional procedure designed to loosen hydrogen and hydrophobic bonds. Its disadvantage is that walls are exposed to mild acid and alkali.

3.	Suspend the preparation successively, by vortexing, in 8 M urea, water, 0.5 M NH_4OH, water, 0.5 M HCOOH and finally water. Each treatment is of ~5 min duration. The final water wash is repeated five times. If the fine structure of the wall is to be preserved, removal of the lipid (steps 1−2) should be followed by a gentler procedure:
4.	Suspend the walls in a slightly alkaline buffer, e.g. 10 mM Tris−HCl or phosphate, pH 8, for ~10 min with occasional agitation. Centrifuge and discard the supernatant.
5.	Repeat this procedure at least five times, but preferably 20.
6.	Wash three times with water.
7.	Suspend in a little water, e.g. 5−10 ml, and freeze-dry. Store cold, preferably below 0°C.

Table 6. Assay of nucleic acid (6).

1.	Take the acid hydrolysate produced in Section 4.1 and, if cloudy, centrifuge at 3000 g for 5 min.
2.	Depending on the type of quartz cell available, but typically 1 × 1 cm, dilute an aliquot of hydrolysate, say 0.5 ml, to 3.0 ml with water and measure the absorbance at 260 nm. It is better to produce an absorption spectrum from 230 to 300 nm and identify the characteristic peak of purines and pyrimidines between 260 and 270 nm. A_{260} is ~1 for 30−40 μg/ml and 1 cm path length.

2.4.2 *Criteria of purity*

The complex structure of cell walls, part of which might be removed in washing or through contaminating hydrolase action, makes it difficult to list positive criteria. Monitoring the removal of membranes by phosphate analysis is of no use since phosphate may be a component of the cell wall, for example mannan. Rather it is the absence of nucleic acid (see *Table 6*) as a cytosolic macromolecular marker, and membranes, seen in electron microscopy, that are better indicators.

However, it is not uncommon to find that criteria for the purity of cell wall preparations are not offered but preparations are characterized as the end-product of a carefully followed procedure.

3. CELL WALL FRACTIONATION

Cell walls have been resolved into their various components with mixed success. Indeed, since it is not clear whether polysaccharides and glycoproteins that comprise the wall should be considered as discrete groups with distinguishable structures or as a continuum of structural variations covalently bound to one another, fractionation procedures are somewhat arbitrary. They are, nonetheless, useful as they provide characteristic patterns for identification and observing change, and must in some measure represent the *in vivo* composition.

Traditionally mild acid, strong alkali and occasionally organic solvents such as dimethyl sulphoxide, have been used for selective solubilization. Indeed the nomenclature of cell wall components, for example alkali-soluble, alkali-insoluble β-glucan, etc. derives from these procedures. More recently, β-D-glucanases of defined specificity have been used to dissect, usually in good yield, the walls of whole cells to produce fragments of wall with molecular weights greater than 30 000, and often in excess of 300 000. Since these digestions can take place in the region of neutrality and at ambient temperatures they currently provide the gentlest conditions in which cell wall fragments can be produced; the products of such digestions presumably conserve the labile bonds destroyed in more rigorous extraction procedures.

3.1 Alkali−acid fractionation

The following procedure is a typical selective solubilization regime. Variations are described at the end.

(i)   Suspend the freeze-dried cell walls at a concentration of 5 mg/ml in 0.75 M (3%) NaOH through which nitrogen has been previously and slowly passed for about 1 h. Bring to 75°C with continued nitrogen purging and close the vessel. For small-scale extractions a screw-topped Universal bottle is ideal. If possible, add a magnetic stirrer bar to the vessel. Slow nitrogen purging should continue but if this is not possible, re-purge the vessel every hour. Every 15−30 min resuspend the cell walls with shaking or, if possible, magnetic stirring. Continue heating for 6 h.

(ii)   Centrifuge the contents at 3000 g for 10 min; store the supernatant and wash the pellet (1:10, v/v) with water by centrifugation. Combine the supernatants.

(iii)   Using the pellet obtained from step (ii) repeat the above process, but use the same volume of 0.5 M (3%) acetic acid at 90°C for 3 h in place of NaOH. A specific gravity of 1 can be assumed for glacial acetic acid; it is easier to dispense by volume than by weight.

(iv)   Repeat the acetic acid extractions twice and combine the supernatants for subsequent isolation of polysaccharides.

(v)   Wash the remaining alkali−acid-insoluble pellet three times with a little water, say 5−10 ml.

(vi)   Neutralize the alkaline extracts with 2 M acetic acid, the acid extracts with 2 M NaOH, and cool to 0°C.

(vii)   With continuous stirring slowly add 3 vol of ethanol, cooled to 0°C, over 3−4 min. Allow the precipitate to settle overnight at 0−4°C.

(viii)   Wash the precipitate by centrifugation at 3000 g for 10 min with ethanol:water (3:1, v/v) and discard the supernatant. Repeat the washing four times.

(ix)   Suspend or dissolve the washed precipitates or alkali−acid-insoluble material in 5−10 ml water and dialyse in washed tubing against 200 ml of stirred water at 4°C. Replace the water after 4 h with a further 200 ml.

(x)   Freeze-dry the contents of the sac.

The mannoprotein in the cell wall is soluble in alkali together with some of the β-glucan. The two can be separated by precipitation of the mannoprotein as a copper complex. The complexing reagent (Fehling's solution) is prepared as two solutions that are mixed prior to use.

(xi)   Prepare the copper solution by dissolving 34.6 g of $CuSO_4.5H_2O$ in 500 ml of

175

water to which two drops of concentrated H_2SO_4 have been added.

(xii)   Dissolve 60 g of NaOH and 173 g of potassium sodium tartrate tetrahydrate in 500 ml of water.

(xiii) Mix equal volumes of the two solutions to produce Fehling's solution which is now complexed with mannoprotein.

(xiv)   Add Fehling's solution to the alkaline extract (steps i and ii) until no more precipitate forms. Allow to settle for $2-3$ h and finally pellet at 3000 g for 10 min.

(xv)   Decant the supernatant solution and isolate the soluble glucan as described in steps vi$-$x.

(xvi)   For 1 min vigorously stir the pelleted copper complex with about 10 vol of ethanol cooled to 0°C and containing 5% (v/v) concentrated HCl. Centrifuge the insoluble mannoprotein at 3000 g for 10 min, discard the supernatant and wash the precipitate by centrifugation with five 10 ml vol of cold ethanol.

(xvii) Follow steps ix$-$x for dialysis and freeze-drying.

The procedure described above for the solubilization of mannan and its subsequent separation from glucan should cause little or no degradation of the polysaccharide moiety. However, the alkali-sensitive bonds of glycosylserine and glycosylthreonine uniting the mannan to the protein, together with the phosphodiester bonds that may occur in the mannan moiety and the disulphide bonds of the protein component, will most certainly be affected. For this reason the isolation of whole mannoprotein should be performed at neutrality, the subsequent resolution of mannoprotein and glucan being achieved by selective precipitation of mannoprotein as a complex with borate and the quaternary ammonium compound Cetavlon (hexadecyl trimethyl ammonium bromide). The following procedure is taken from Lloyd (7) and Shibata *et al.* (8,9).

(xviii) Stir freeze-dried cell walls (10 g) in 100 ml of 50 mM phosphate buffer, pH 7.0, at 100°C for 2 h. Cool to room temperature and centrifuge at 3000 g for 10 min.

An alternative extraction, used by Peat *et al.* (10), is to autoclave the cell wall preparation in 20 mM citrate buffer, pH 7.0, at 140°C for 2 h.

(xix)   Decant the supernatant solution and extract the residue twice more using the same procedure. Combine the supernatant extractions.

(xx)   Dissolve sodium acetate (5.4 g) in the extract (300 ml) and add ethanol (900 ml) with stirring over a period of $4-5$ min. Allow the precipitated polymer to settle overnight, preferably at 4°C. Decant or aspirate the supernatant from the residue and wash it three times, by centrifugation at 3000 g for 10 min, with ethanol:water (3:1, v/v).

(xxi)   Dissolve or suspend the washed precipitate in water (50 ml) and dialyse against stirred water (21) at 4°C. Replace the water after 4 h and continue for another 4 h. Freeze-dry the contents of the sac.

(xxii) To separate the mannoprotein from other extracted polymers, dissolve freeze-dried extract (5 g) in water (100 ml) at room temperature and, with constant stirring, add a solution of Cetavlon (4 g dissolved in 50 ml of water) over $4-5$ min. Store for 2 h at room temperature then remove any precipitate that has formed by centrifugation at 3000 g for 10 min. This precipitate may be discarded or stored for future investigation. The mannoprotein should remain in solution.

(xxiii) Add a solution of (1%) boric acid (100 ml) with stirring to the complexed mannoprotein solution, adjust the mixture to pH 8.8 with 2 M NaOH, and allow
the precipitate to settle at room temperature for 1 h.

(xxiv) Collect the precipitate by centrifugation at 3000 *g* for 10 min, store the supernatant solution, which may contain other cell wall components and wash the pellet
by centrifugation with two 50 ml vol of 0.5% sodium acetate solution adjusted
to pH 8.8.

(xxv) Dissolve the washed pellet in 2% acetic acid (50 ml) at room temperature and
add, with stirring over 4−5 min, 150 ml of ethanol containing sodium acetate
(1 g). Store for 1 h, then wash the precipitate by centrifugation with 4 × 200 ml
of ethanol containing 2% acetic acid.

(xxvi) Dissolve the washed precipitate in water (50 ml), adjusting the pH, if necessary,
to 7 with 1 M NaOH. Dialyse against water (500 ml), replacing the water after
4 h and continuing for another 4 h. This is the mannoprotein fraction.

(xxvii) Whilst stirring the retentate add 3 vol of ethanol, containing 1% sodium acetate,
over 4−5 min. Allow the precipitate to settle for 1 h then wash with ethanol
(3 × 100 ml). After the final wash, invert the centrifuge tube to drain traces
of ethanol and dry the pellet over P_2O_5 at room temperature. This is the glucan
fraction.

Using this method the majority of mannoprotein (>90%) should be precipitated.
Small amounts may be left in solution after step xxv.

A more elegant method of resolving β-D-glucans from mannoprotein, especially
where analysis of cell wall components is on a small scale, for example less than 200 mg,
is to exploit the binding characteristics of lectins. Concanavalin A (Con A) binds to
α-D-glucosyl and α-D-mannosyl residues, but has no affinity for β-D-glucosyl groups.
Fractionation on a column of Con A immobilized to Sepharose (Con A−Sepharose)
provides a fast and easy method of separating glucan from mannoprotein which is adsorbed, although any glycogen present is also bound. Digestion with purified amylo-
glucosidase before or after column chromatography, followed by dialysis, will remove
the α-glucan.

The following is a typical separation procedure.

(i)   Pour 10 ml of Con A−Sepharose 4B gel, normally provided by the supplier
        in a swollen state, into a column approximately 1 × 20 cm and wash with 100 ml
        of 0.02 M Tris−HCl, pH 7.4; 0.5 M NaCl to equilibrate the column and remove
        preservative.

(ii)   Apply 1 ml of solubilized cell wall components, obtained from 1 g of wet weight
        of cells and dialysed against elution buffer.

(iii)   Elute with buffer until all non-bound carbohydrate (β-D-glucan) has been washed
        through. This will be 2−3 bed vol, that is 20−30 ml. The phenol−sulphuric
        acid procedure (Section 4.2) can be used to monitor the effluent for carbohydrate.

(iv)   Desorb the mannoprotein using at least 20−30 ml elution buffer containing 0.5 M
        α-methyl mannoside. Dialyse the two eluted glucan and mannoprotein fractions
        each against two changes of water (IL).

The binding of glycosyl groups to Con A requires the presence of both Ca^{2+} and
Mn^{2+} ions. Once chelated, these cations are normally firmly bound to the lectin at

neutral pH. If their presence is in doubt, resuspension of Con A−Sepharose 4B in a solution of 1 mM $MnCl_2$ and $CaCl_2$ pH 4.5 will ensure the replacement of these ions before initial washing and equilibration (step i above) is carried out.

There are variations of the alkaline extraction protocol using different times and temperatures and stronger or weaker bases. An alternative extraction is in 0.5 M (3%) NaOH at 4°C for 6 days (11). A two-stage base extraction recommending 1 M (5.6%) KOH at 60°C for 20 min followed by 10 M (40%) NaOH at 100°C for 1 h interposed with extraction by dimethyl sulphoxide at 22°C for 16 h has been reported (12). Sometimes the extraction process is repeated before proceeding to the next stage. Pre-treatment with very mild acid, for example acetate buffer pH 5, for 3 h at 75°C is reported to improve alkaline extraction (13). Sonic vibration is said to assist alkaline solubility (13). Treatment with nitrous acid, depolymerizing de-acetylated chitin, is reported to render soluble previously intractable glucan (12). The chitin must have undergone a suitably vigorous exposure to alkali, for example 10 M at 100°C for 1 h for efficient de-acetylation. The procedure is as follows:

(i)   To an aqueous suspension (1 ml) of water-washed alkali-treated cells, add 1.5 ml of freshly prepared 2 M $NaNO_2$ solution followed by 0.5 ml of 2 M HCl. Store for 90 min in a stoppered tube with occasional shaking.

(ii) Centrifuge the cells at 3000 g for 10 min. Neutralize the supernatant solution with 2 M NaOH and precipitate with ethanol, following steps vi−x above.

3.2 Enzymic digestion of whole cells

A number of commercial enzymes are available that specifically hydrolyse β-1,3-D-glucosidic bonds in yeast cell walls. These preparations sometimes contain proteases and mannanase but do not hydrolyse β-1,6-D-glucosyl linkages. Moreover, a propor-tion of 1,3-linkages in the cell wall remain resistant to the enzyme with the consequence that large pieces of wall can be solubilized. The method described is for the small-scale isolation of mannoproteins and glucans from whole cells. It is derived from reference (14) and has been applied to *Saccharomyces cerevisiae* and *Candida albicans*.

(i)   Suspend washed cells at a concentration of 5 mg dry weight/ml in 4 ml of 50 mM Tris−HCl buffer, pH 7.5 and add glucanase to a final concentration of 0.5−1.0 mg/ml. Maintain the cells in suspension at 30°C using a shaking water bath, rotary shaker, or roller tube device. Useful commercially available glucanases are those derived from *Arthrobacter luteus* and include Zymolyase (Miles) and Lyticase (Sigma).

(ii) Monitor the extent of solubilization by centrifuging 0.1 ml aliquots of the digest, taking 50 μl of supernatant for total carbohydrate assay (Section 4.2).

(iii) When no more carbohydrate is released, centrifuge the remainder of the digest at 3000 g for 10 min, remove the supernatant and heat it at 100°C for 5 min to inactivate the glucanase. Should any precipitation occur, re-centrifuge the solution.

(iv) Remove water by rotary evaporation or freeze-drying and dissolve the residue in about 1 ml of water.

(v)   Apply the solution to a column (1 × 40 cm) of Bio-Gel P-30 (Bio-Rad Laboratories) equilibrated in water containing 5 mM sodium azide as an antimicrobial agent. An alternative eluant is 0.1 M acetic acid. Elute at a flow-rate of about 3−4 ml/h, collecting 1 ml fractions.

(vi) Monitor carbohydrate content of fractions by taking 100 μl samples for the phenol−sulphuric acid procedure (Section 4.2). Large molecular weight glucans and mannoprotein are eluted between 8 and 15 ml; small oligosaccharides between 26 and 35 ml. A broad band of small molecular weight polysaccharides often appears between these two peaks.

(vii) Remove sodium azide or acetic acid by dialysis against two changes of 30 volumes of water. Alternatively acetic acid can be removed by freeze-drying, provided a cold-trap is fitted in the vacuum system.

4. CHARACTERIZATION OF CELL WALL FRACTIONS

The simplest characterization of cell wall fractions comprises a gross analysis of component sugars, identifying their nature and proportion. In this manner the ratios of glucan, mannoprotein, chitin, etc. are established. Further characterization is usually required since a number of different glucans are probably present. A useful procedure is to examine their susceptibility to periodate oxidation. Both the production of acid and reduction of periodate per anhydroglucose residue are indicators of the type of linkages present. Further characterization requires methylation analysis and enzyme dissection which are considered to be beyond the scope of this chapter but for which references are provided.

4.1 **Acidic hydrolysis of fractions**

(i) Weigh into a thick-walled or centrifuge tube of about 10−15 ml capacity, preferably fitted with a ground-glass stopper, about 2−3 mg of sample. Add 1 ml of 90% formic acid and loosely close the top with a stopper or a glass marble. Heat in a boiling water bath for about 90 min. Concentrations of polysaccharide greater than 0.3% may produce unwanted condensation reactions. If it is suspected that the sample contains more than 50% of protein, the concentration of polysaccharide should be dropped to 0.1%.

(ii) Remove formic acid by rotary evaporation, add 2 ml of water and again evaporate to dryness. A ground-glass joint on the tube is useful as an adaptor can directly couple it to the evaporator. Alternatively, a short sleeve of plastic tubing produces an effective union.

(iii) Add 2 ml of 1 M H_2SO_4, replace the stopper or marble and heat in a boiling water bath for a further 3 h.

(iv) Cool the contents to 0°C in melting ice and add portions of solid barium carbonate to neutralize the acid. Vortex the mixture to ensure rapid mixing. CO_2 is evolved and a heavy white precipitate of barium sulphate appears. Neutrality can be judged by dipping small slivers of indicator paper into the liquid. The final solution is slightly alkaline. It is therefore desirable to maintain the temperature in the region of 0°C to avoid degradation of the released sugars. Add 2 ml of ethanol and stir.

(v) Centrifuge the contents of the tube for 3−4 min at 3000 g and remove the supernatant. With stirring, wash the precipitate by centrifugation with three 2 ml volumes of cold 50% aqueous ethanol and combine the supernatants. Use of 50% aqueous ethanol reduces the slight opalescence experienced with water washings.

(vi)   Evaporate to dryness or freeze-dry and store below 0°C.

The resolution and quantitative determination of individual sugars can be carried out using HPLC, GLC after derivatization, for example to alditol acetates, or by individual assay of each sugar using chemical or enzymic methods. The method using HPLC is described. A number of suitable columns are commercially available.

(i)   Dissolve the sample in glass double-distilled water and pass through a pre-filter such as AP25 (Millipore) and 0.45 μm filter (Millipore or Oxoid).

(ii)   Inject 20 μl on to a carbohydrate analysis column (e.g. Waters Associates) using an eluant of acetonitrile−water (85:15 v/v) at a flow-rate of 1 ml/min and ambient temperature. Sugars are detected by a differential refractometer.

A note of caution must be added concerning the assay and total acidic hydrolysis of chitin. Although extensive hydrolysis occurs in the above conditions, they may not be sufficient. 4 M HCl for 4 h at 100°C is considered better, although release of sugars into these conditions will cause some small percentage to be degraded. This can be assessed by examining glucosamine levels measured at 4, 6 and 8 h under these conditions and determining the rate of loss. HCl is readily removed by rotary evaporation followed by storage over NaOH pellets in an evacuated desiccator to remove the last traces of acid. Glucosamine and galactosamine are resolved on conventional amino acid analysis columns.

The enzymic hydrolysis of chitin presents a number of problems which must be remembered when using chitinase. The fibrous and highly insoluble nature of the substrate, especially if lodged in cell wall complexes, requires that sufficient time be allowed for digestion. The alkaline extraction regimes may have de-acetylated the polymer, partially or wholly, to chitosan. Re-acetylation will be required before enzymic hydrolysis is possible. Furthermore, the purer preparations of chitinase may lack chitobiase that hydrolyses the released disaccharide to *N*-acetyl glucosamine. The colorimetric assay (18) of *N*-acetyl glucosamine by the Morgan and Elson method will not detect the disaccharide.

The release of reducing sugars from a polymer can be monitored by the copper reductometric procedure of Nelson and Somogyi (5), a typical range being 5−100 μg glucose. Glucose can be specifically assayed using glucose oxidase (15), the range being 5−50 μg. Use of the chromogen 2,2′-azino-di-[3-ethyl]-benzthiazoline sulphonate (ABTS) in place of *O*-dianisidine is recommended as it is more sensitive (3−20 μg glucose) and avoids the possible carcinogenic nature of the anisidine. Hexosamines (5−30 μg) can be estimated using the Elson and Morgan procedure (16) modified by Blix (17) and *N*-acetyl hexosamines by the method of Reissig *et al.* (18). A typical range for the latter is 5−50 μg of *N*-acetyl glucosamine or 10−150 μg of *N*-acetyl galactosamine.

4.2 Assay of carbohydrate; phenol−sulphuric acid procedure (19)

This assay, a trifle disconcerting when first used, is a fast, sensitive method for measuring most, but not all sugars whether they be mono-, oligo- or polysaccharides. It will not, for instance, detect glucosamine. It is useful for monitoring eluates from column chromatography (Section 3.2). Great care must be taken in using this procedure as the hot reagents are very corrosive. It is essential to wear eye protection and have running water available in case of spillage. The procedure is best carried out in a fume cupboard with the interposed screen well down.

(i)   In a thick walled Pyrex glass test-tube, about 1.5×15 cm, place 1.0 ml of sample, 1.0 ml of 5% phenol solution and mix well.

(ii)   Using a syringe dispensor, rapidly deliver 5.0 ml of analytical grade concentrated H_2SO_4 into the contents of the tube already in place in a rack. The mouth of the tube must be pointing upright and away from the operator. A great deal of heat is generated and short-lived 'boiling' occurs. Briefly vortex the contents to ensure thorough mixing, allow to stand and, when cool, read the absorbance at 490 nm. The measurements should be made no longer than 15 min after reaction. The assay is linear to about 80 μg glucose to give an absorbance of about $0.6-0.7$ for a 1 cm path length. The glucose used for the standard solution should be dried at 40°C for 18 h *in vacuo* over P_2O_5.

It is advisable to use a teat and Pasteur pipette for transfer between tube and cuvette.

4.3 Periodate oxidation

This procedure is useful where the ratio of β-1,3- to β-1,6-linked glucans is required for characterization of a cell wall fraction.

(i)   Weigh about $10-20$ mg of material into a stoppered tube of about 15 ml capacity. Suspend or dissolve in 8.0 ml of water. Alternatively dispense an equivalent amount of suspended or dissolved material kept stored as frozen stock preparations. Sonication often assists the production of a fine suspension that can be reproducibly sampled.

(ii)   Prepare a fresh solution of 0.12 M sodium metaperiodate (mol. wt 214) by dissolving 1.28 g in 50 ml water and add 2.0 ml to the polysaccharide solution or suspension.

(iii) Replace and firmly secure the stopper, exclude light by wrapping aluminium foil around the tube and store at 4°C.

(iv)   If the polysaccharide undergoing oxidation, is insoluble, then slowly rotate, roll, or in some manner agitate the tube to keep the material in suspension. Hence the requirement for a stoppered tube. Kept at 4°C, the oxidation is allowed to proceed for a minimum of 60 h, but preferably for 100 h.

(v)   To follow the course of oxidation, remove 30 μl samples, dilute to 10.0 ml with water and measure the absorbance at 223 nm. Suggested sampling times are 0, 6, 24, 48, 72 and 96 h. This will measure the remaining metaperiodate. Calibrate the assay by measuring A_{223} of 30 μl of 24 mM metaperiodate diluted to 10.0 ml and by a similar measurement derived from 10 ml of 24 mM metaperiodate to which 100 μl of re-distilled ethane diol (ethylene glycol) has been added. Allow 30 min to elapse for the diol to reduce completely the metaperiodate before taking the 30 μl sample. The difference in absorbance at 223 nm is a measure of 24 mM metaperiodate.

It is advisable to measure oxidation of standardized compounds to check the fidelity of operations. Since resistance to oxidation is good evidence for a $1-3$ glycosidic linkage, this negative experimental observation must be complemented by a positive indication of oxidation. α-Methyl glucoside is often used, reducing 2 mol of metaperiodate per mol of sugar.

The acid liberated can be titrated with 5 mM HCl using methyl red as indicator. An alternative method, providing a better end-point, is to reduce completely excess metaperiodate to iodate with ethane diol, add an excess of iodide ion and titrate the liberated iodine with thiosulphate. One mol of proton is equivalent to 1 mol of thiosulphate.

(vi)   To a 1.0 ml sample of oxidation mixture placed in a stoppered tube add 0.1 ml of ethane diol and store closed in the dark for 30 min at room temperature.

(vii) Add 1 ml of 0.6 M (10%) KI solution, at which point the brown colour of iodine may be seen.

(viii) As soon as possible, titrate the iodine with 5 mM sodium thiosulphate (commercially available as a standardized solution) using $1-2$ drops of a 1% starch glycollate or starch solution as indicator. The end-point is the complete discoloration of the blue starch complex.

A comparison of carbohydrate content (Sections 4.1, 4.2) with metaperiodate reduced and acid produced will allow an assessment of the susceptibility of the fractions to periodate oxidation. As an example, internally located unbranched 1,6-linked glucosyl residues will reduce 2 mol of metaperiodate and liberate 1 mol of formic acid per residue; a similarly located 1,3-linked residue is completely resistant.

To produce a practical guide for the detailed characterization of cell wall polymers, or a comprehensive survey of analytical techniques, requires another chapter. Sufficient characterization procedures have been described to provide a reasonable indication of the cell wall composition and structure. Laboratory procedures determining sugars as tri-methyl silyl and alditol acetate derivatives by GLC $(20-22)$, the ratio and type of glycosidic linkage using methylation analysis (23) and probing structures by GLC$-$MS $(21-22)$ and ^{13}C-NMR (24) can be found in the appropriate volumes of *Methods in Enzymology* $(20-24)$.

5. ACKNOWLEDGEMENTS

I would like to thank Susan Scott, Alison Ramsay, Christoper Mulloy, John Houston and Robert Sturgeon for helpful discussions and reports on procedures unfamiliar to the author.

6. REFERENCES

1. Cabib,E., Roberts,R. and Bowers,B. (1982) *Annu. Rev. Biochem.*, **51**, 763.
2. Bartnicki-Garcia,S. (1968) *Annu. Rev. Microbiol.*, **22**, 87.
3. Sietsma,J.H. and Wessels,J.G.H. (1979) *J. Gen. Microbiol.*, **114**, 99.
4. Elorza,M.V., Murgui,A. and Sentandreu,R. (1985) *J. Gen. Microbiol.*, **131**, 2209.
5. Nelson,N. (1944) *J. Biol. Chem.*, **153**, 375.
6. Horikoshi,K. and Iida,S. (1964) *Biochim. Biophys. Acta*, **83**, 197.
7. Lloyd,K.O. (1970) *Biochemistry*, **9**, 3446.
8. Shibata,N., Ichikawa,T., Tojo,M., Takahashi,M., Ito,N., Okubo,Y. and Suzuki,S. (1985) *Arch. Biochem. Biophys.*, **243**, 338.
9. Okubo,Y., Shibata,N., Ichikawa,T., Chaki,S. and Suzuki,S. (1981) *Arch. Biochem. Biophys.*, **212**, 204.
10. Peat,S., Whelan,W.J. and Edwards,T.E. (1961) *J. Chem. Soc.*, **29**.
11. Fleet,G.H. and Manners,D.J. (1976) *J. Gen. Microbiol.*, **9**, 180.

12. Sietsma,J.H. and Wessels,J.G.H. (1981) *J. Gen. Microbiol.*, **125**, 209.
13. Bacon,J.S.D., Farmer,V.C., Jones,D. and Taylor,I.F. (1969) *Biochem. J.*, **114**, 557.
14. Tkacz,J.S. (1984) In *Microbial Cell Wall Synthesis and Autolysis*. Nombela,C. (ed.), Elsevier, Amsterdam, p. 287.
15. Lloyd,J.B. and Whelan,W.J. (1969) *Anal. Biochem.*, **30**, 476.
16. Elson,L.A. and Morgan,W.T.J. (1933) *Biochem. J.*, **27**, 1824.
17. Blix,G. (1984) *Acta Chem. Scand.*, **2**, 467.
18. Reissig,J.L., Strominger,J.L. and Leloir,L.F. (1955) *J. Biol. Chem.*, **217**, 959.
19. Dubois,M., Gilles,K.A., Hamilton,J.K., Rebers,P.A. and Smith,F. (1956) *Anal. Chem.*, **28**, 350.
20. Laine,R.A., Esselman,W.J. and Sweeley,C.C. (1972) In *Methods in Enzymology*. Ginsburg,V. (ed.), Academic Press, New York, Vol. 28, p. 159.
21. Nilsson,B. and Zopf,D. (1982) In *Methods in Enzymology*. Ginsburg,V. (ed.), Academic Press, New York, Vol. 83, p. 46.
22. McNeil,M., Darvill,A.G., Åman,P., Franzen,L. and Albersheim,P. (1982) In *Methods in Enzymology*. Ginsburg,V. (ed.), Academic Press, New York, Vol. 83, p. 3.
23. Lindberg,B. and Lönngren,J. (1978) In *Methods in Enzymology*. Ginsburg,V. (ed.), Academic Press, New York, Vol. 50, p. 3.
24. Barker,R., Nunez,H.A., Rosevear,P. and Serianni,A.S. (1982) In *Methods in Enzymology*. Ginsburg,V. (ed.), Academic Press, New York, Vol. 83, p. 58.

7. APPENDIX

Listed below are addresses of suppliers in the UK that stock appliances referred to in Section 2.2

Braun Homogenizer, FT Scientific Instruments Ltd, Station Road Industrial Estate, Bredon, Tewkesbury GL20 7HH, UK.

X-Press, Biotec Ltd, LKB House, 232 Addington Road, South Croydon CR2 8YD, UK.

Sorvall Omnimixer, Du Pont (UK) Ltd, Wedgewood Way, Stevenage, Herts SG1 4QW, UK.

Bead-Beater, Life Sciences Laboratories Ltd, Sarum Road, Luton, Beds LU3 2RA, UK.

French Press, V.A.Howe, 12−14 St Anne's Crescent, London SW18 2LS, UK.

Micronizing Mill, McCrone Research Associates Ltd, 2 McCrone Mews, Belsize Lane, London NW3 5BG, UK.

CHAPTER 9

Yeast mitochondria

DAVID RICKWOOD, BERNARD DUJON and VICTOR M.DARLEY-USMAR

1. INTRODUCTION

It is remarkable that the structure and functions of mitochondria are so similar in a very wide range of eukaryotic organisms. Thus besides their intrinsic interest it has also been possible to use yeast mitochondria as a model for the mitochondria of other types of organisms. For such studies with yeast the ability of yeast to survive without functional mitochondria has been a particularly powerful tool not only in studying the biogenesis of mitochondria but also in dissecting the functional complexes of the respiratory chain.

In order to assess any of the properties of mitochondria it is important to be able to purify them free of contaminants that may interfere with the activity of interest. However, the 'purity' of mitochondria depends on the area of interest and this also determines the method of assay used to determine the purity. For example, if one is interested in electron transport processes then usually one is not interested whether or not the mitochondria are contaminated with nuclear DNA, while such contamination could be very problematical for studies of the transcription of mitochondrial DNA (mtDNA).

This chapter describes in detail the methods used for the genetic analysis of mitochondrial DNA. The isolation of mitochondria from the yeast *Saccharomyces cerevisiae* and the methods that can be used to assess the integrity and purity of the isolated mitochondria are also given. These methods are likely to be applicable to a wide range of other yeasts with a minimum of changes. This chapter also describes the ways of assaying the respiratory activity of yeast mitochondria and briefly touches on some of the important methods used in studying the molecular biology of yeast mitochondria. However, the reader must appreciate that within the space limitation of a single chapter it has not been possible to cover these areas in great detail and neither has there been room to include a description of the methods used to isolate the individual components of the respiratory chain. If further details are needed then the reader should consult the volume in the Practical Approach series which is devoted to the methods for studying mitochondria (1).

2. STRUCTURE AND FUNCTIONS OF MITOCHONDRIA

2.1 Mitochondrial membranes

All mitochondria consist of a double membrane structure (*Figure 1*) which thus form two compartments, the matrix enclosed by the inner membrane and the inter-membrane

**Mitochondrial
compartments**

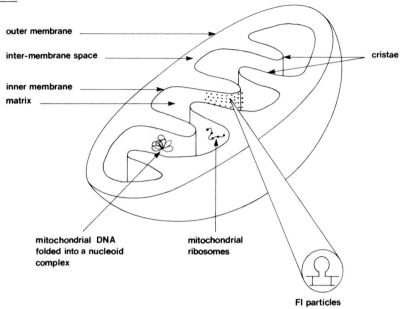

Figure 1. Schematic representation of the structure of a mitochondrion.

space between the two membranes. The compositions of each membrane are very different and reflect the different functions of each. The outer membrane is very similar to the endoplasmic reticulum in terms of composition but it appears to have a number of specific receptors which are involved in the import of mitochondrial proteins synthesized on the cytoplasmic ribosomes. The inner mitochondrial membrane has an unusual composition in that it has a high ratio of protein to lipid of about 3:1 and an unusual lipid composition in that it is enriched in cardiolipin and contains almost no cholesterol.

2.2 Mitochondrial compartments

Both mitochondrial membranes, the matrix and the inter-membrane space can each be thought of as individual compartments in that each is associated with a set of enzymes that reflects the various metabolic processes associated with that compartment of the mitochondrion (2). The components of the respiratory chain form an integral part of the inner membrane (*Table 1*) and the two sides of the inner membrane are associated with different proteins. The inner membrane is impermeable to ions and the pH gradient that forms across the membrane is essential for the generation of ATP. The inner membrane surrounds the matrix of the mitochondrion, this is a soluble fraction containing the enzymes of the citric acid cycle and the oxidation of fatty acids (*Table 1*). The matrix also contains the mitochondrial ribosomes involved in translation of the mRNA transcribed from the mtDNA which is also attached to the inside of the inner membrane.

Table 1. Compartments of the mitochondrion and their associated metabolic functions (2).

Outer membrane

Function	*Associated enzyme(s)*
Oxidation of neuroactive aromatic amines	Monoamine oxidase
Cardiolipin biosynthesis	e.g. glycerol phosphate acyl transferase
Transport of nuclear coded and cytoplasmically synthesized proteins	Not yet identified
Electron transfer	NADH cytochrome c reductase (rotenone-insensitive). The function of this protein is not yet fully defined.

Intermediate space

Function	*Associated enzymes*
Maintenance of adenine nucleotide balance	Adenylate kinase
	Nucleoside diphosphokinase
	Nucleoside monophosphokinase
Electron transfer from complex III to complex IV of the respiratory chain	Cytochrome c
Processing of proteins imported from cytoplasm	Not yet isolated

Inner membrane

Function	*Associated macromolecules*
Oxidative phosphorylation	The phospholipid bilayer is essential to maintain the proton gradient.
	Four electron transfer complexes, three of which couple electron transfer to formation of a proton gradient, and a proton-driven ATP synthetase
Transport of pyridine nucleotides	ADP/ATP translocase
Ca^{2+} ion transport	e.g. Ca^{2+} ATPase
Transport of metabolites	Pyruvate carrier, $H_2PO_4^-/OH^-$ antiport, carnitine shuttle.

Matrix

Function	*Associated macromolecules*
Oxidation of pyruvate to acetyl CoA	The pyruvate dehydrogenase complex
Oxidation of ketone bodies	e.g. 3-ketoacid CoA transferase
Oxidation of amino acids	e.g. glutaminase, glutamate dehydrogenase, aspartate aminotransferase, α-ketoglutarate transaminases
Part of the urea cycle	Carbamylphosphate synthetase, ornithine transcarbamylase
Oxidation of fatty acids to acetyl CoA	Fatty acyl-CoA dehydrogenase, enoyl hydratase, β-hydroxyacyl-CoA dehydrogenase, β-ketoacyl-CoA thiolase
Protection against oxidative stress	Superoxide dismutase, catalase, glutathione peroxidase, glutathione reductase
Processing of proteins imported from the cytoplasm	Proteases for specific signal peptides of imported proteins
Inheritance of genes coding for mitochondrial RNA and some proteins	Mitochondrial DNA, DNA polymerase and primase
Synthesis of 7 membrane components of the proteins of respiratory chain and oxidative phosphorylation	Ribosomes and apparatus for transcription and translation

187

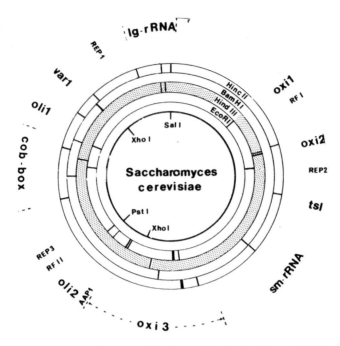

Figure 2. Map of the mitochondrial genome of *Sacch. cerevisiae* derived from ref. 5. The map shows the position of the principal genes and the restriction nuclease digestion patterns.

2.3 **Yeast mitochondrial DNA**

2.3.1 *General features of mitochondrial DNA*

Sacch. cerevisiae is a facultative aerobe and so it has the ability to survive mutations to the mtDNA or even its complete loss, events that would be fatal for obligate aerobic species. Mutations of the mtDNA of normal respiratory-competent cells (RC cells) result in respiratory-deficient cells (RD cells). The mutations that can occur are varied in their sites of action but all result in *Sacch. cerevisiae* being unable to grow on non-fermentable carbon sources such as glycerol. The phenotype of respiratory deficiency is referred to as the *petite* phenotype due to the small size of the colonies which grow on solid nutrient agar. As described in Section 3 *petite* mutations are of various types and have been invaluable in constructing the genetic map of the *Sacch. cerevisiae* mtDNA genome by deletion and recombination mapping.

Fungal mitochondrial genomes show considerable variation in size and gene organization. *Schizosaccharomyces pombe* has a mtDNA of approximately 19 kb (3) and is therefore similar in size to the compact mitochondrial genomes of higher eukaryotic species which are about 16 kb. However, mtDNAs of the species *Brettanomyces sp.* can be as large as 108 kb (4). *Sacch. cerevisiae* mtDNA varies in size from 78 kb to 85 kb depending on the specific strain (5). *Figure 2* shows a map of the yeast mitochondrial genome. The strain-specific variation in genome size can be attributable to a combination of factors that include intergenic insertion/deletions. This is combined with the presence or absence of facultative introns and mini-inserts in specific genes.

Basically, all fungal mtDNAs encode the same complement of mRNAs and have the translational machinery to express these mitochondrial genes. However, large intergenic regions containing A + T-rich sequences are present in many yeast mtDNA genomes and are tentatively considered to be non-coding regions.

2.3.2 *Genetic aspects of mitochondrial DNA*

Mutations in the mitochondrial genome of *Sacch. cerevisiae* (reviewed in ref. 6) belong to three general classes, the specific properties of which imply that different experimental procedures may be used to manipulate them. Firstly, point mutations (typically single base substitutions) or short deletions within one of the structural mitochondrial genes encoding components of the respiratory chain, of the ATP synthase complex or of the mitochondrial protein synthesis machinery. Such mutations confer an RD phenotype and will be referred to in this chapter as *mit⁻* mutants. Note that this class encompasses mutations originally termed *mit⁻* as well as mutations originally termed *syn⁻*, *pho⁻* or *gin⁻*. The *pho⁻* mutants, with mutations affecting the ATP synthase genes, are not physiologically RD, yet they cannot grow on non-fermentable carbon sources. For this reason, they can be considered to be like other *mit⁻* mutants for most genetic tests. Secondly, point mutations within one of the structural mitochondrial genes as above but conferring an RC phenotype with resistance to a specific inhibitor of the mitochondrial function concerned (typically an antibiotic resistance or a drug resistance). They will be referred to in this chapter as *ant*^{R} mutants even though this class encompasses resistance to drugs other than antibiotics. Thirdly, very large deletions in the mitochondrial genetic map accompanied by reiteration of the conserved DNA segment of, in the extreme case, complete loss of all of the mtDNA. They obviously all confer an RD phenotype and are non-revertible. They will be referred to in this chapter as either *rho⁻* mutants if a segment of mtDNA is still present or *rho⁰* mutants for those cells devoid of mtDNA.

A normal cell of *Sacch. cerevisiae* contains 20−35 molecules of mtDNA (7). The actual figure may be slightly variable from strain to strain and varies according to growth conditions, but in all cases a few dozen mtDNA molecules are present in a cell and are all active at the same time. The distribution of mtDNA molecules between individual mitochondria is certainly important for the physiology of the cell but can be largely disregarded in the genetic methods described here because the molecules of mtDNA are apparently redistributed in an efficient manner during the time of a genetic cross. An explanation for this fact is probably that, in rapid growth conditions, mitochondria tend to fuse with one another, forming a very small number of large and branched mitochondria or even only one giant mitochondrion (8). The multiplicity of the mtDNA molecules results in a cell being in either one of two states, namely either *homoplasmic*, in which case all of the mtDNA molecules are identical in terms of the genetic markers they carry, or *heteroplasmic*, in which different mtDNA molecules carrying different genetic markers are present simultaneously. This definition is theoretical but, in practice, successive subcloning experiments as described in Section 3.2.1 provide an excellent operational definition of homoplasmic and heteroplasmic cells that is most useful in all mitochondrial genetics of *Sacch. cerevisiae*.

189

Table 2. Growth media used for the isolation and characterization of mitochondrial mutants.

Type of medium	Composition
YPglu	2% D-glucose, 1% yeast extract, 1% Bactopeptone
N0	YPglu buffered with 50 mM phosphate buffer pH 6.24
YP10	10% D-glucose, 1% yeast extract, 1% Bactopeptone
YPdif	0.1% D-glucose, 2% (v/v) glycerol, 1% yeast extract, 1% Bactopeptone
YPgal	2% D-galactose, 1% yeast extract, 1% Bactopeptone
YPgly	2% (v/v) glycerol, 1% yeast extract, 1% Bactopeptone
N3	YPgly buffered with 50 mM phosphate buffer pH 6.24
N1	2% (v/v) ethanol, 1% yeast extract, 1% Bactopeptone buffered with 50 mM phosphate buffer pH 6.24
W0	2% D-glucose, 0.67% Yeast−Nitrogen Base without amino acids
W10	10% D-glucose, 0.67% Yeast−Nitrogen Base without amino acids
G0	Minimal medium see *Table 2* of Appendix to this chapter
Ggal	G0 containing 2% D-galactose instead of glucose
N3C	N3 containing 0.4% chloramphenicol adjusted to pH 6.5
N3E	N3 containing 0.5% erythromycin adjusted to pH 6.24
N3S	N3 containing 0.5% spiramycin adjusted to pH 6.24
N3P	N3 containing 0.2% paromomycin adjusted to pH 6.5
N3O	N3 containing 3 mg/l of oligomycin, 0.5% methanol
N3V	N3 containing 1 mg/l of venturicidin, 0.5% methanol
N3oss	N3 containing 2 mg/l of ossamycin, 0.5% methanol
N1A	N1 containing 0.1 mg/l antimycin A, 0.1% ethanol
N1F	N1 containing 5 mg/l funiculosin, 1% ethanol
N3D	N3 containing 35 mg/l diuron, 2% acetone
N3M	N3 containing 0.3 mg/l mucidin, 0.1% ethanol
N3myx	N3 containing 2 mg/l myxothiazol, 2% ethanol

3. BASIC GENETIC TECHNIQUES FOR YEAST MITOCHONDRIA

3.1 Isolation of mitochondrial mutants

3.1.1 *Induction and isolation of rho⁻ and rho⁰ mutants*

Rho⁻ and *rho⁰* mutants are frequent. Cultures of *rho⁺* strains routinely contain about 1% of such mutants under normal conditions. The frequency of mutants can reach up to 100% after mutagenesis. *Table 2* summarizes the types of media used for the isolation and characterization of mutants; full details of these media are given in the Appendix to this chapter.

Isolation of rho⁰ mutants. Rho⁰ mutants form a specific subclass of RD mutants which completely lack all mtDNA. They are particularly useful in two aspects. Firstly, they provide a null control against which other mitochondrial mutations can be tested by crosses and secondly they can be used in coordination with the *kar1* mutation to in-

Table 3. Long-term storage of yeast mitochondrial mutants.

Storage medium

Mitochondrial mutants can be stored indefinitely at −70°C in the following medium
    1% (w/v) yeast extract (Difco)
    1% (w/v) Bactopeptone (Difco)
    2% (w/v) D-glucose
   25% (v/v) glycerol
Prepare this medium, distribute 0.5−1 ml aliquots in small screw-cap tubes and autoclave.

Preparation of strains for storage

This is most important since cells should be cultivated in the appropriate medium to eliminate undesirable mutants or revertants. For ant^R mutants use the corresponding antibiotic- or drug-containing medium. If several ant^R mutations are present simultaneously, use either the N3 medium or the antibiotic or drug corresponding to the less stable mutations (i.e. the mutation conferring the strongest selective disadvantage on N3 medium). Check purity of culture as in *Figure 3*. For *mit⁻* mutants, use YP10. If the strain is *pet9*, verify the absence of *mit⁺* revertants by the qualitative replica cross to a *PET9 rho⁰* strain as in *Figure 4*. If the strain is *PET*, verify the absence of RC revertants by plating on N3 and check the purity of the culture by crossing subclones to a *rho⁻* mutant containing the wild-type *mit⁺* allele, using the qualitative replica cross method. Clones with original *mit⁻* genotype will give rise to RC diploids while spontaneous *rho⁻* mutants (which tend to accumulate in *PET9 mit⁻* cultures) will not. For *rho⁻* mutants, use YPglu or YP10 and test the purity of the culture as in Section 3.2.2.

Freezing and thawing of cells

If culture is on solid medium, dissociate a loopful of cells into storage medium. If the culture is liquid, add 0.1−0.2 ml of culture directly into storage medium. Do not grow cells in this medium. Freeze immediately at −70°C.
    To use the stocks, thaw and inoculate into YPglu (*rho⁻*), YP10 (*mit⁻*), N3 or antibiotic-containing medium (*ant^R*). Incubate for 1−2 days at 28°C.

troduce a specific mitochondrial genotype within a known nuclear background (see Section 3.2.3). A specific induction of *rho⁰* mutants can be obtained by growing cells in the presence of ethidium bromide as follows.

(i)   Grow the cells of the desired haploid strain (with appropriate auxotrophic markers) in N3 medium (if RC) or YPglu medium (if RD).

(ii)   Pellet the cells by centrifugation at 1000 *g* for 5 min, wash and resuspend in sterile water at a density of 10^7 cells/ml (use a haemocytometer or calibrated spectrophotometer).

(iii)   Inoculate 0.05 ml of the cell suspension into 5 ml of N0 medium. Add 0.01 ml of an ethidium bromide stock solution at 10 mg/ml in sterile water (to obtain a final concentration of 20 μg/ml of ethidium bromide). Wrap the tube or flask containing the culture in aluminium foil and incubate at 28°C for 24 h with shaking to prevent cell sedimentation.

(iv)   Transfer 0.05 ml of the previous culture into 5 ml of fresh N0 medium containing 20 μg/ml of ethidium bromide and incubate as in step (iii).

(v)   Repeat the transfer once again as in step (iv). Note that one culture in the presence of ethidium bromide is usually sufficient to convert all of the cells into *rho⁰* cells. The purpose of three successive cultures in the presence of ethidium bromide is to ensure that all cells are indeed *rho⁰*.

Table 4. Criteria for determining the mitochondrial inheritance of yeast mutations.

Verification of a rho^0 genotype

Apply the following criteria to verify that an RD clone is a rho^0 mutant and not an undesirable rho^-, mit^- or *pet* mutant.

1. Check for the total absence of mtDNA by running total yeast DNA in a CsCl analytical gradient (see Section 6.2.1); mtDNA is lighter than nuclear DNA. This is the original definition of rho^0 mutants (9). The procedure is, however, time consuming and not every laboratory is equipped with an analytical ultracentrifuge.
2. Spot ~ 1 μg of total yeast DNA from a minilysate (see Section 6.2.2) on a nitrocellulose sheet, bake at 80°C under vacuum and hybridize using CsCl-purified rho^+ mtDNA (Section 6.2.3) as a probe. Include spots of 1 μg total DNA from a rho^+ and from a known rho^0 mutant as controls.
3. Cross the presumed rho^0 clone with a series of mit^- tester strains representative of the various mitochondrial genes and loci (same procedure as for testing rho^- mutants, see *Figure 4*). A rho^0 clone will produce *only* RD diploids whatever the mit^- tester used.
4. Cross the presumed rho^0 mutant to a rho^+ tester and verify that zygotic suppressiveness is zero to within the limits of statistical significance given by the total number of colonies scored (see Section 3.5).

ant^R mutants

Apply the following criteria to distinguish between mitochondrial mutations and undesirable nuclear mutations.

1. Cross the ant^R mutant to a rho^+ tester strain with the corresponding ant^S allele and possessing another (non-allelic) ant^R mitochondrial mutation, used as a control. Use the quantitative random cross method only (Section 3.3.2). Verify that the mitotic segregation of the ant^R/ant^S phenotype occurs among diploids and that the percentage transmission of the ant^R allele is coordinated with that of the control allele (see *Table 7*).
2. Cross the ant^R mutant to a rho^0 tester strain, using the quantitative random cross method. Verify that all diploid clones are ant^R.
3. Isolate a rho^0 derivative of the ant^R mutant and cross it to a rho^+ tester strain, using the quantitative random cross method. Check that all of the diploid clones are ant^S.
4. Sporulate one homoplasmic ant^R diploid clone from the cross (step 1) and check that all spores of tetrads are ant^R.

mit^- mutants

Apply the following criteria to distinguish between mitochondrial mutations and undesirable rho^- (or rho^0) or *pet* mutations.

1. Cross the presumed mit^- mutant to a rho^+ mit^+ tester strain, using the quantitative random cross method of Section 3.3.2. Check the mitotic segregation of the RC/RD phenotype among the diploids derived from the cross.
2. Cross the presumed mit^- mutant to a rho^0 tester strain and check that the entire diploid population is RD.
3. Cross the presumed mit^- mutant to a series of well-characterized rho^- mutants representing, together, every part of the mitochondrial genome. Check that RC diploids are formed with at least one rho^- mutant.

rho^- mutants

Apply the following criteria to distinguish between a rho^- mutant and an undesirable *pet* or rho^0 mutant.

1. Cross the presumed rho^- mutant to a rho^0 PET tester strain and check for the complete absence of RC cells in the diploid progeny.
2. Cross the presumed rho^- mutant to a series of mit^- tester strains representative of the various mitochondrial genes of loci. A rho^- mutant will either give rise to RC diploids in the progeny *of some but not all* crosses or will only give rise to RD diploids in all crosses. In the latter case, use criteria 3 and 4 to distinguish from a rho^0 mutant.
3. Cross the presumed rho^- mutant to a rho^+ tester strain and measure the zygotic suppressiveness as described in Section 3.5.
4. Prepare a minilysate from the presumed rho^- mutant, digest with restriction endonuclease, electrophorese and hybridize using purified mtDNA from a rho^+ strain as probe. Check for the presence of mtDNA and the existence of large deletions in the mitochondrial genome as compared with the rho^+ strain.

(vi) Pellet the cells, wash and resuspend in sterile water. Count the cell density and dilute to 10^3 cells/ml. Plate 0.1 ml aliquots on YPglu and incubate the plates at 28°C for 2−3 days. Note that only one dilution is necessary since the ethidium bromide treatment does not reduce cell viability.

(vii) Replicate the YPglu plate onto YPgly or N3 medium to ensure that all the clones are RD (this is always the case).

(viii) Pick up a single colony on the YPglu plate, grow in 5 ml of YPglu and store mutant (*Table 3*).

If *rho*^{0} mutants are needed from a large number of strains (such as when testing the mitochondrial inheritance of *ant*^{R} mutants, see *Table 4*) the following method may provide a quicker alternative.

(i) Place the strains to be mutagenized in grids on N3 medium (e.g. a 5 × 5 grid) and incubate at 28°C for 2−3 days.

(ii) Replicate on N0 medium containing 40 μg/ml of ethidium bromide. Incubate at 28°C in darkness for 2 days.

(iii) Transfer by replica plating onto fresh N0 medium containing 40 μg/ml of ethidium bromide. Incubate as in step (ii).

(iv) Repeat the transfer three times in succession.

(v) After the fourth passage on ethidium bromide, replica plate on YPglu and N3 media. Incubate at 28°C for 3 days.

(vi) Check for the complete absence of growth on N3 and use the YPglu plate for further tests, as necessary.

Caution: ethidium bromide is mutagenic. Wear gloves and dispose of wastes properly. Only add ethidium bromide to the medium after autoclaving; the ethidium bromide stock solution can be filter-sterilized easily.

Isolation of independent spontaneous rho$^-$ mutants (primary clones) from a rho$^+$ strain. Because *rho*$^-$ mutants are frequent, it is not necessary to induce them, so long as the number of mutants needed is not too large. The major difficulty with spontaneous mutants, however, is to ensure that they are of independent origin. The following method can be used for *rho*$^-$ mutants.

(i) Grow the *rho*$^+$ strain in N3 until stationary phase in order to eliminate all preexisting *rho*$^-$ mutants.

(ii) Dilute to 10^3 cells/ml in sterile water.

(iii) Plate 0.1 ml aliquots on YPdif and incubate at 28°C for at least 5−6 days (prepare enough plates for the number of *rho*$^-$ mutants desired, the frequency of spontaneous mutants is usually between 0.5 and 5%).

(iv) Score the plates and pick up RD colonies (*petite*) individually. These constitute primary clones and need subcloning (see the following note on subcloning).

Induction of a range of rho$^-$ mutants (primary clones) from a rho$^+$ strain. Rho$^-$ mutants can be induced efficiently by a wide variety of chemicals or treatments (see ref. 6 for a review). Most studies on the induction of *rho*$^-$ mutations (or *rho*^{0} mutations since the two were rarely distinguished) have focused on UV irradiation, ethidium bromide mutagenesis or berenil mutagenesis. Below is the method used to determine the kinetics of induction of *rho*⁻ mutants by ethidium bromide and to obtain a collection of independent mutants (10).

(i)   Grow the desired haploid strain (with appropriate auxotrophic markers) in N3 medium until stationary phase (remember that *rho*⁻ diploid cells never sporulate so that it is useless to induce a complete collection of *rho*⁻ mutants from a diploid strain).

(ii)   Pellet the cells by centrifugation at 1000 g for 5 min, wash and resuspend in sterile water at a density of 10^7 cells/ml.

(iii)   Mix the following in a sterile tube or conical flask wrapped in aluminium foil: 3.85 ml of sterile water; 0.5 ml of 0.5 M sodium potassium phosphate buffer, pH 6.5; 0.5 ml of cell suspension; 0.1 ml of 500 μg/ml cycloheximide stock solution in sterile water and 0.05 ml of 500 μg/ml ethidium bromide stock solution in sterile water.

(iv)   Incubate at 28°C with gentle shaking to prevent sedimentation.

(v)   Take 0.1 ml aliquots at various time intervals [e.g. immediately after mixing (zero time) and every hour thereafter until 6 h]. Dilute in sterile water to 10^3 cells/ml and plate 0.1 ml aliquots on YPdif medium. Note that the kinetics of induction may vary from strain to strain. When using a given strain for the first time it may be advisable to take aliquots of the mutagenesis mixture for longer time periods.

(vi)   Incubate YPdif plates at 28°C for at least 5–6 days.

(vii)   Score the plates and draw the kinetics of induction as the log of the frequencies of RC colonies remaining as a function of time of treatment. A straight line should be found over a long time interval.

(viii) Pick up the RD (*petite*) colonies individually at a time corresponding to a moderate induction (this reduces the proportion of undesirable *rho*⁰ mutants). These constitute primary clones and need to be subcloned (see the following note on subcloning).

The following method provides a quicker alternative if it is not necessary to study the kinetics of induction.

(i)   Grow the desired haploid strain (with appropriate auxotrophic markers) in N3 medium until stationary phase.

(ii)   Pellet the cells by centrifugation at 1000 g for 5 min, wash and resuspend in 5 ml of N0 medium at a density of 10^7 cells/ml. Add 0.01 ml of an ethidium bromide stock solution at 10 mg/ml (this gives a final concentration of 20 μg/ml ethidium bromide). Wrap the tube in aluminium foil and incubate at 28°C for 60 min (this corresponds to a moderate induction for most strains. Do not extend the incubation time to ensure that mutants remain independent).

(iii)   Pellet the cells as in step (ii), wash and resuspend in sterile water at a cell density of 10^7 cells/ml.

(iv)   Plate 0.1 ml aliquots on YPdif and incubate at 28°C for at least 5–6 days.

(v)   Score the plates and pick up RD colonies individually. These constitute primary clones and need to be subcloned (see the following note on subcloning).

Subcloning of rho⁻ cells. RD colonies picked out by any of the above methods represent primary clones (i.e. clones arising from the mutagenized cells themselves). Primary clones are very often mixtures of rho^0 and rho^- cells, the latter themselves being mixtures of different rho^- mutations. It is therefore necessary to subclone each primary clone at least once prior to using it as a pure rho^- mutant of one particular genotype. Primary clones can be either subcloned individually or mixed and subcloned altogether (see below). Test all clones (primary or secondary) as described in Section 3.3.1.

Direct isolation of large collections of secondary clones of rho⁻ mutants. Because subcloning of individual rho^- primary clones is time consuming, the following procedure is recommended to obtain a large collection of secondary clones directly.

(i)   Pick out RD colonies from one of the above procedures, taking care that equivalent numbers of cells are taken from each colony to limit any bias. Dissociate in a separate tube containing 10 ml of sterile water.

(ii)   When a sufficient number of primary clones has been picked up (use moderate induction to limit the frequency of rho^0 clones), mix thoroughly. This suspension constitutes a random mixture of primary clones. Do not grow such a mixture to avoid bias. Count the cell density, dilute to 10^3 cells/ml and plate 0.1 ml aliquots on YPglu.

(iii)   Incubate for 2 or 3 days at 28°C. Each colony represents a secondary clone. Secondary clones usually arise from homoplasmic cells and are pure within the limit of stability of the particular rho^- mutant considered (see Section 3.2.2).

(iv)   Test the secondary clones by the qualitative replica cross procedure (Section 3.3.1).

Deletion of additional mtDNA from a characterized rho⁻ mutant. Rho⁻ mutants are susceptible to further deletion and can be treated for this purpose essentially like rho^+ strains. Spontaneous deletions may occur with a significant frequency in some rho^- mutants of low stability. In other cases, or if you need a large collection of different deletions, use ethidium bromide mutagenesis (as described previously) or manganese mutagenesis (see Section 3.1.3). Note that firstly, for most rho^- mutants, the kinetics of loss of mtDNA fragments is slower than for rho^+ strains, and secondly mutant clones are indistinguishable by their phenotype from non-mutant clones. Plate the mutagenized cultures on YPglu or YP10 only (YPdif is useless for this!). The mutant clones picked up are primary clones and need subcloning.

3.1.2 Isolation of ant^R mitochondrial mutants

Mitochondrial point mutations conferring an ant^R phenotype are rare (of the order of $10^{-7}-10^{-8}$/cell/division). However, because they remain RC and, in addition, become resistant to a specific inhibitor, ant^R mutants can easily be selected for. Thus, there is no major need for a mutagenesis to increase their frequency (although manganese mutagenesis as described in Section 3.1.3 can be used).

The following method can be used to select a collection of independent mitochondrial ant^R mutants.

(i)   Grow the desired haploid strain with at least one auxotrophic requirement in N3 medium (remember that it is necessary that the parental strain be RC to select ant^R mutants: use only a rho^+ mit^+ *PET* strain).

(ii)   Dilute in sterile water to 10^3 cells/ml and plate 0.1 ml aliquots on N3 or streak the culture on N3 plates for single colonies. Incubate at 28°C for 3 days.

(iii) Pick up single colonies (subclones) and inoculate individually in 5 ml of N3 medium. Incubate at 28°C for 2 days. Prepare as many subclones as you eventually want mutants.

(iv)   Pellet the cells by centrifugation at 1000 *g* for 5 min and resuspend in 1 ml of sterile water.

(v)   Plate 0.1 ml of each suspension on N3 or N1 medium containing the appropriate antibiotic or drug.

(vi)   Incubate the selective plate at 28°C for at least a week to allow appearance of resistant colonies over the lawn of inhibited cells. Some antibiotics require longer incubation times (e.g. N3C or N3M plates require up to 3 weeks of incubation). Note the gradual appearance of colonies by examining the plates at regular intervals (e.g. every other day). Prevent desiccation during prolonged incubation by placing the plates in a humidity-controlled incubator or put the plates in a container with wet pads in a Petri dish.

(vii) Pick up a single mutant colony per original subclone to ensure that mutants are of independent origin.

(viii) Inoculate in 5 ml of N3 medium and incubate at 28°C for 2 days.

(ix)   Dilute in sterile water to 10^3 cells/ml and plate 0.1 ml on the N3 medium containing the antibiotic or drug (or streak out for single colonies). This subcloning step ensures that the colony picked up on the selective plate was actually due to a resistant mutation and not a physiological adaptation of sensitive cells to a prolonged incubation on the selective medium at high cell density.

(x)   Pick up a single subclone from each mutant, inoculate in N3 or N1 medium containing the antibiotic or drug, incubate at 28°C for 2−3 days and store the mutants (*Table 3*).

Although this procedure is extremely simple, the following remarks will help to ensure a successful isolation of ant^R mitochondrial mutants.

Firstly, for several antibiotics (e.g. chloramphenicol, spiromycin, paromomycin) the concentrations of antibiotics necessary to inhibit the wild-type sensitive strains can be relatively close to the concentrations used to select resistant mutants. Hence, a residual growth of the sensitive cells on the selective plates may occur. The level of resistance of wild-type cells varies considerably from strain to strain. It is advisable to select strains with the lowest level of resistance before isolating mitochondrial mutants or before using them as tester strains in crosses.

Secondly, although most ant^R mitochondrial mutants are healthy RC cells, some show growth rates slower than the wild-type on N3 or YPgly media (e.g. chloramphenicol-resistant or mucidin-resistant cells). It is recommended to grow ant^R mutants always in the appropriate antibiotic-containing medium prior to using them for genetics or molecular tests in order to eliminate possible spontaneous revertants.

Thirdly, not all resistant mutants isolated by the above procedure are due to mutations in the mitochondrial genome and it is *absolutely necessary* that the mutants are tested genetically to determine if the mutation is mitochondrial or nuclear prior to further consideration (see *Table 4*). Note that mitochondrial mutants are generally resistant to much higher concentrations of antibiotics or drugs than their nuclear counterparts; hence the use of high concentrations of antibiotics in media. Resistance to high concentrations is *not*, however, sufficient to prove that the mutation is mitochondrial and not nuclear.

3.1.3 *Isolation of mit⁻ mitochondrial mutants*

Point mutations in the mitochondrial genome conferring an RD phenotype are very rare as compared with the production of rho^- or rho^0 mutants. It is therefore necessary to increase their frequency by mutagenesis and to eliminate the rho^- or rho^0 mutants. This is best achieved by combining manganese mutagenesis (11) with the use of the *op1* (*pet9*) nuclear mutation which is lethal if combined with a rho^- or rho^0 mutation (12) but not if combined with a *mit⁻* mutation. The following method has been developed (13).

(i) Grow a rho^+mit^+pet9 haploid strain with at least one auxotrophic marker in YP10 medium.

(ii) Pellet the cells by centrifugation at 1000 *g* for 5 min, wash and resuspend in sterile water, count the cells and dilute in water to 10^7 cells/ml.

(iii) Inoculate 0.1 ml of the cell suspension in 5 ml of YP10 medium. Add 0.1 ml of a 350 mM $MnCl_2$ sterile solution (to a final concentration of 7 mM $MnCl_2$ in the culture medium). Note that Mn^{2+} is toxic, so do not exceed a final concentration of 10 mM.

(iv) Incubate at 28°C for 24 h with shaking to prevent cell sedimentation.

(v) Count the cells and dilute in sterile water to 2×10^3 cells/ml. Plate 0.1 ml aliquots on YP10 plates. Remember that *mit⁻* mutants are rare even after induction, so make sure that you plate enough Petri dishes to obtain several thousands of colonies in total.

(vi) Incubate at 28°C for 3 days. The population of colonies appearing is composed of non-mutants and of the desired *mit⁻* mutants (rho^- and rho^0 mutants, which are also induced by Mn^{2+}, are eliminated). However, all colonies are RD, due to the *pet9* mutation (in fact *pet9* does not confer an actual RD phenotype, but cells grow so extremely slowly on N3 or YPgly that they do not form visible colonies).

(vii) Pick up colonies individually and place in 8 × 8 grids on YP10. Incubate at 28°C for 3 days.

(viii) Replicate the grids on YP10, N3, W0 media and on W0 plates covered with a lawn of *pet9* rho^0 tester strain of opposite mating type and auxotrophic requirements (see Section 3.3.1). Incubate at 28°C for 3 days.

(ix) Check for the absence of growth of haploids on the N3 and W0 replicas and the confluent growth of diploids on W0. Replica plate the diploids on N3. Incubate the plates at 28°C for 3 days.

(x) Score diploid colonies that fail to grow on the N3 replica and pick up the corresponding haploid clone on the YP10 grid.

(xi) Dissociate and dilute to 10^3 cells/ml in sterile water. Plate 0.1 ml aliquots on YP10. Incubate and repeat steps (vii)−(x) on subclones.

(xii) Pick out a single subclone from each mutant. Inoculate in YP10 and incubate at 28°C for 2 days. Store the mutant subclone (*Table 3*).

3.1.4. *Isolation of conditional mit⁻ mitochondrial mutants*

A number of temperature-sensitive mitochondrial mutants have been described. Those conferring a conditional RD phenotype occur in many, if not all, mitochondrial genes. Temperature-sensitive *mit⁻* mutants can be treated as other *mit⁻* mutants at their restrictive temperature but as RC strains at their permissive temperature (although their growth rate on N3 medium may be affected). For the isolation of such mutants one must first define the restrictive and permissive temperatures according to ones require-ments (e.g. 35°C and 25°C, respectively if heat-sensitive mutants are desired or 18°C and 28°C, respectively if cold-sensitive mutants are required) and then apply either of the following methods.

If the pet9 mutation is used. Steps (i)−(viii) are the same as in Section 3.1.3.

(ix) Replica plate diploids on N3 medium in duplicate. Incubate one N3 plate at the restrictive temperature and the other at the permissive temperature.

(x) Score the diploid colonies that fail to grow on the N3 replica at the restrictive temperature but do grow at the permissive temperature. Pick out the correspond-ing haploid clone on the YP10 grid.

(xi) and (xii) Same as in Section 3.1.3.

If a direct phenotypic screening is used.

(i) Grow a *rho⁺mit⁺pet9* haploid strain with at least one auxotrophic mutation in N3 medium.

(ii)−(vii) Same as in Section 3.1.3 except that the population of colonies is RC.

(viii) Replicate the grids on N3 medium in duplicate. Incubate one plate at the restric-tive temperature and the other at the permissive temperature.

(ix) Score haploid colonies that grow on N3 at the permissive temperature but not at the restrictive temperature. Pick out the corresponding clone on YP10 (do not worry about *rho⁻* or *rho⁰* mutants as they will never be conditional mutants!).

(x) Cross with a *rho⁰* tester strain and verify that the diploids grow on N3 at the permissive temperature but not at the restrictive temperature (this step eliminates undesirable nuclear *pet⁻* mutations).

(xi) and (xii) Same as in Section 3.1.3.

3.2 Segregation tests

3.2.1 *Determination of the homoplasmic/heteroplasmic state of cells and the purity of cultures*

Heteroplasmic cells may originate from crosses, mutagenesis or spontaneous mutations (mainly *rho⁻* mutations which are frequent and unstable, see below). It is of primary importance in mitochondrial genetics to know the homoplasmic/heteroplasmic state of a cell. This can only be determined *a posteriori* from the pure/mixed character of the clone derived from that cell, after mitotic segregation of the mtDNA molecules has taken place. The mechanism by which segregation takes place is not entirely clear (see ref. 6 for review) but it does appear that, under normal conditions, the process is rapid, that is heteroplasmic cells always give rise to mixtures of different homoplasmic cells during formation of a visible colony. A single subcloning is therefore usually sufficient to isolate homoplasmic cell lines from heteroplasmic situations. Rare cases of persistent heteroplasmons resulting from local peculiarities of mtDNA (e.g. a duplication) have been reported. However, even under such exceptional circumstances the formation of a visible colony is sufficient to result in mixed clones composed of numerous homoplasmic cells in addition to the persistent heteroplasmic type.

The purity of cultures can be determined by plating out diluted aliquots on a non-selective medium (e.g. YPglu or YP10 for haploids or W0 or W10 for diploids) and by replicating the subclones on selective media (e.g. N3 for distinguishing between RC and RD cells or antibiotic-containing media for distinguishing between *ant*[R] and *ant*[S] cells). The pure/mixed character of the culture is demonstrated by the homogeneity/ heterogeneity of the subclones (it is necessary to examine a sufficient number of subclones for statistical significance), the purity of each subclone itself being determined by subsequent subcloning and/or by the appearance of replica colonies on the selective media (i.e. confluent growth, sectored growth, papillate growth or complete absence of growth). *Figure 3* shows an example of this.

For correct interpretation of the appearance of the replica colonies, it is necessary to note that, from an average size colony, $2-5 \times 10^5$ cells are deposited by the velvet on the replica. This figure sets the lower limit of detection of the replica method (i.e. absence of growth means a clone in which the frequency of cells able to grow on the selective medium is below $2-5 \times 10^{-6}$). This figure is particularly convenient for mitochondrial genetics because firstly it is well below the frequency of the rarer homoplasmic cell type segregated out of a heteroplasmic cell, and secondly it is at least one order of magnitude above the spontaneous mutation rate for point mutations or revertants. Complete absence of growth of the replica trace on a selective medium is therefore demonstrative of a pure clone derived from a homoplasmic cell. Confluent growth of the replica, on the contrary, may indicate either a pure clone (i.e. entirely composed of cells able to grow on the selective medium) or a mixed clone in which the frequency of such cells is sufficiently high. Sectors or papillate growth are obviously indicative of mixed clones derived from heteroplasmic cells.

Now, the classical distinction between RC cells whose mitochondrial genotype can be determined directly from their phenotype and RD cells whose mitochondrial genotype can only be determined after marker rescue obviously holds true for the above analysis of subclones.

3.2.2 *Methods to check the purity of and/or purify a rho⁻ culture*

All *rho⁻* mutants are intrinsically unstable and spontaneously undergo secondary mutation, deletions, rearrangements or complete loss of their mtDNA. Because all *rho⁻* (and *rho⁰*) mutants have the same RD phenotype there is no direct method to eliminate

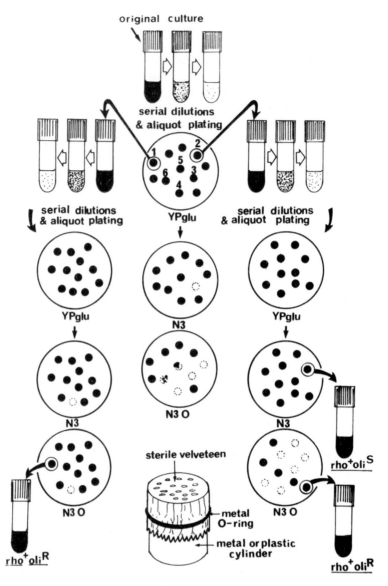

Figure 3. Subclonings and replica tests determine the homoplasmic/heteroplasmic state of cells and the purity of cultures. The example shows the test by replica and the isolation of homoplasmic subclones from a hypothetical *rho*[+] population containing homoplasmic and heteroplasmic cells for the *oli*[R] mutation. **Clone 1** is a pure *rho*[+] *oli*[R] clone originated from a homoplasmic cell. Subcloning reveals 100% of clones of the same type (except for the spontaneous *rho*[−] mutants). **Clone 2** is a mixed clone originated from a heteroplasmic cell containing simultaneously *rho*[+] *oli*[R] and *rho*[+] *oli*[S] mtDNA molecules. Subcloning permits the isolation of pure *rho*[+] *oli*[R] subclones and pure *rho*[+] *oli*[S] subclones. **Clone 3** is either a pure *rho*[−] (or *rho*[0]) clone or a mixed clone containing various *rho*[−] mutants. Subcloning and qualitative replica crosses are necessary to distinguish between these possibilities. **Clone 4** is a homoplasmic *rho*[+] *oli*[S]. **Clone 5** is a mixed clone as clone 2 but in which the proportion of *rho*[+] *oli*[R] cells is lower. **Clone 6** is a mixed clone as clone 5 but in which the proportion of *rho*[+] *oli*[R] cells is even lower. Note that the conclusions drawn about the homoplasmic/heteroplasmic state of the parental cell of each clone are valid only if each colony originates from a single cell. Clumpy strains make the analysis more difficult and should be avoided whenever possible.

such undesirable mutations and the purity of each culture needs to be determined careful-ly. The probability that the original genotype remains unaltered is generally constant per cell division (this is the intrinsic stability of the *rho*⁻). Consequently, the overall frequency of cells of the original genotype in a culture (the purity of the culture) shows an exponential decrease relative to the total number of cell generations of the culture. The intrinsic stability varies to a large extent from one *rho*⁻ mutant to the next and is also influenced by the nuclear genotype of the strain used. In practice, it is impossi-ble to work with *rho*⁻ mutants with a stability lower than 0.95/cell/generation (in this case only 30% of the cells have the original genotype after the growth of a visible col-ony!). But even with *rho*⁻ mutants of high stability the proportion of cells with the original genotype rapidly decreases in the population during cultivation (e.g. *rho*⁻ mutants with stability values equal to 0.99 still give rise to more than 20% of cells with altered genotypes after the growth of a visible colony). For this reason it is strongly recommended not to grow a *rho*⁻ mutant from a previous culture without checking the purity of that culture and the number of generations that it has undergone; in ad-dition it is advisable to store aliquots of the tested cultures at −70°C (see *Table 3*) for subsequent use.

The purity of a *rho*⁻ culture can be determined by a qualitative replica cross and/or by minilysate analysis.

Replica cross test.

(i)   Dilute the culture in sterile water to 10³ cells/ml, store the undiluted culture in the refrigerator at 5°C.

(ii)   Plate 0.1 ml of diluted culture on YPglu, incubate at 28°C for 3 days.

(iii)   Place colonies on grids on YPglu (~ 200 − 300 colonies in total are needed if statistical significance of the composition of the population is desired) and incu-bate for 3 days.

(iv)   Test colonies by the qualitative replica cross using appropriate *rho*⁺ tester strains as shown in *Figure 4*.

Minilysate analysis.

(i) and (ii) As for the replica cross.

(iii)   Pick up colonies one by one and inoculate into 5 ml of YPglu.

(iv)   Prepare minilysates as in Section 6.2 and test the mtDNA by restriction digest and hybridization using CsCl-purified mtDNA as a probe.

If purity is satisfactory, aliquot the original culture and store aliquots at −70°C (*Table 3*). Always prepare enough aliquots of your clone for the foreseeable future. It is less work to prepare more aliquots than to repeat the testing of cultures. Use one aliquot to inoculate the culture needed for each experiment.

If purity is not satisfactory, pick up colonies with the required genotype on original YPglu plates and inoculate into 10 ml of YPglu or YP10 medium. Incubate at 28°C for 2 days. Test the purity of the subclones as before.

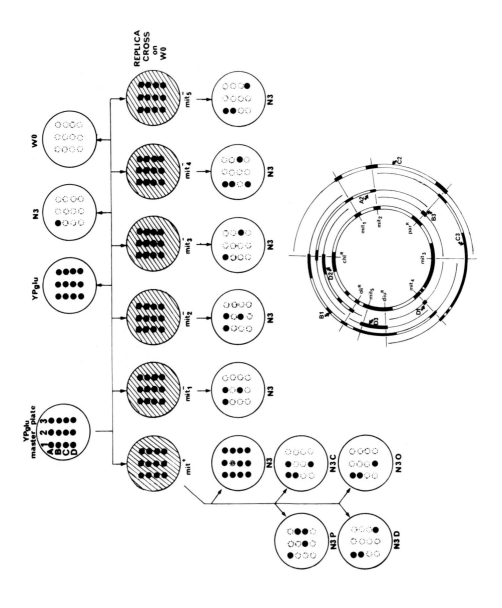

3.2.3 *How to transfer a known mitochondrial genotype into a new nuclear genetic background*

This can be performed by either of the following methods but the results are not always equivalent.

Method A: sporulation. This method can only be applied to rho^+mit^+ genotypes. Cross your strain with the mitochondrial genotype of interest to a rho^0 strain with opposite mating type and different auxotrophic requirements. Do a quantitative random cross using Method A steps (i)−(iv) of Section 3.3.2. Pick out the diploids and sporulate as described in *Table 5*. Note that all haploid segregants have the same mitochondrial genotype but different nuclear genotypes (unless the rho^0 tester was already isogenic to your strain of interest except for mating type).

Method B: cytoduction. This method can be applied to all types of mitochondrial mutations. It is based on the property of the *kar1* mutant to delay nuclear fusion after cytoplasmic fusion during mating. As a result, a fraction of zygotic buds are haploids with the nucleus of one parent but the cytoplasm of both. Combination of the *kar1* mutation with the rho^0 mutation offers, therefore, a simple method to transfer a known mitochondrial genome into a new genetic background (25).

(a) For rho^+mit^+ strains.

(i)   Cross your strain with the mitochondrial genotype of interest to a *kar1* rho^0 strain with opposite mating type and different auxotrophic requirements (*Table 6*). Use synchronous mating as in Method B of Section 3.3.2, steps (i)−(vi).

(ii)   When zygotes are formed take aliquots at various time intervals (e.g. every 2 h), dilute to 10^3 cells/ml and plate 0.1 ml aliquots on N3. Incubate at 28°C for 2−3 days.

Figure 4. The qualitative replica cross. The example shows the genetic analysis of individual rho^- clones by crosses to a series of well-defined tester strains (all are α *HIS1 MET1 ade2 pet9* rho^+ and either mit^+ or mit^-) and qualitative analysis of the resulting diploids by replica. **Clone A1** is the rho^+ control (use the parental strain from which the collection of rho^- mutants is derived). In the hypothetical example shown the genotype of the parental strain is *a his1 met1 ADE2 PET* rho^+ chl^R par^R diu^R mit^+. **Clone A3** is a rho^0 control (derived from the same parental strain as the collection of rho^- mutants tested). Complete genotypes of other clones can be deduced as follows:
clone A2, rho^- chl^R oli^0 par^0 diu^0 $mit1^+$ $mit2^+$ $mit3^0$ $mit4^0$ $mit5^0$
clone B1, rho^- chl^R oli^R par^0 diu^R $mit1^0$ $mit2^0$ $mit3^0$ $mit4^+$ $mit5^+$
clone B2, rho^- HIS
clone B3, rho^- chl^0 oli^0 par^R diu^0 $mit1^0$ $mit2^0$ $mit3^0$ $mit4^0$ $mit5^0$
clone C1, either rho^- chl^0 oli^0 par^0 diu^0 $mit1^0$ $mit2^0$ $mit3^0$ $mit4^0$ $mit5^0$ or rho^0
clone C2, rho^- chl^0 oli^0 par^R diu^0 $mit1^+$ $mit2^+$ $mit3^0$ $mit4^0$ $mit5^0$
clone C3, rho^- chl^0 oli^0 par^R diu^0 $mit1^0$ $mit2^0$ $mit3^+$ $mit4^+$ $mit5^0$
clone D1, rho^- chl^0 oli^0 par^0 diu^0 $mit1^0$ $mit2^0$ $mit3^0$ $mit4^+$ $mit5^0$
clone D2, rho^- chl^R oli^R par^0 diu^0 $mit1^0$ $mit2^0$ $mit3^0$ $mit4^0$ $mit5^0$
clone D3, rho^- chl^0 oli^0 par^0 diu^R $mit1^0$ $mit2^0$ $mit3^0$ $mit4^0$ $mit5^+$.
The insert gives a schematic representation of the mitochondrial map of *Sacch. cerevisiae* (major genes in full) and places the segment retained by each rho^- clone with its maximum limits (thin lines). Such a map can only be drawn if rho^- clones tested are secondary clones or clones of subsequent orders. For primary clones, there is no indication if the different genetic markers are retained on the same mtDNA molecules of the same cells. Only pick out the interesting rho^- clones on original YPglu plates (other plates have diploid cells issued from the cross, not the desired rho^- mutant cells).

Table 5. Sporulation of mitochondrial mutants of *Saccharomyces cerevisiae*.

Sporulation medium

0.25% (w/v) yeast extract (Difco)
0.1% (w/v) D-glucose
0.1 M potassium acetate
Prepare this medium and autoclave.

Sporulation and separation of ascospores

1. Inoculate desired homoplasmic diploid clone or mating mixture into 1−2 ml of sporulation medium placed in a 100 ml conical flask (sporulation is better if a very small volume is used). Incubate at 28°C for 2 or 3 days with shaking. Remember that only RC cells sporulate (see text).
2. Monitor tetrad formation under a microscope.
3. Take 0.1 ml of sporulated culture, add 0.1 ml (10 000 U) of β-glucuronidase from *Helix pomatia* (Industrie Biologique Française or Sigma) and incubate at 28°C for 60 min. Verify ascus disruption under microscope.
4. Either separate ascospores by micromanipulation on YPglu or dilute in 10 ml sterile water and vortex thoroughly.
5. In the second case, count the cell density with a haemocytometer and dilute to 10^4 and 10^3 cells/ml. Plate 0.1 ml aliquots of each dilution on YPglu. Incubate at 28°C for 2−3 days.
6. Replicate on W0 and either N3 or antibiotic-containing media (if homoplasmic RC diploids are sporulated) or YP10 and N3 (if heteroplasmic *mit⁺/mit⁻* zygotes are sporulated). Incubate at 28°C for 3 days.
7. Pick out either RC auxotrophic clones on N3 or antibiotic medium (for *ant*R mutations) or RD auxotrophic clones on YP10 (for *mit⁻* mutations). Dissociate in water and test auxotrophic requirements by dropping onto W0 plates supplemented with various combinations of nutrients (as required for the markers present in the diploid, see *Table 2* of Appendix). Incubate for 3 days at 28°C.
8. Test the mating type of the clones with the desired nuclear genotype by crossing to **a** and α tester strains with complementary auxotrophy on W0 plates.
9. In the case of sporulation of heteroplasmic zygotes, check the mitochondrial genotype of RD haploid segregants by crossing to *rho⁻* tester strains as in Section 3.3.1 (this step is necessary to eliminate undesirable spontaneous *rho⁻* mutants that may appear during sporulation and subsequent steps and to distinguish between the required *mit⁻* mutants from the *mit⁺ pet9* strains if *pet9* is used).

(iii) Replicate on W0 and W0 agar supplemented for the auxotrophic requirements of the *karl rho⁰* parent. Incubate at 28°C for 3 days.

(iv) Score the RC clones with auxotrophic requirements identical to the *karl rho⁰* parent. They are cytoductants with the nucleus of the *karl* parent and the mtDNA of your strain.

(v) If you want to transfer your mitochondrial genotype of interest into a specific non-*karl* nucleus, cross this cytoductant to a *KAR rho⁰* strain with the desired nuclear genetic background as in steps (i) and (ii) above, then replicate on W0 and W0 agar supplemented for the auxotrophic requirements of the *KAR rho⁰* parent. Pick out the RC clones with auxotrophic requirements identical to those of your *KAR* parent. These are the final cytoductants with both mitochondrial and nuclear genotypes of interest.

(b) For *rho⁻* mutants.

(i)−(iii) Same as in Method (a) except that aliquots are plated on YPglu instead of N3 medium.

(iv) Pick up RD clones with auxotrophic requirements identical to the *karl* parent.

Table 6. Standard yeast strains for mitochondrial genetics (14).

KL14-4A: a *his1 trp2 rho$^+$ome$^+$chl$^R_{321}$oli$^R_1$par$^R_{454}$*
KL14-4A/I21: *leu2* derivative of KL14-4A
Both strains are used as standard *rho$^+$* strains for crosses and mapping purposes. Large collections of *rho$^-$* mutants are available from both strains. Parental strains of several original *ant^R* mutations (15,16).

D273-10B/A: α *met rho$^+$ ome$^+$*
Standard *rho$^+$* strain for crosses and mapping purposes. Large collections of *rho$^-$* mutants and of *mit$^-$* mutants available from this strain. (17)

777-3A: α *ade1 pet9 rho$^+$*
Standard *pet9* strain (*rho$^-$* lethal). Large collection of *mit$^-$* mutants available from this strain (13).

AB1-4A/8: a *his4C ADE1 PET9* derivative of 777-3A *rho$^+$*
AK51-17A: α *met ADE1 PET9* derivative of 777-3A *rho$^+$*
Both strains are used for crosses and/or transfer of *mit$^-$* mutations from 777-3A to isogenic *PET9* nuclear backgrounds (18).

MH32-12D: a *ade2 his1 rho$^+$ chl$^R_{321}$ ery$^R_{221}$ oli$^R_{144}$ par$^R_{454}$*
MH41-7B: a *ade2 his1 rho$^+$ chl$^R_{321}$ ery$^R_{514}$ oli$^R_{145}$ par$^R_{454}$*
Large collections of *rho$^-$* mutants available from these strains (19).

J69-1B: α *ade1 his rho$^+$*
Parental strain of numerous *mit$^-$* mutants (especially in the ATP synthase genes) (20).

ID41-6/161: a *ade lys rho$^+$ome$^-$ chl$^R_{321}$ oli$^R_{1-4}$ par$^R_1$*
Parental strain of numerous *mit$^-$* mutants (21).

55R5-3C: a *ura1 rho$^+$ome$^-$*
Parental strains of numerous original *ant^R* mutants. Standard *ome$^-$* strain (22).

DP1-1B: α *his1 trp1 rho$^+$ome$^+$*
Parental strain of several original *ant^R* mutants. Standard *ome$^+$* strain (22).

IL828-3C. a *his1 ura1 ome$^+$ chl$^R_{321}$ ery$^R_{221}$ oli$^R_1$*
IL828-5D. α *his1 ome$^+$ chl$^R_{321}$ ery$^R_{221}$ oli$^R_1$*
IL871-1D. a *ura1 trp1 ome$^-$ chl$^R_{321}$ ery$^R_{221}$ oli$^R_1$*
IL871-1A. α *his1 trp1 ome$^-$ chl$^R_{321}$ ery$^R_{221}$ oli$^R_1$*
Standard *ome* tester strains (15,23,24).

JC8/55: a *leu1 can^R kar1 rho^0* (25)
CK50A/50: α *leu2 trp1 tyr6 phe can^R kar1 rho^0* [(Perrodin and Slonimski, unpublished). Both strains are used to transfer mitochondrial mutations into new nuclear backgrounds by cytoduction.]

Distinguish the desired *kar1 rho$^-$* cytoductant from non-mated *kar1 rho^0* parent by either the qualitative replica cross using a known *mit$^-$* tester strain (Section 3.3.1) or the minilysate procedure (Section 6.2.2). In all cases verify the identity of mtDNA of the cytoductant selected with the original *rho$^-$* mutant, using the minilysate procedure, to eliminate the possibility of secondary mutations of the *rho$^-$* mtDNA.

(v) Same as in Method (a) except that the RD clones are picked out and screened as in step (iv).

(c) For *mit$^-$* mutants.

(i)−(iii) Same as Method (a), except that aliquots are plated on YP10 instead of N3 medium.

(iv) Pick up RD clones with auxotrophic requirements identical to the *kar1* parent.

Distinguish the desired *kar1 mit⁻* cytoductant from the non-mated *kar1 rho⁰* parent by the qualitative replica cross method using a known *rho⁻* tester strain (Section 3.3.1).

(v)   Same as in step (iv).

Note that the frequency of cytoductants is variable from cross to cross. Plate $200-300$ colonies from the mating mixture at each time interval until $6-8$ h after zygote formation.

3.3 Tests for recombination

3.3.1 *Qualitative replica crosses*

Because all *rho⁻* (or *rho⁰*) mutants have the same RD phenotype, determination of their mitochondrial genotype requires rescuing their alleles, by recombination, into an RC cell (26). The rationale is as follows.

mit⁺ alleles: When crossed to a given *mit⁻* mutant, each *rho⁻* mutant which retains the fragment of mtDNA containing the corresponding *mit⁺* allele will give rise to at least some RC diploids. On the contrary, each *rho⁻* mutant that has lost the fragment of mtDNA containing the corresponding *mit⁺* allele will give rise to RD diploids only. In this case the genotype of the *rho⁻* mutant for that particular allele is written *mit⁰*. Because the test of diploids is qualitative, the frequencies of RC recombinants generated are not taken into account so long as they are above the limit of detection of the replica method (see Section 3.2.1).

antᴿ alleles: The same principle applies to rescuing *antᴿ* mutations carried by the *rho⁻* mutants except that crosses are performed to a *rho⁺* tester strain and that the diploids are tested for the *antᴿ/antˢ* phenotype instead of the RC/RD phenotype. *Rho⁻* mutants that have lost the fragment containing the *antᴿ* allele are indicated *ant⁰*.

Determination of the markers retained or lost by rho⁻ mutants (Figure 4)

(i)   Place the *rho⁻* mutants in grids on YPglu medium (grids 8×8 are convenient for this purpose) and incubate at 28°C for $2-3$ days. Remember to incorporate the *rho⁺* parental strain and its *rho⁰* derivative on each grid as controls.

(ii)   Inoculate *pet9 rho⁺ mit⁻* tester strains (with opposite mating-type and auxotrophic requirements) in 10 ml of YP10 medium. Prepare a set of tester strains with *mit⁻* mutations representative of the various genes or loci for which you want to determine the loss/retention in the *rho⁻* tested (see *Tables 6* and *7*). If the *rho⁻* mutants tested originate from a *rho⁺* strain which has *antᴿ* mutations, also inoculate a *pet9 rho⁺ mit⁺* tester strain in 10 ml of YP10 medium. Incubate at 28°C for $2-3$ days without agitation.

(iii)   Spread 0.1 ml of the YP10 cultures on W0 medium to form a lawn (plating should be very regular for good mating efficiency).

(iv)   Replica plate the YPglu grids onto W0, N3 and YPglu media first, then onto the lawns of tester strains (remember to use a new velvet for each lawn to avoid cross-contamination; a unique master plate can tolerate $5-10$ successive impressions if good quality velvets are used). Incubate at 28°C for 3 days. Keep the master plates in the refrigerator at 5°C.

Table 7. A set of yeast mitochondrial mutations useful for mapping. Mutations are listed clockwise on map (see ref. 6).

Gene	Mutations		Phenotype	Approximate map location (units on KL14-4A map)
21S rRNA[a]		cs901	conditional *syn*[−]	95−97%
		ery[R]$_{221}$	erythromycin resistance	97−98%
		chl[R]$_{321}$	chloramphenicol resistance	0%
Asp tRNA		ts170	conditional *syn*[−]	5−6%
Tyr tRNA		tsm8	conditional *syn*[−]	6−7%
Sub II cyt. ox. (*oxi1* gene)		M13-249	respiratory deficient	15−17%
Sub III cyt. ox. (*oxi2* gene)		G199	respiratory deficient	20−22%
16S rRNA		*par*[R]$_{454}$	paromomycin resistance	35−38%
Sub I cyt. ox. (*oxi3* gene)		G20	respiratory deficient	45−58%
		G922	respiratory deficient	
Sub 6 ATPase (*oli2* gene)		*pho*[−]$_1$	phosphorylation deficient	60−65%
		oli[R]$_{144}$	oligomycin resistance	
Cyt.*b* (cob-box gene)	exon 1	G706	respiratory deficient	
		ana[R]$_{1-32}$	antimycin resistance	
		diu[R]$_{2-732}$	diuron resistance	
		muc[R]$_{1-771}$	mucidin resistance	
	intron 2	G1370	respiratory deficient	
	exon 4	G1988	respiratory deficient	72−80%
		ana[R]$_{2-25}$	antimycin resistance	
		diu[R]$_{1-731}$	diuron resistance	
	intron 4	G1659	respiratory deficient	
		M7832	respiratory deficient	
	exon 6	M6-200	respiratory deficient	
		muc[R]$_{2-772}$	mucidin resistance	
Sub 9 ATPase (*oli1* gene)		*oli*[R]$_1$	oligomycin resistance	82−83%

[a]Under the control of the *ome* locus.

(v)   Sort the W0 replica cross plates. Check for the absence of growth of replicas on N2 and W0 plates. Check for the confluent growth of diploids over the replica colonies and for the absence of growth of the lawn elsewhere (a slight residual growth of the auxotrophic lawn may be visible; it is due to the very high density of the inoculum but is not a problem so long as a *pet9* strain is used).

(vi)   Replica plate the diploids on N3 medium alone (for the *mit*[−] testers) and/or on N3 containing antibiotic (for the *mit*[+] tester). Incubate the new replicas at 28°C for 3 days. Store all other plates in the refrigerator at 5°C.

(vii)   Sort all the plates and score the result of each *rho*[−] mutant individually (see example in *Figure 4*).

This method indicates which alleles are present in each *rho*[−] clone but does not demonstrate that the segments between retained alleles are present. Always apply the minilysate procedure or purify the mtDNA to check continuity of the fragments retained as compared with the *rho*$^+$ mtDNA.

Mapping of mit$^-$ *mutants using petite deletion mapping.* This is the reciprocal of the previous method.

(i) Place the *mit*$^-$ to be mapped in grids on YP10 plates. Incubate at 28°C for 2−3 days.

(ii) Inoculate *rho*$^-$ testers (with opposite mating type and auxotrophic requirements) into 10 ml of YP10 medium. Prepare a set of *rho*$^-$ mutants with overlapping fragments of mtDNA that cover either the entire genome or the segment in which the *mit*$^-$ mutations are expected. Incubate at 28°C for 2−3 days.

(iii)−(vii) Same as for the previous method.

Each *mit*$^-$ mutant that gives rise to RC diploids with a given *rho*$^-$ tester is a mutation that maps within the fragment retained by that *rho*$^-$. Each *mit*$^-$ mutant that gives rise to RD diploids only with a given *rho*$^-$ tester is a mutation that maps outside of the fragment retained by that *rho*$^-$. Use an appropriate set of *rho*$^-$ mutants with overlapping fragments for a quick and precise mapping of the *mit*$^-$ mutation. For significant results it is critical that your set of *rho*$^-$ mutants has been submitted to precise restriction mapping (or, better, to complete DNA sequencing). Any *rho*$^-$ mutant whose retained fragment might be rearranged as compared with the *rho*$^+$ map should be avoided. Remember that, since the *petite* deletion mapping is a recombination test and not a complementation test, it is not necessary for functional segments of the mtDNA to be retained in the *rho*$^-$ mutants. However, any limitation or inhibition of recombination will result in a lack of RC recombinants in diploids which may be misleading. Check the consistency of the results, using different *rho*$^-$ testers, to deduce the final mapping assignments.

3.3.2 *Quantitative random cross*

This is the standard type of cross for the quantitative determination of the percentage transmission of mitochondrial alleles and of the recombination frequencies between pairs of alleles. It is based on the fact that both recombination and segregation of mtDNA molecules are rapid and efficient. The cross can be made either by placing the two strains in contact directly on the minimal medium selective for diploids or by synchronous mating in complete medium, followed by selection of diploids in minimal medium. The first method is quicker but results in a severe counter-selection of RD cells due to their slow growth on W0 medium. This method is therefore reserved for crosses between *ant*^{R} and *ant*^{S} cells. The second method must be used for all crosses involving *mit*$^-$ or *rho*$^-$ mutations in which a quantitative measurement of the frequency of RD cells in the progeny is needed.

Method A: crosses involving RC cells only.

(i) Grow the parental strains in N3 medium (if *ant*^{S}) or in the antibiotic-containing medium (if *ant*^{R}). Remember that the two parents should have opposite mating type and auxotrophic requirements.

(ii) Dilute each culture 10 times in sterile water.
(iii) Mark a W0 plate with three points (a, x, b) about 2 cm from each other. Place one drop of culture of one parent strain over points a and x, wait for complete absorption, then place one drop of the other parent strain over points b and x. Wait for complete absorption.
(iv) Incubate the W0 plate at 28°C for 3 days.
(v) Check for the confluent growth of prototrophic diploids over the x mark and the complete absence of growth of both haploid parents over the positions of a and b.
(vi) Pick up the entire diploid patch, using a sterile loop, and dissociate it in 10 ml of sterile water (picking up the entire patch ensures that the number of zygotic clones is as high as possible such that the interclonal variance, which is high, is properly averaged).
(vii) Dilute in sterile water to 10^3 cells/ml and plate 0.1 ml aliquots on W0 plates. Prepare 5 – 10 plates to obtain enough total colonies for statistical significance.
(viii) Incubate the plates at 28°C for 3 days.
(ix) Replica plate on YPglu, N3 and antibiotic-containing media as appropriate for the ant^R mutations involved in your cross. Incubate at 28°C for 3 days (or longer for some antibiotic-containing media, check the different levels of growth of resistant and sensitive colonies at regular intervals). Keep all the plates in the refrigerator until growth of resistant colonies is visible on the slowest growing plates incubated.
(x) Sort the plates. Score the mitochondrial genotype of each colony individually.
(xi) Calculate and interpret as described in Section 3.3.3.

Method B: crosses involving one (or two) RD parents.

(i) Grow the parental strains in N3 or antibiotic-containing medium, for the RC parent, and in YP10, for the RD parent(s).
(ii) Plate 0.1 ml of each culture on W0 plates to check for the absence of prototrophic revertants. Incubate at 28°C for 3 days.
(iii) Pellet the cells in the remainder of the cultures, wash and resuspend in YP10 at a density of 10^7 cells/ml.
(iv) Mix 5 ml of the suspension from each parent and incubate at 28°C for 1.5 h with shaking.
(v) Pellet the cells by centrifugation at 1000 *g* for 5 min. Let the tube stand for 15 min.
(vi) Gently resuspend the pellet in the same medium taking care not to disrupt cell aggregates. Incubate further at 28°C for 2 – 3 h. Monitor the appearance of zygotes under the microscope.
(vii) When the zygotes have formed, centrifuge as before, wash and resuspend the pellet in 10 ml of sterile water.
(viii) Inoculate 0.5 ml of the suspension into 10 ml of W10 and incubate at 28°C for 2 days *without* shaking.
(ix) Inoculate 0.1 ml of the previous culture into 10 ml of W10 and incubate again at 28°C for 2 days *without* shaking.

(x)   Dilute to 10^3 cells/ml in sterile water and plate 0.1 ml aliquots on W0 medium. Incubate at 28°C for 3 days.

(xi)   Replica plate onto YPglu, N3 and antibiotic-containing media as appropriate for the mutations studied. Incubate at 28°C for 3 days (or longer for some antibiotic-containing media; check the different levels of growth of the resistant and sensitive colonies at regular intervals). Keep all of the plates in the refrigerator until growth of resistant colonies is visible on the slowest growing incubated plates.

(xii)   Sort the plates. Score the total number of RD colonies, then score the genotype of each RC colony individually.

(xiii)   Calculate and interpret the results as described in the next section.

3.3.3 Calculation of linkage between mitochondrial genetic markers

Carry out a quantitative random cross (Methods A or B as appropriate). Never use the results from individual zygotic clones because the interclonal variance is high for reasons unrelated to genetic distances. Interpret the results according to the type of cross as follows.

(i) *Crosses involving only RC cells and in which the different ant^R mutations determine distinguishable phenotypes.* This is the simplest case since all genotypes can be directly determined from the phenotypes of the homoplasmic diploid clones issued from the cross. Crosses can be multifactorial but, in practice, three point crosses are more convenient to score and provide adequate information. An example is given in *Table 8*.

In order to deduce properly the genetic linkage from the number of recombinant clones

---

Table 8. How to compute transmissions, recombination and allelic distributions in three-factor mitochondrial crosses of *Saccharomyces cerevisiae*.

First case

Crosses involving RC parents only and in which the different ant^R loci determine distinguishable phenotypes:
(e.g. $rho^+mit^+ant^R_1ant^S_2ant^S_3 \times rho^+mit^+ant^S_1ant^R_2ant^R_3$)

Types found among progeny:

	ant1	ant2	ant3	designation
	R	S	S	P1 (parental)
	R	S	R	R1
	R	R	S	R2
RC	R	R	R	R3 (recombinants)
	S	S	S	R4
	S	S	R	R5
	S	R	S	R6
	S	R	R	P2 (parental)

RD (corresponding to spontaneous rho^- mutants, if any)

Calculations

1.   % transmission of alleles from parent 1:
       %tra(ant1) = (P1+R1+R2+R3)/Total RC × 100
       %tra(ant2) = (P1+R1+R4+R5)/Total RC × 100
       %tra(ant3) = (P1+R2+R4+R6)/Total RC × 100

Transmissions of all alleles of the same parent should be equal within limits of statistical significance independently of their genetic location (except for alleles linked to *ome*). This parameter measures the parental contribution and is characteristic of a given cross: this is the *coordinated transmission*.

2. % recombinants between allelic pairs = first type + second type

$$\%\text{rec}(ant1\text{-}ant2) = (R2 + R3)/\text{Total RC} \times 100 + (R4 + R5)/\text{Total RC} \times 100$$
$$\%\text{rec}(ant1\text{-}ant3) = (R1 + R3)/\text{Total RC} \times 100 + (R4 + R6)/\text{Total RC} \times 100$$
$$\%\text{rec}(ant2\text{-}ant3) = (R1 + R5)/\text{Total RC} \times 100 + (R2 + R6)/\text{Total RC} \times 100$$

The two reciprocal recombinant types between two allelic pairs should be equal within limits of statistical significance (except for alleles linked to *ome*).

3. Distribution of third allelic pair among recombinants between the two others and conclusions (valid only for alleles unlinked to *ome*):

	if dis > 1	if dis < 1
dis (*ant3*) = R2/R3 + R5/R4	*ant3-ant1* linked	*ant3-ant2* linked
dis (*ant2*) = R3/R1 + R4/R6	*ant2-ant3* linked	*ant2-ant1* linked
dis (*ant1*) = R1/R2 + R6/R5	*ant1-ant2* linked	*ant1-ant3* linked

If *dis* not significantly different from 1: absence of linkage.

Second case

Crosses involving a *mit⁻* parent (e.g. $rho^+ mit_1^- ant^S_2 ant^S_3 \times rho^+ mit_1^+ ant^R_2 ant^R_3$)

Types found among progeny:

	ant2	ant3	designation
	S	S	R4
RC	S	R	R5 (recombinants)
	R	S	R6
	R	R	P2 (parental)

RD corresponding to all other types

Calculations

1. % transmission of alleles from parent 1:

$$\%\text{tra}(mit) = \text{RD}/\text{Total RC} + \text{RD} \quad \text{(valid only if Method B of Section 3.3.2 properly used)}$$

2. % recombinants between allelic pairs (absolute figures)

$$\%\text{rec}(mit1\text{-}ant2) = 2 \times (R4 + R5)/\text{Total RC} + \text{RD} \times 100$$
$$\%\text{rec}(mit1\text{-}ant3) = 2 \times (R4 + R6)/\text{Total RC} + \text{RD} \times 100$$

3. Linkage calculations:

if (R4 + R5)/(R4 + R6) > 1 then *mit1-ant3* linked

if (R4 + R5)/(R4 + R6) < 1 then *mit1-ant2* linked

if (R4 + R5)/(R4 + R6) not significantly different from 1: absence of linkage

counted, it is necessary to take into account the parental contribution. This figure is a complex parameter that depends simultaneously upon the number of mtDNA molecules contributed by each parent to the zygote, the kinetics with which such parental mtDNA molecules segregate out into homoplasmic buds, the efficiency with which mtDNA molecules of the two parents find each other to make recombination possible, the bud positions and, perhaps, other cellular factors as well. Such phenomena are not quantitatively equivalent in strains having the same mitochondrial mutations but different nuclear genetic backgrounds. As a result the parental contribution varies from cross to cross. In most instances, the parental contribution is such that the final frequency of recombinants between two unlinked allelic pairs does not exceed 10 − 15% for each reciprocal recombinant type. Genetic linkage is demonstrated for frequencies of recombinants significantly below the $10-15\%$ limit so long as the parental contribution in the particular cross studied is not extremely biased in favour of one or the other parent.

Examination of the distribution of alleles of a third allelic pair among recombinants between the two others eliminates the problem of the parental contribution, hence the interest of three point tests. In this case an equal distribution of the two alleles of the third pair is expected when the third allelic pair is unlinked to either of the other two. An unequal distribution of the third allelic pair reflects linkage to one or the other (or even both) of the first two. The distribution favours the combination which remains parental for the two alleles brought in *cis* in the cross (see *Table 8*). The greater the bias, the stronger the linkage and the shorter the genetic distance.

(ii) *Crosses involving RC cells in which two different ant^R mutations determine the same phenotype.* In this case, only the wild-type recombinant between the two allelic pairs determining the same phenotype can be distinguished from the two parental types. Distribution of alleles of the third pair among such recombinants can be used to determine genetic linkage between this pair and the other two, as explained previously.

(iii) *Crosses involving both mit^+/mit^- and ant^R/ant^S allelic pairs.* In this case, a fraction of the progeny will be composed of RD clones, the mitochondrial genotypes of which cannot be determined. Distribution of the alleles of each ant^R/ant^S allelic pair among the remaining RC subpopulation simultaneously depends upon the genetic linkage between the mit^+/mit^- and ant^R/ant^S allelic pairs and the parental contribution in this particular cross.

If the quantitative random cross has been performed using the best non-selective conditions of Method B, then the parental contribution can be estimated by calculating the percentage of RD clones out of the total number of clones. However, even in this case, it is better to use an additional ant^R/ant^S allelic pair, unlinked to the first one, as a control. The distributions of both ant^R/ant^S allelic pairs among the mit^+ progeny will be equal in the absence of linkage. Genetic linkage between the mit^+/mit^- pair and either one of the ant^R/ant^S pairs is demonstrated if the two distributions differ significantly (see *Table 8*).

(iv) *Crosses involving two mit^+/mit^- allelic pairs.* In this case, only the wild-type mit^+mit^+ recombinant will be RC. For the same reasons as above, the percentage of this recombinant is limited to $10-15\%$. Genetic linkage is established for the percentage of RC recombinants significantly below this limit so long as the parental contribution in the particular cross is not extremely biased in favour of one or the other parent. In this case, however, the parental contribution cannot be estimated independently since both parents are indistinguishable RD.

(v) *Crosses involving a rho$^-$ mutant.* In this case, a fraction of the progeny will be RD and if the quantitative random cross has been performed using the best non-selective conditions of Method B then the percentage of RD clones among the total represents the parental contribution of the rho^- mutant. Sometimes this percentage is referred to as a delayed suppressiveness and should be clearly distinguished from the zygotic suppressiveness described in Section 3.5 even though both are obviously not independent. Distribution of ant^R/ant^S alleles among the rho^+ subpopulation can be (and have been) examined but this results in complex data beyond the scope of this chapter (see ref. 6 for review).

(vi) *Important aspects of the analysis of genetic crosses.* Firstly, conclusions about genetic linkage drawn above are valid only if precautions are taken in the cross not to select against any particular genotype during growth of the diploid cells issued from that cross. This is generally the case for most ant^R mutations using Method A. If there is any doubt as to a possible selective advantage under such conditions, then use Method B.

Secondly, remember that the observable percentage of recombinants between any two allelic pairs is low (upper limit $10-15\%$) and becomes even lower in the case of genetic linkage or when the parental contribution is excessively biased. Always score a total number of colonies large enough to obtain statistically significant data for each recombinant type counted. (Finding five colonies of one type and three of the other does not demonstrate that the ratio is 5/3. In fact, the ratio is almost indeterminate from such data. Precision increases if you count 50 colonies of one type and 30 of the other and improves further if you count more colonies.)

Thirdly, mitochondrial recombination is very active and genetic linkage is only found for relatively short physical distances (e.g. shorter than $2000-3000$ bp or even less). For longer distances use petite deletion mapping (Section 3.3.1).

3.3.4 *Determination of the omega allele of a rho$^+$ strain*

The optional intron of the 21S rRNA gene encodes a double-strand specific endonuclease that cleaves the intron-less forms of that gene at a specific site as a prerequisite for the insertion of a copy of that intron. This molecular phenomenon exerts a strong effect on the recombination of flanking genetic markers in crosses between intron-plus strains (ome^+) and intron-minus strains (ome^-). Consequently, the quantitative random cross can be used to determine if a strain is ome^+ or ome^- as follows.

(i) Cross the rho^+ strain in parallel with two tester strains (one ome^+ reference and one ome^- reference) containing, at least, the two intron-linked mutations chl^R_{321} and ery^R_{221} (see *Table 6*).

(ii) For each cross analyse the ratio of the two reciprocal recombinants, that is $chl^R ery^S/chl^S ery^R$.

(iii) If this ratio is high (usually $30-50\%$, at least) with the ome^- tester and close to 1 with the ome^+ tester, then the strain is ome^-. If this ratio is close to 1 with the ome^+ tester and very low (usually $0.02-0.03\%$ or lower) with the ome^- tester, then the strain is ome^+. Rare instances may occur in which both ratios are close to 1 with the two testers. If this is the case the strain is a mutant of the ome system and it is not possible to deduce the presence/absence of that intron by this genetic test.

The presence of the intron can, obviously, also be determined by molecular analysis of mtDNA. In particular, one can use the minilysate procedure to screen large numbers of strains or clones. Detect the presence of the intron by hybridization with a specific intron probe (available upon request from B.Dujon).

3.3.5 *Isolation of mitochondrial recombinants*

Because genetic recombination between mtDNA molecules is an efficient process, the formation of recombinants, even between linked mutations, is easily obtained by crosses between haploid strains containing the mitochondrial mutations of interest. However, the screening and the isolation of such recombinants into strains able to mate (needed for subsequent crosses) may, depending upon their phenotype, become time consuming. The following cases should be distinguished.

(i) *Recombinants between ant^R mutations conferring different phenotype.* This is the simplest case since the recombinants are RC and phenotypically distinguishable. Use the quantitative random cross method to obtain homoplasmic diploid clones, pick out the desired recombinant and sporulate as in *Table 5*. Since mitochondrial mutations do not sporulate during meiosis of a homoplasmic diploid, tetrad analysis is not necessary (except if you want to use it as a criterion of mitochondrial inheritance of your *ant^R* mutations). Haploid segregants can be used directly for subsequent experiments.

(ii) *Recombinants between ant^R mutations conferring the same phenotype.* This is a laborious procedure since the double mutant is indistinguishable from the parents. Use the quantitative random cross method. Pick out homoplasmic diploids of the resistant phenotype at random and mix them (pick out enough clones to ensure that at least one recombinant clone is present). Sporulate the mixture as in *Table 5*. Haploid segregants may be either single *ant^R* mutants of one type or the other or the desired double *ant^R* mutant. Check the mitochondrial genotypes by back-crosses of haploid segregants to each single mutant. Use the quantitative random cross method. Examine the appearance of sensitive recombinants in the progenies (the double mutant will not produce any sensitive recombinant in both crosses, score enough diploids for statistical significance, knowing the frequency of recombinants between the two *ant^R/ant^S* allelic pairs).

(iii) *Recombinants involving mit$^-$ mutations.* The same difficulty exists in constructing double *mit$^-$* recombinants or *mit$^-$ ant^R* recombinants, namely that the homoplasmic diploid is RD and does not sporulate. To circumvent this difficulty, zygotes are sporulated immediately after their formation before segregation of mtDNA molecules produces homoplasmic cells. Use the quantitative random cross Method B from step (i) to (vi). When zygotes are formed, inoculate into sporulation medium and proceed as in *Table 5*; it is essential to include step (ix). Verify the mitochondrial genotypes of segregants by the qualitative random cross method as follows. If a double *mit$^-$* mutant is being constructed, cross to each single mutant and check for the absence of RC recombinants in both crosses by replica plating on N3 medium. If a *mit$^-$ ant^R* recombinant is being constructed, cross to a *mit$^+$ ant^S* strain and verify the formation of *mit$^+$ ant^R* recombinant by replica plating on the antibiotic-containing medium.

(iv) *Methods using the mating (MAT) system.* Mutations of the MAT system allow the formation of diploids able to mate, hence the possibility of constructing mitochondrial recombinants without the need for sporulation. To do this cross the mitochondrial mutant (an **a** strain wild-type for the MAT system) to a haploid strain with the following nuclear genotype: *hml a⁻ mat a⁻ HMRα marl* (mate as an α strain). Use the quantitative random cross method and score phenotypes of diploid clones as normal. Diploids can now be used directly as an **a** parent in crosses. In particular, the qualitative replica cross method can be used to determine their mitochondrial genotype.

3.4 Complementation tests

Functional complementation between mitochondrial mutations can be examined during the transient heteroplasmic stage that follows zygote formation and precedes the appearance of recombinants. Complementation can be assayed either by measuring the oxygen consumption of cells arising from crosses between different *mit⁻* mutants or by examining the mitochondrial translation products of cells arising from crosses between a *rho⁻* mutant and a wild-type cell.

3.4.1 *Restoration of respiration in crosses between mit⁻ mutants*

(i)   Grow the desired *mit⁻* mutants of opposite mating type and auxotrophic requirements in YP10 medium (note that mutants with mutations affecting the ATP synthetase genes cannot be used as they are themselves able to consume oxygen).

(ii) Centrifuge at 1000 *g* for 5 min to pellet the cells, wash and resuspend in YPgal at a cell density of 10^7 cells/ml.

(iii) Add 50 ml of each suspension to a 500 ml conical flask. Incubate at 28°C for 2 h with shaking.

(iv) Pellet the cells by centrifugation at 1000 *g* for 5 min. Let the tube stand for 15 min, then gently resuspend the pellet in the same medium.

(v)   Incubate at 28°C for a further 1−3 h. Monitor zygote formation under the microscope.

(vi) Take 10 ml aliquots at regular time intervals (e.g. every 30 min) immediately after the first zygotes start to appear.

(vii) Immediately dilute 0.1 ml of each aliquot in 10 ml of sterile water, then dilute again 1 ml in 10 ml. Plate 0.1 ml of each dilution on W0 and N3 media. Incubate the plates at 28°C for 5 days.

(viii) Immediately centrifuge the remaining 9.9 ml of culture, wash and resuspend the cells in 4 ml of 50 mM phthalate buffer, pH 4.0 containing 1% (v/v) ethanol. Incubate at 30°C for 5 min and vortex vigorously (for maximum oxygenation).

(ix) Place 1.8 ml of the aerated suspension into a Clark electrode (see Section 5.2.1). Add 5 μl of 5 mM dinitrophenol and record polarographic measurements of oxygen consumption at 30°C, using an automated oxygraph (e.g. Gilson).

(x)   Repeat the same measurement on a second 1.8 ml sample after addition of dinitrophenol and of 40 μl of 0.4 M KCN (great care required with cyanide solution). Subtract values from the previous measurement.

(xi) Count the number of colonies on W0 plates and calculate the number of prototrophic cells in the suspension placed in the Clark electrode. Plot as a function of time.

(xii) Count the number of colonies on N3 plates and calculate the number of cells able to give rise to RC colonies as before. This is a measure of the appearance of recombinants (a cell can form a colony on N3 if it contains recombined wild-type mtDNA molecules at the time of plating or if it is able to produce such molecules from its heteroplasmic pool during its survival time on the N3 medium). Plot the ratio of recombinants over the prototrophic cells as a function of time.

(xiii) Calculate respiration as nanomoles of KCN-sensitive oxygen consumption/h/10^6 prototrophic cells. Plot as a function of time. Complementation is positive if a significant rate of respiratory activity (approaching that of the wild-type) is observed during the time period before wild-type recombinants are found ($\sim$4–8 h after zygote formation in most cases). Partial complementation (with lower respiratory activity but significantly above background) is sometimes observed.

3.4.2 *Appearance of new translation products in rho*$^-$ × *rho*$^+$ *crosses*

(i) Grow the desired *rho*$^-$ mutant in YPglu and the *rho*$^+$ tester strain (with opposite mating type and auxotrophic requirements) in N3 medium. Note that the expected new translation product formed as a result of complementation should be distinguishable from the translation products of the *rho*$^+$ parent.

(ii)–(v) Same as in Section 3.4.1 but use a ratio of 3:1 of the *rho*$^-$ to *rho*$^+$ parental cells for mating.

(vi) Take 5 ml aliquots at regular time intervals and analyse the translation products as described in Section 6.4.

(vii) Dilute 0.1 ml of each aliquot in 10 ml of sterile water and dilute again 1 ml in 10 ml. Plate 0.1 ml aliquots of each dilution on W0 medium. Incubate at 28°C for 3 days. Score the plates and calculate the percentage of prototrophic cells relative to the *rho*$^+$ haploid parent.

(viii) Compare the frequency of appearance of new translation product with the previous ratio.

3.5 Quantitative and qualitative tests for suppressiveness

Suppressiveness was originally defined as the percentage of zygotic clones composed entirely of *rho*$^-$ cells arising from a cross between *rho*$^-$ mutant and a wild-type (26). This is a characteristic figure for a given *rho*$^-$ mutant and is quantitatively reproducible so long as the purity of culture is maintained.

3.5.1 *Quantitative determination of zygotic suppressiveness*

(i) Grow the desired *rho*$^-$ mutant as well as a *rho*^{0} (derived from the same strain) in YPglu or YP10 medium until stationary phase.

(ii) Grow the *rho*$^+$ tester strain in N3 medium until stationary phase.

(iii) Plate 0.1 ml aliquots of each culture on W0 medium to confirm the absence of prototrophic cells, incubate at 28°C for 3 days.

(iv) Dilute an aliquot of the *rho*$^-$ culture to 10^3 cells/ml and plate on YPglu to determine the purity of the culture as described in Section 3.2.2.

(v) Centrifuge the remainder of the culture at 1000 *g* for 5 min to pellet the cells, wash and resuspend in YP10 medium at a density of 10^7 cells/ml.

(vi) Prepare two sterile 100 ml conical flasks and place 5 ml of the *rho*$^+$ suspension in each. Add 5 ml of the *rho*$^-$ suspension in one flask and 5 ml of the *rho*^{0} suspension in the other. Incubate at 28°C for 90 min with shaking.

(vii) Pellet the cells by centrifugation at 1000 *g* for 5 min. Allow the tube to stand for 15 min.

(viii) Gently resuspend the pellet in the same medium taking care not to disrupt cell aggregates. Incubate at 28°C for a further 1−3 h. Monitor the appearance of zygotes under the microscope.

(ix) When the first zygotes appear, stop the incubation by placing the mating mixtures at 0°C. Never allow the first buds to separate from zygotes prior to plating.

(x) Dilute in sterile water to 10^3 zygotes/ml and plate 0.1 ml aliquots on W0 medium (if the percentage of visible zygotes is insufficient, prepare serial dilutions and plate all dilutions).

(xi) Incubate the plates at 28°C for 3−4 days.

(xii) Replicate on YPglu and N3 plates and incubate at 28°C for 2−3 days. [This step may be omitted since *rho*$^+$ *and rho*$^-$ clones can usually be distinguished on W0 medium by their size and colour (*rho*$^+$ clones are large, thick and white while *rho*$^-$ clones are small, flat and slightly translucent). The appearance of colonies on W0 is not, however, definite proof that the colony is *rho*$^-$.]

(xiii) Score the number of zygotic clones entirely composed of RD cells and compare with the total number of zygotic clones examined. Score at least 300 clones in total to obtain sufficient statistical significance.

(xiv) Score the purity of the culture. If purity is satisfactory proceed to step (xv), if not subclone and repeat the whole test as there are no means of predicting the effect of the secondary mutations on the suppressiveness.

(xv) The zygotic suppressiveness (S) of the *rho*$^-$ mutant is given by

$$S = \frac{S1 - S0}{100 - S0} \times 100$$

where S1 and S0 are the percentage of RD zygotic clones in crosses of the *rho*$^-$ mutant and of the *rho*^{0} mutant, respectively.

3.5.2 *Qualitative determination of zygotic suppressiveness (27)*

This procedure allows one to distinguish rapidly between *rho*$^-$ mutants with extremely high zygotic suppressiveness (hypersuppressive mutants) and other *rho*$^-$ mutants. Hypersuppressive mutants (S > 95%) form a distinct class of *rho*$^-$ mutants in which one origin of replication is retained and highly amplified.

(i) Place the *rho*$^-$ clones in grids (8 × 8 grids are convenient, see *Figure 5*) on YPglu medium and incubate at 28°C for 3 days.

(ii) Grow a *rho*$^+$ *pet9* strain of opposite mating type (and appropriate auxotrophic requirements) in YP10 medium.

(iii) Plate 0.1 ml of the YP10 culture on W0 medium (plating of the lawn should be very regular for good mating efficiency).

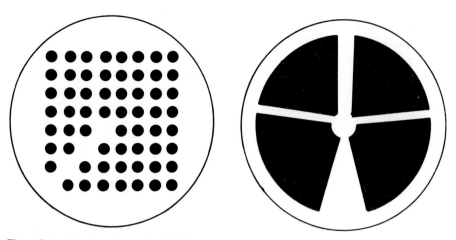

Figure 5. Popular dispositions of haploid clones for the qualitative replica cross. **Left:** 8 × 8 grid. **Right:** quadrant. Both dispositions are self-oriented.

(iv)   Replicate the YPglu grids on YPglu, W0 and N3 media first, then onto the lawn. Incubate all replica plates at 28°C for 3 days.

(v)   Check that there is no growth of replicas on N3 and W0 plates. Check for the confluent growth of prototrophic diploids on each replica over the lawn and for the complete absence of growth on the N3 and W0 plates.

(vi)   Replicate the diploids on N3 medium. Incubate at 28°C for *only* 2 days.

(vii) Score the growth of diploid colonies on N3 plates. Confluent growth indicates the presence of all types of *rho⁻* mutants except hypersuppressive. The appearance of discrete colonies of RC cells growing over the inhibited RD cells indicates a hypersuppressive *rho⁻* mutant.

(viii) Pick out the desired *rho⁻* clone on the YPglu replica plate.

3.5.3 *Semi-quantitative determination of zygotic suppressiveness (quadrant test) (28)*

This test is based on the same principle as before except that the colonies of diploids are made larger making it easier to distinguish between the different classes of suppressiveness.

(i)   Place the *rho⁻* clones on YPglu medium in quadrants (as shown in *Figure 5*) using a sterile loop (ensure even inoculation over the quadrant). Incubate at 28°C for 3 days. Keep the original culture in the refrigerator at 5°C.

(ii)–(iv) Same as in Section 3.5.2.

(v)   Score the growth of diploid colonies on N3 medium. Confluent growth indicates *rho⁻* mutants of low or moderate suppressiveness (lower than 50%); numerous mini-colonies of RC cells over the inhibited RD cells indicate high suppressiveness (~50–90%) while rare mini-colonies of RC cells indicate hypersuppressiveness (>95%).

(vi)   Pick out the desired *rho⁻* clone from the original culture.

218

4. ISOLATION OF MITOCHONDRIA

4.1 Introduction

It is more difficult to prepare pure mitochondria from fungal cells than from animal cells because of the presence of the tough cell wall. Most work has centred on *Sacch. cerevisiae* although the methods described here can be adapted to most types of yeast. The major problem in all cases is to break open the cells without disrupting the integrity of the cell organelles. There are two approaches to this, either the cell wall can be digested enzymatically to give protoplasts or broken physically by using various grinding methods.

4.2 Preparation of mitochondria

4.2.1 Lysis of cells by protoplasting

The disadvantages of enzymatic protoplasting of cells are that it involves prolonged incubation under adverse conditions (e.g. high concentrations of cells and thiols) using a crude extract from snail gut (see Section 3.2 of Chapter 6) or mushrooms (*Table 9*) and the conversion to protoplasts may be incomplete. However, a number of workers do prefer to use the protoplasting technique for preparing mitochondria from yeast for a variety of applications (see Sections 6.2.1–3).

4.2.2 Lysis of cells by grinding with glass beads

In the authors' experience grinding is a rapid method of breaking open cells but usually it is necessary to compromise between maximizing the degree of breakage and avoiding fragmentation of cellular structures (29) which in turn can contaminate the mitochondrial preparation. In the authors' experience of *Sacch. cerevisiae*, grinding with glass beads can give rapid preparations of pure mitochondria. The method used by the authors is as follows and is based on the method of Lang *et al.* (30) (see also Chapter 8, Section 2.2.5).

(i) Harvest the yeast cells by centrifugation at 1000 *g* for 20 min at 5°C, then wash them by resuspension in cold distilled water followed by re-centrifugation as before.

(ii) Suspend each gram of cells in 2 vol of cold lysis medium containing 0.6 M sorbitol, 1 mM EDTA, 10 mM Tris–HCl, pH 7.4, put into a screw-top 500 ml bottle and then add 3 g of acid-washed glass beads (diameter 0.5–0.75 mm) for each ml of cell suspension. As a rough guide it is best not to put more than

Table 9. Preparation of a mushroom extract for protoplasting yeast cells (28).

1. Wrap 500 g of ripe mushrooms (*Agaricus campestris*, preferably, or *Agaricus bisporus*) in muslin and pulverize with a pestle and mortar.
2. Collect the juice, dilute with 10 ml of distilled water and pulverize the tissue again.
3. Repeat this procedure and leave all the pooled juice extracts to stand overnight at 4°C.
4. Centrifuge the juice at 5000 r.p.m. for 20 min and collect the supernatant.
5. Filter the extract through a series of Millipore filters to sterilize it and keep it frozen at −22°C.

75 ml of yeast cell suspension in this size of bottle.

(iii) Shake the cells through a distance of about $80-100$ cm twice a second (2 Hz) for 30 sec and then cool in ice. Repeat this shaking procedure until a high proportion of the cells are lysed, usually this needs a total shaking time of 3 min. The degree of lysis can be checked microscopically after staining with Trypan Blue.

(iv) Pour off the supernatant from the glass beads and wash the beads with the lysis medium once or twice to recover the remainder of the lysed cells.

Depending on the skill of the person, between $40-90\%$ of the cells should be lysed using this method. If you wish to prepare mitochondria on a large scale then it is best to use a grinding mill in either batch or continuous mode; in the latter case it may well be possible to link the grinding mill directly to a continuous flow centrifuge such as a Sharples. Grinding mills are extremely effective in lysing cells but care must be taken to ensure that the mill does not cause excessive damage and fragmentation to the subcellular organelles; not only may the mitochondria become damaged but also fragmented cellular components may become bound to the outside of the mitochondria (29).

4.2.3 *Isolation of mitochondria from yeast cell lysates*

(i) Centrifuge the cell lysate at 2500 *g* for 10 min at 5°C to pellet unbroken cells, cell wall debris and nuclei.

(ii) Very carefully pour off the supernatant; avoid disturbing the pelleted material. If you think that there may be some contamination of the supernatant by the pellet then repeat step (i).

(iii) Centrifuge the supernatant at 15 000 *g* for 10 min at 5°C to pellet the mitochondria. Discard the supernatant.

(iv) Gently resuspend the mitochondria in 0.8 M sucrose, 1 mM EDTA, 10 mM Tris−HCl, pH 7.4 and 0.1% bovine serum albumin (BSA) prior to loading them onto a $1-2$ M sucrose continuous gradient containing 10 mM Tris−HCl, pH 7.4, 1 mM EDTA and 0.1% BSA.

(v) Centrifuge the gradients in a swing-out rotor at 80 000 *g* for 90 min at 5°C.

(vi) The intact mitochondria will be found as a brown band at a density of about 1.18 g/ml; damaged mitochondria will band lighter (~ 1.14 g/ml) or denser (~ 1.22 g/ml) depending on whether both membranes or only the outer membranes are damaged.

(vii) Remove the band of intact mitochondria using a Pasteur pipette with its tip bent at an angle of 90° to facilitate collection of the band.

The use of sucrose gradients gives a very pure population of mitochondria with only minor contamination by co-banding lysosomal material. It has been found that mitochondria from cells lysed by grinding can be contaminated by adhering chromatin from fragmented nuclei; such contamination can be minimized by avoiding excessively harsh grinding methods. For studies of transcription and replication any minor contamination of mitochondria with nuclear DNA can be removed by digesting the crude mitochondrial pellet at 0°C for 30 min in isotonic sucrose solution containing 10 mM $MgCl_2$ at a ratio of 300 Kunitz units/mg of mitochondrial protein. After digestion the mitochon-

Table 10. Gradients for the isopycnic purification of mitochondria.

Gradient medium	Concentration range	Centrifugation conditions g/time /min)	Density of mitochondria (g/ml)
Sucrose	1.0–2.0 M	80 000/120	1.19
Percoll[a]	25–60% (v/v)	40 000/20	1.10
Metrizamide	20–50% (w/v)	80 000/60	1.16
Nycodenz	20–50% (w/v)	80 000/60	1.17

[a]The gradient also contains 0.25 M sucrose as the osmotic balancer.

dria must be washed free of Mg^{2+} with buffered isotonic sucrose containing 5 mM EDTA since otherwise these divalent cations adversely affect the subsequent purification of mitochondria on the sucrose gradient.

One feature of yeast is that some strains have respiratory defective mitochondria as a result of changes in the nuclear or mtDNA (see Sections 2.3 and 3). Such mitochondria can be purified in the usual way because they band at a density similar to normal mitochondria although the appearance of the band especially in terms of colour may be different because of a lack of cytochromes which are responsible for the brown colour of normal mitochondria. In the case of yeast grown anaerobically there is some evidence for slight changes in density of the pro-mitochondria but even so similar purification protocols can be used (31).

Other gradient media besides sucrose have been used for the isopycnic gradient purification of mitochondria including colloidal silica media and iodinated gradient media (*Table 10*) but none has been proved to be really superior to sucrose at least for purifying yeast mitochondria.

Instead of isopycnic gradient centrifugation some workers prefer to attempt to purify mitochondria by repeated washing of the crude mitochondrial pellet obtained at step (iii) with isotonic sucrose solution. Depending on the type of experiment this method may produce sufficiently pure mitochondria, however it is important to be aware that such washing procedures are inherently much less efficient than sucrose gradients. Thus although washing methods tend to produce higher yields of 'mitochondria', care is necessary to ensure that such preparations do not include artefacts.

4.3 Preparation of mitoplasts

In some cases it is desirable to remove the outer membrane of mitochondria to remove co-sedimenting contaminants adventitiously bound to the outside of the mitochondria or to remove from the mitochondria those enzyme activities associated with the outer membrane. The formation of mitoplasts by removing the outer membrane is possible because of the different composition of the inner and outer membranes. Originally mitoplasts were prepared using a technique of swelling and shrinking (32) but now it is more usual to take advantage of the different composition of the inner and outer membranes by selective solubilization of the outer membrane with digitonin (33). To prepare mitoplasts suspend the sucrose gradient-purified mitochondria in 0.6 M mannitol, 10 mM Tris−HCl, pH 7.0, 0.1% BSA and add a 2% solution of re-crystallized digitonin to give a final ratio of digitonin to mitochondrial protein of 0.4:1.0. Mix well, leave in

Table 11. Marker enzymes for subcellular fractions.

Subcellular fraction	Marker activity
Mitochondria	
inner membrane	Succinate dehydrogenase
outer membrane	Monoamine oxidase, kynurenine hydroxylase
inter-membrane space	Adenylate kinase
matrix	Fumarase, isocitrate dehydrogenase
Lysosomes	Aryl sulphatase, acid phosphatase
Peroxisomes	Catalase
Endoplasmic reticulum	Glucose-6-phosphatase
Golgi bodies	Galactosyl transferase
Plasma membrane	5'-Nucleotidase
Nuclei	Nuclear DNA

ice for 1 min and pellet the mitoplasts by centrifugation at 20 000 *g* for 10 min at 5°C. It is particularly important not to expose the mitochondria to too great an amount of digitonin as otherwise the mitoplasts are damaged also. Using this method of preparation the mitoplasts show normal, coupled respiratory activity (33).

5. CHARACTERIZATION OF MITOCHONDRIA

5.1 Determination of the purity of mitochondria

5.1.1 *Analysis of purity on the basis of marker enzymes*

As indicated in *Table 1*, the mitochondrion is the site of a number of metabolic pathways not found in other organelles of the cell. Similarly a number of other organelles also show enzymatic activities not found elsewhere in the cell. These so called marker enzyme activities can be used to determine if one type of organelle is contaminated by another. *Table 11* lists some of the marker enzyme activities associated with the different types of cell organelle.

The other important uses of the mitochondrial marker enzyme activities are that they can be used not only to determine if the outer membrane is present (33) but also to gauge the intactness of the mitochondria. The inner membrane is impermeable to a wide range of enzyme substrates and so if one measures the apparent enzyme activity of an enzyme such as succinate dehydrogenase in a mitochondrial suspension and compares it with the total activity of the suspension after disruption of the membranes by, for example, lysis with 1% Triton X-100 then the greater the latency of the activity the more intact is the preparation of mitochondria.

5.1.2 *Analysis of purity on the basis of DNA*

For the many studies of the molecular biology of mitochondria, especially of the sequence and expression of mtDNA, the purity of the mitochondrial preparations has been judged by the analysis of the mtDNA. In yeast the buoyant density of the mtDNA in isopycnic CsCl gradients is less than that of the nuclear DNA and moreover this difference can be accentuated by centrifuging the DNA in the presence of specific DNA-binding dyes such as bisbenzamide as described in Section 6.2.1. An alternative approach is to digest the mtDNA isolated from purified mitochondria (see Section 4) with appropriate restriction nucleases. If the mitochondria are pure then one should obtain a defined number of restriction fragments the number of which depends on the restriction nuclease used. These restriction fragments give sharp bands when separated by electrophoresis on agarose gels; any nuclear contamination of the mitochondria is revealed as a background smear of bands while smearing of the bands of mtDNA indicates heterogeneity as a result of degradation of the DNA.

5.2 Functional characterization of intact mitochondria

The mitochondrial electron transport chain conserves the energy of electron transfer in the form of a proton gradient which is then used to drive the synthesis of ATP via the ATP synthetase. This function is critically dependent on the inner membrane being impermeable to protons. For a simple scheme of oxidative phosphorylation showing this relationship between electron transfer and ATP production see *Figure 6*. A more

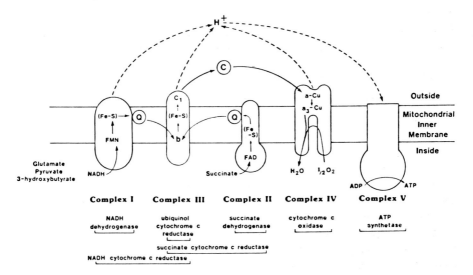

Figure 6. The proteins of oxidative phosphorylation. The electron transport proteins (complexes I−IV) and the ATP synthetase (complex V) are shown in their probable orientation in the mitochondrial inner membrane. The lipid soluble electron shuttle co-enzyme Q is depicted as Q and the water-soluble electron transfer protein cytochrome c as C. Co-enzyme Q is shown here, for simplicity, as transferring electrons from complexes I and II to complex III in a linear fashion. In reality the situation is more complex and the reader is referred to ref. 34 for a more complete explanation of electron transfer in this part of the chain. The site of proton translocation at 'coupling sites' I (complex I), II (complex III) and III (complex IV) are shown. The activities measured from the chain can include the entire sequence and in coupled mitochondria the ATP synthetase. This is the case when the O_2 electrode is used for assay. For example if NADH is used then 'NADH oxidase' activity is measured. Artificial electron acceptors can be added to pick up electrons earlier in the chain and their reduction monitored spectrophotometrically. Exogenous cytochrome c is often used for this purpose. Some of the typical activity measurements made and the parts of the chain they span are also shown.

detailed discussion of energy conservation is given elsewhere (34). In mitochondria with an intact inner membrane, electron transfer can only proceed if the proton gradient is continuously dissipated by the flow back into the mitochondrion via the ATP synthetase. In practical terms this means that in intact mitochondria, electron transfer, which can be measured by O_2 consumption, can only occur if ADP is also present. This phenomenon is described as 'coupling'. The extent of coupling in a mitochondrial preparation can be easily measured as the respiratory control ratio (RCR) and is the most rapid and effective method for determining mitochondrial functional integrity. The consumption of oxygen in the presence of substrate and ADP (state 3 respiration) is measured and compared with the rate of respiration after the ADP has been consumed (state 4 respiration). Mitochondria which show no difference in state 3 and state 4 respiration are 'uncoupled' a state which can also be induced by some reagents (e.g. dinitrophenol) appropriately termed uncouplers. In addition to measures of mitochondrial function, structural criteria for purity can be assessed by isopycnic centrifugation or electron microscopy. Mitochondria isolated from facultative anaerobes such as the yeast *Sacch. cerevisiae* change in respect to their composition depending on conditions of oxygen concentration and other culture conditions (35).

5.2.1 *The oxygen electrode*

Mitochondrial respiration is most conveniently and quickly measured using a 'Clark' type oxygen electrode. This consists of a silver/silver chloride reference anode surrounding (generally) a platinum cathode. These electrodes are immersed in saturated KCl solution and separated from the reaction vessel by a thin Teflon membrane that is permeable to oxygen but which prevents electrode poisoning. The electrodes are polarized at a voltage of 0.6 V. At the platinum cathode electrons reduce oxygen molecules to water.

$$4H^+ + 4e^- + O_2 \rightarrow 2H_2O$$

The chloride anions migrate to the anode and release electrons.

$$4Ag + 4Cl^- \rightarrow 4AgCl + 4e^-$$

The overall result is that transfer of electrons from the cathode to the anode occurs causing a current to flow between the two electrodes which can be measured in an external circuit. The current is proportional to the partial pressure of oxygen in the sample and, as the response is linear, only two calibration points are necessary.

The current which flows for air-saturated water at 20°C is a few microamps. The current generated is very temperature-dependent and it is therefore important to operate the electrode at constant temperature.

It can be appreciated from the above equations that as oxygen is consumed by the electrode, the solution close to the electrode becomes progressively anaerobic. It is therefore necessary to stir the solution vigorously to ensure that the bulk solution under study is continuously brought into contact with the electrode. The most convenient way to do this is by a magnetic stirrer. Small magnetic stirrer bars can easily be made by sealing pins or pieces of paper clip in capillary tubing.

A suitable circuit for polarizing the electrodes and allowing measurements to be made is shown in *Figure 7*. For output to a chart recorder, R is chosen such that:

$$R = 2 \times \text{span of recorder (in mV)} \times 10^3 \ \Omega$$

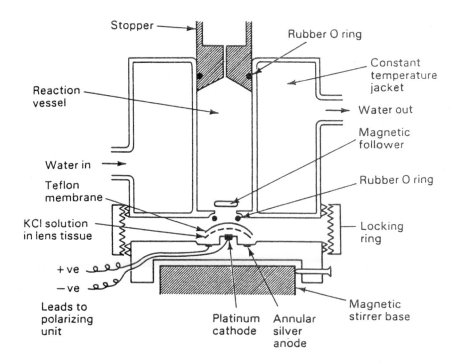

Figure 7. Oxygen electrode. A suspension of mitochondria (2.0−3.0 ml) in buffer is introduced into the water-jacketed reaction chamber which is then closed by a tight-fitting stopper. This stopper is gently pushed down to remove the air space completely and care is taken to exclude all air bubbles from solution, the conical depression in the base of the stopper helping in this task. Additions of substrates, uncouplers etc. are made through the narrow part in the stopper using a syringe capable of delivering microlitre quantities. The solution is stirred continuously with a magnetic stirrer. The current flowing in the external, polarizing circuit is proportional to the dissolved oxygen concentration in the solution in the reaction chamber.

The electrodes may conveniently be enclosed in a small chamber of approximately 3 ml volume. This chamber is isolated from contact with the atmosphere by a close-fitting cap. Solutions are introduced into the chamber and their oxygen concentrations measured directly once the apparatus has been calibrated. (Suitable electrodes, polarizing and measuring circuits may be purchased from Rank Brothers, High Street, Botisham, Cambridge, UK).

5.2.2 *Calibration of the oxygen electrode*

(i)   Pipette 3 ml of distilled water into the electrode chamber, add a few crystals of sodium dithionite ($Na_2S_2O_4$). The oxygen concentration rapidly falls to zero. The output of the electrode to the chart recorder also falls and the position of the pen on the chart recorder is set to zero.

(ii) Wash out the electrode chamber several times with distilled water, previously equilibrated with air at 20°C into the electrode chamber, itself thermostatted at 20°C.

(iii) Set the chart recorder to 90% of its deflection by suitable choice of sensitivity. This level now corresponds to 260 μM dissolved oxygen. This provides two known oxygen concentrations, that is 0 and 260 μM from which other concentrations can be determined.

5.2.3 *Measurement of respiratory control ratio (RCR)*

The conditions described below are suitable for the oxygen electrode described previously and assume a final volume of 2.5 ml.

(i)   Add isolation buffer [0.6 M sucrose, 10 mM KH_2PO_4, 12 mM, EDTA, 10 mM Tris−HCl, pH 7.4, 0.1 ml of BSA (fraction 5, 50 mg/ml)].

(ii)   Add mitochondria to give a final concentration of 0.2−0.5 mg/ml and measure the background rate for 1−2 min.

(iii) Add substrate. For glutamate/malate, to a final concentration of 2.5 mM from stock solution of 80 mM adjusted to pH 7.4 with KOH, and for succinate a final concentration of 5 mM from a stock solution of 160 mM. Measure the rate of oxygen uptake for 1 min.

(iv)   Add ADP to a final concentration of 180 μM from a stock solution of 11.6 mM. Measure the oxygen uptake for 10 min (state 3 respiration; state 4 respiration occurs when the ADP has been consumed).

(v)   Calculate the RCR by dividing the rate of oxygen uptake in state 3 by the rate of oxygen uptake in state 4. For the P/O ratio calculate the oxygen consumed from the start of state 3 respiration to its end. The P/O ratio is then the ratio of the nmoles of ADP added, divided by the nanograms of atoms of oxygen utilized during state 3 respiration. The RCR values for intact yeast mitochondria prepared by protoplasting are 3.5 with both succinate- and NADH-linked substrates (36). Other differences also exist in the early part of the respiratory chain when yeast and mammalian mitochondria are compared. For example the yeast mitochondria do not contain a phosphorylation site for NADH-linked substrates and so the P/O ratio is in the range of 1.5−1.8 (37). For succinate the P/O ratio is 1.2−1.6 (38).

5.3 Spectral analysis of cytochromes

The electron transport chain has four cytochromes which can be distinguished from each other by visible-light absorption spectroscopy. Cytochromes b and c_1 are components of complex III and are in the ratio of 2:1. Cytochrome c is a small, water-soluble protein, which transfers electrons from complex III to complex IV, and has a characteristic α-peak absorbance at 550 nm. Complex IV or cytochrome c oxidase contains four redox centres: two Cu atoms and two haem a groups. The haem a groups, which absorb in the visible region of the spectrum, are in distinct structural environments with cytochrome a forming part of the electron gate of this molecule and cytochrome a_3 part of the oxygen binding site. (For a more detailed review of these proteins see refs 39 and 40.) It is often useful to analyse the visible absorption spectra of mitochondria, for example, to determine concentration of components, characterize respiratory chain mutants or monitor purification procedures. Described below is a simple method for taking the reduced−oxidized difference spectrum of a sample of mitochondria. The cytochromes are redox proteins in which the oxidized and reduced proteins have different absorption spectra: a fact that can be exploited to overcome the problems associated with spectroscopy of scattering solutions such as suspensions of mitochondria. In one cuvette put the oxidized mitochondria (as prepared) and in the sample cuvette the reduced sample (using dithionite as a reductant). At the same concentration of mitochondria the light scattering should be the same for both samples and so it is possible to make absorbance spectra with stable baselines.

(i) Suspend the mitochondria at a concentration of 4 mg/ml and then divide equally into two stoppered cuvettes.

The next stage depends on the type of spectrophotometer you have available. If you have a double-beam instrument proceed to step (ii), if you have a single-beam instrument, with a facility for memorizing baseline, proceed to step (iv).

(ii) If you are using a double-beam spectrophotometer, to record the baseline, place the cuvettes in the reference and sample compartments and record the spectrum from 500 to 650 nm with a full scale absorbance of 0.2 absorbance units. Since this is a difference spectrum and may therefore record negative values the baseline should be positioned in the centre of the chart recording paper. The baseline you have recorded should be flat. If the signal is unduly noisy then increase the slit width and increase the scan time.

(iii) To one cuvette, the sample, add a few grains (~ 50) of dithionite. It is important not to add too many as this will acidify the solution and the protein will precipitate. The other cuvette is the reference. Record the spectrum as for step (ii).

(iv) If you are using a single-beam instrument with a facility for memorizing the

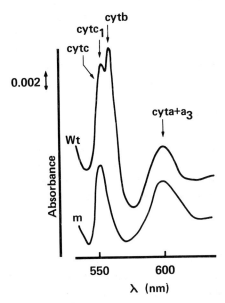

Figure 8. Visible absorption spectra of wild-type (Wt) and mutant (m) yeast. The reduced minus oxidized difference spectrum for the wild-type and mutant yeast mitochondria (derived from ref. 38). The mutant is deficient in the absorbance band at 562 nm characteristic of cytochrome b. Cytochrome $a + a_3$ are shown as absorbing in the 600–605 nm region but in fact this absorbance peak is predominantly due to cytochrome a with an approximately 20% contribution of cytochrome a_3.

baseline, place the reference cuvette in the machine and memorize the absorption spectrum from 500 to 650 nm. Replace the reference cuvette with the dithionite-reduced sample [step (iii) above] and record the spectrum.

Typical spectra for mitochondria of a wild-type yeast and a mutant are shown in *Figure 8*, which also shows the positions of the characteristic absorbance peaks of the cytochromes.

6. ISOLATION AND CHARACTERIZATION OF NUCLEIC ACIDS AND PROTEINS

6.1 Introduction

As can be gauged from the section on the genetics of mitochondria, the biogenesis of mitochondria involves complex interactions between the nuclear and mitochondrial genomes. The mtDNA exhibits both regulated transcription and a pattern of replication which is quite clearly distinct from that occurring in the nucleus. In addition, the formation of functional mitochondria is dependent on protein synthesis both in the cytoplasm and within the mitochondrion, each of which is quite distinct from the other. It is both beyond the scope of this chapter and inappropriate to review the complexities of the replication, transcription and translation in mitochondria, however these topics have been covered in much greater detail elsewhere (1). In this part of the chapter an attempt will be made to give the reader an insight into the methods used for studying these processes in mitochondria.

6.2 Preparation and analysis of yeast mtDNA

The mitochondrial genome of *Sacch. cerevisiae* is relatively small and has been extensively sequenced. Detailed analysis of the mtDNA provides insights into the processes of both DNA replication and transcription as well as providing useful information on its genetic content. Furthermore, the minilysate and colony hybridization methods are quick procedures allowing rapid genetic screening of numerous clones.

6.2.1 Purification of mtDNA

Several methods have been published for the preparation of mtDNA but most of them involve isolation of the DNA from whole cells rather than purified mitochondria, since it is very easy to separate the mtDNA from the nuclear DNA on CsCl gradients and it also removes the possibility that the mtDNA may become degraded during the isolation of the mitochondria. The following method is used by one of the authors and it gives a good yield of mtDNA.

(i) Grow the cells of the desired strain until early stationary phase in $1-3$ l of N3 medium or antibiotic-containing medium (if RC) or YP10 medium (if RD). Keep an aliquot to determine the purity of the culture.

(ii) Pellet the cells by centrifugation at 1000 g for 10 min, rinse the cells in water and centrifuge again. Weigh the pellet of cells. The following procedure is for 15 g of cells.

(iii) Resuspend the cells in 100 ml of 50 mM sodium phosphate buffer, 25 mM ED-TA, 1% (v/v) β-mercaptoethanol, pH 7.5 and add 15 mg of Zymolyase 60 000 (Seikagaku Kogyo, Tokyo). Incubate at 37°C until cells lyse (medium becomes viscous, usually 5–15 min).

(iv) Immediately add 100 ml of 0.2 M Tris–HCl, 80 mM EDTA and 1% SDS, pH 9.5. Incubate at 65°C for 30 min.

(v) Cool in ice, then add 50 ml of 5 M potassium acetate. Incubate at 0°C for 45 min.

(vi) Centrifuge at 5000 g for 10 min at 5°C. Keep the supernatant and add 150 ml of isopropanol. Mix and freeze at −20°C for 60 min.

(vii) Centrifuge at 5000 g at 0°C for 15 min. Discard the supernatant. Add 20 ml of 1 mM EDTA, 10 mM Tris–HCl, pH 7.8 and dissolve the pellet (this may be difficult, if so use a glass Teflon Potter-Elvehjem homogenizer if necessary).

(viii) Centrifuge at 15 000 g for 15 min at 5°C. Keep the supernatant, adjust the volume with buffer as necessary for ultracentrifugation tubes (usually 30 ml is convenient). To each 1 ml of solution add 24 μl of a 10 mg/ml stock solution of bisbenzimide (Hoechst 33258) and 1.16 g of CsCl.

(ix) Centrifuge at 100 000 g at 15°C until equilibrium. Visualize the DNA bands under UV light (blue fluorescence). The upper band is the mtDNA, the lower intense band is nuclear DNA. Carefully pipette out the mtDNA. Place in a second (smaller) ultracentrifuge tube and centrifuge again at 100 000 g until equilibrium.

(x) Collect the pure mtDNA band (upper). Extract the bisbenzimide with three successive extractions with isopropanol (*previously equilibrated with CsCl solution*). Add 4 vol of 0.3 M sodium acetate. Mix and add 3 vol of ethanol. Freeze at −70°C for 5 min and centrifuge at 10 000 g for 10 min.

(xi) Discard the supernatant and dissolve the pellet in 300 μl of 0.3 M sodium acetate. Add 1 ml of ethanol, freeze and centrifuge.

(xii) Rinse the pellet with ethanol and dry under vacuum. Re-dissolve the pellet in 500 μl of 1 mM EDTA, 10 mM Tris–HCl, pH 7.8.

One of the authors has found that a modification of this method that works well is to break open the yeast cells by grinding the cells with glass beads as described in Section 4.2.2 instead of protoplasting the cells as described in step (iii).

6.2.2 *Quick, small-scale preparation of yeast mitochondrial DNA (minilysates)*

Several methods have been published for the rapid preparation of total DNA out of which the mtDNA can be analysed directly by the Southern blotting technique using specific probes. The method described here is frequently used in the authors' laboratory and gives reliable results. It permits the analysis of numerous clones with limited effort (typically up to 100 clones can be analysed simultaneously), hence its usefulness in screening *rho⁻* mutants, polymorphic variants or recombinants.

(i) Grow the cells of the desired clone in a small volume (usually 3 ml are sufficient) of either YPglu medium (if RD) or N3 medium (if RC) until late log phase. It is not necessary to measure the optical density of the culture; an experienced eye is sufficient in this instance.

(ii)   Take a 1.5 ml sample of the culture with a sterile pipette and transfer to a micro-centrifuge tube. Store the remainder of the culture in the refrigerator until complete analysis of the mtDNA is finished.

(iii) Spin the tube at 12 000 *g* for about 10 sec (the pelleting of yeast cells in a microcentrifuge is almost immediate).

(iv)   Discard the supernatant by inverting the tube and quickly drain on a clean tissue (do not leave tubes inverted for a long time as the pellet may start to drip down the side of the tube).

(v)   Resuspend the pellet in 1 ml of water, vortex and centrifuge again as before.

(vi)   Discard the water by inverting the tube and quickly drain on a clean tissue.

(vii) Resuspend the pellet in 0.2 ml of 25 mM EDTA, 1% (v/v) β-mercaptoethanol, 50 mM sodium phosphate buffer, pH 7.5. Vortex and add 0.1 ml of a 1 mg/ml Zymolyase 60 000 solution.

(viii) Incubate at 37°C until the cells lyse (usually a few minutes is sufficient). Lysis can be easily monitored as the medium becomes highly viscous and a quick spin at 12 000 *g* no longer results in the formation of a pellet.

(ix)   Immediately after the lysis, add 0.2 ml of 80 mM EDTA, 1% (w/v) SDS, 0.2 M Tris−HCl, pH 9.5. Mix and incubate at 65°C for 30 min.

(x)   Add 0.1 ml of 5 M potassium acetate, mix and incubate at 0°C for 45 min.

(xi)   Centrifuge at 10 000 *g* for 10 min.

(xii) Pipette out the supernatant and transfer to a fresh microcentrifuge tube.

(xiii) Add 0.25 ml of 7.5 M ammonium acetate and fill the tube with ethanol. Mix and freeze at −70°C for 5 min (do not use a longer precipitation step since impurities will also precipitate after longer periods).

(xiv) Centrifuge at 12 000 *g* for 5 min at 0°C. Discard the supernatant and re-dissolve the pellet in 0.2 ml of 0.3 M sodium acetate. If solubilization is difficult, use a vortex mixer or micropipette. Ensure that solubilization is complete (if not NH_4^+ ions may be trapped in the pellet and interfere with subsequent utilization of the DNA). After solubilization, fill the tube with ethanol and freeze at −70°C.

(xv)   Centrifuge at 12 000 *g* for 5 min at 0°C, discard the supernatant and fill the tube with ethanol to rinse the pellet.

(xvi) Centrifuge, discard the ethanol supernatant and dry under vacuum.

(xvii) Re-dissolve the pellet in 100 μl of 1 mM EDTA, 10 mM Tris−HCl, pH 7.8. There should be enough DNA from such a preparation for 10 gel loadings.

(xviii) Restrict the DNA with appropriate restriction endonuclease(s) (see *Figure 1*) and analyse fragments by electrophoresis on a 1% agarose gel. If a minilysate is made from a *rho*⁺ strain or from a *rho*⁻ mutant containing a long fragment of the mitochondrial genome, blot onto nitrocellulose and hybridize using specific probes (if possible use recombinant plasmids with inserts of mtDNA as probes or use only highly purified mtDNA since any nuclear DNA contamination will result in a high background or even false hybridization). If a minilysate is made from a *rho*⁻ mutant retaining only a short segment of the mitochondrial genome (e.g. <5000 bp), amplification is usually sufficient for the mtDNA bands to be clearly visible over the background of nuclear DNA after staining the gel with 0.1 μg/ml of ethidium bromide.

6.2.3 *Analysis of colonies by DNA hybridization*

(i) Dilute the appropriate culture to 10^3 cells/ml and plate 0.1 ml on YPglu medium (if RD) or N3 medium (if RC). Incubate at 28°C for 2 days (YPglu) or 3 days (N3).

(ii) Place in grids on the same medium, picking colonies out one by one with sterile toothpicks. Prepare grids in duplicate in order to save viable colonies on one plate. Incubate at 28°C for 2 or 3 days.

(iii) Gently lay a Whatman 541 filter disc (9 cm diameter) on one plate of each duplicate. Store the other plate in the refrigerator.

(iv) Allow a few minutes contact with colonies, then slowly peel off the filter (only manipulate the filter with smooth forceps).

(v) Flip the filter disc onto Whatman 3MM sheets saturated with a solution of 25 mM EDTA, 1% (v/v) β-mercaptoethanol, 50 mM sodium phosphate, pH 7.5, containing 0.1 mg/ml of Zymolyase 60 000 (Seikagaku Kogyo, Tokyo). Incubate at 37°C for 1−2 h in a closed container to prevent desiccation.

(vi) Transfer the filter disc onto Whatman 3MM sheets saturated with 0.5 M NaOH and incubate for 10 min at room temperature.

(vii) Drain the filter disc on clean tissue and transfer onto Whatman 3MM sheets saturated with 1 M Tris−HCl, pH 7.5. Incubate for 10 min at room temperature. Repeat once.

(viii) Transfer the disc onto Whatman 3MM sheets saturated with 2.5 M NaCl, 0.5 mM Tris−HCl, pH 7.5. Incubate for 2 min.

(ix) Drain the filter on a clean tissue and air dry.

(x) Bake the filters in a vacuum oven at 80°C for 2 h.

(xi) Float the baked filters on 6 × SSC (Appendix IV) until they are thoroughly wetted, then submerge the filter in this solution for 5 min.

(xii) Pre-hybridize the filter in a plastic bag using 0.5 ml/cm^2 of filter in pre-hybridization medium containing 50% de-ionized formamide, 0.1% SDS, 5 × Denhardt's solution, 5 × SSPE and 100 μg/ml of heat-denatured sonicated calf thymus DNA. Incubate the filter with gentle agitation at 42°C for 2 h. Denhardt's solution is prepared as a 50-fold concentrated solution containing 1% Ficoll, 1% polyvinylpyrrolidone (PVP) and 1% BSA while SSPE is prepared as a 20-fold concentrated solution containing 3 M NaCl, 20 mM EDTA and 0.2 M sodium phosphate buffer, pH 7.4; both of these stock solutions are made in advance and can be stored at −20°C. Formamide is deionized by stirring it with Biorad AG501X8(D) mixed-bed resin (10 g/100 ml) for 4 h at room temperature and then it is also stored at −20°C.

(xiii) Add the denatured mtDNA probe, labelled with [^{32}P]phosphate by nick-translation either by using one of the commercially available kits (e.g. from Amersham International) or the procedure outlined in *Table 12*. The specific activity of the probe should be in the range of $5 \times 10^7 - 5 \times 10^8$ c.p.m./μg and the final concentration in the hybridization solution should be approximately 5 ng/ml.

(xix) Re-seal the bag and mix the probe into the solution by gentle squeezing and inversion of the bag and then incubate the filter for 18 h at 42°C.

(xv) After hybridization, carefully remove the filter from the plastic bag and wash

Table 12. Nick-translation of mtDNA.

1.	Dissolve 200−250 ng of DNA in 10 μl of 1 mM EDTA Tris−HCl, pH 8.0.
2.	Add 4 μl of nucleotides/buffer solution containing three of the four dNTPs 100 μM of each in 50 mM Tris−HCl, pH 7.8, 10 mM MgSO₄, 0.1 mM DTT.
3.	Add 2−5 μl of the other [α-³²P]dNTP (10 mCi/ml, 370 MBq/ml).
4.	Add 2 μl of enzyme solution (1 unit of DNA polymerase I, 20 pg of DNase I in a buffer containing 20 mM Tris−HCl, pH 7.5, 10 mM MgSO₄, 10% glycerol and 50 μg/ml BSA).
5.	Make up to 20 μl with distilled water, mix gently and incubate at 15°C for 75 min.
6.	Stop the reaction by adding 2 μl of 0.5 M EDTA, 0.25% bromophenol blue (the dye makes it possible to follow the purification of the labelled DNA by gel filtration).
7.	Load the reaction mixture onto a Sephadex G-75 column (prepared in a siliconized Pasteur pipette) equilibrated with 1 mM EDTA, 10 mM Tris−HCl, pH 8.0, elute with the same buffer, collect four-drop fractions and measure the radioactivity of each fraction by Cerenkov counting. DNA is eluted first then the peak of deoxynucleotides and the dye.

in approximately 200 ml of SSC containing 0.1% SDS at room temperature for 15 min. Then wash the filter three or four times in the same solution at 65°C each for 20−30 min.

(xvi) Dry the filter on a piece of Whatman 3MM filter paper, cover the filter with Saran wrap and press the covered, dried filter onto X-ray film (Hyperfilm MP Amersham International or equivalent).

(xvii) After development, scan the spots corresponding to the standards and the unknown samples using a spectrophotometer and deduce the relative concentrations in the unknown samples from the relative areas under the peaks.

6.2.4 *Methods for studying DNA replication*

(i) *In vivo studies of DNA replication. Sacch. cerevisiae* is unusual in that it is not possible to label the DNA with thymidine: the usual salvage pathway is not available because cells lack thymidine kinase (41). DNA can be labelled by growing cells in defined medium in the presence of radioactively labelled adenine; such labelling can be greatly enhanced if it is possible to use a mutant which is auxotrophic for adenine. However, unlike thymidine, adenine is incorporated extensively into both RNA and DNA and thus there is no really specific labelling of the DNA. Alternatively, cells can be grown in the presence of [³²P]phosphate but this labels lipids and proteins as well as both RNA and DNA and so these types of labelling procedures have not been widely used for studying the replication of the mtDNA. However, *in vivo* labelling combined with isotope dilution methods has been used to demonstrate the fact that the relative amounts of nuclear DNA and mtDNA in yeast are constant irrespective as to whether yeast cells are subjected to glucose repression (42).

(ii) *In vitro studies of DNA replication.* The DNA polymerase of *Sacch. cerevisiae* has been fully characterized by the work of Wintersberger's laboratory (43−45). The activity of the DNA polymerase can be studied in isolated mitochondria (*Table 13*) or alternatively its properties can be analysed after purification of the enzyme (45). Since this work most of the studies of DNA replication in mitochondria have focused on the role of the *ori* sequences which appear to represent initiation sites for replication of the mtDNA. This topic is extremely complex and the reader should consult one of the reviews of this topic for further details (e.g. ref. 43).

Table 13. Assay of mitochondrial DNA polymerase activity.

1.	Prepare mitochondria by grinding yeast cells and purify the mitochondria by isopycnic centrifugation (see Section 4.2).
2.	Add 20 μg of mitochondrial protein[a] to give a final volume of 0.25 ml of reaction medium containing:

> 25 μg activated DNA[b]
> 50 mM Tris−HCl, pH 7.5
> 10 mM MgCl$_2$
> 1 mM EDTA
> 50 μM each of dATP, dCTP and dGTP
> 50 μM [^{3}H]dTTP (40 μCi/μmol; 1.5 MBq/μmol).

3.	Incubate the reaction mixtures at 35°C for 15 min (remember to prepare a zero time, non-incubated control).
4.	Terminate the incubation by cooling in ice and the addition of 0.5 ml of 15% TCA containing 1% sodium pyrophosphate.
5.	Collect the acid-precipitable material by filtration onto Whatman GF/C filters and wash the filters successively with 5% TCA containing 1% sodium pyrophosphate, 5% TCA and ethanol:ether (1:1).
6.	Dry the filters in an oven at 45°C and measure the radioactivity of each in a liquid scintillation counter.

[a]Whole mitochondria or mitochondrial extracts can be used; in the latter case it is best to use a higher concentration of MgCl$_2$ (44,45).
[b]Activated DNA is DNA that has been nicked by incubation with very small amounts of pancreatic DNase. To activate the DNA add 0.02 μg of pancreatic DNase to 15 mg of native DNA in 20 mM Tris−HCl, pH 7.5 and 10 mM MgCl$_2$ and incubate at 35°C for 15 min prior to use.

6.3 Preparation and analysis of mitochondrial RNA

6.3.1 *Isolation of RNA from mitochondria*

For a number of techniques it is necessary to isolate the RNA prior to analysis. During the isolation procedure it is of paramount importance to ensure that the RNA is not degraded. Contamination with exogenous nucleases can be minimized by baking all glassware at 200°C for 2 h and by autoclaving all plasticware and solutions at 120°C for 20 min. Solutions that do not contain compounds with free amino groups can also be treated with diethylpyrocarbonate (DEPC, 20 μl/100 ml) prior to autoclaving; this inactivates proteins and when heated it decomposes into CO_2 and ethanol. Often endogenous nucleases are also present in the transcription mixture, especially if one is using a system derived from whole mitochondria, and so it is very important to inactivate all of the proteins and to separate them from the RNA after *in vitro* incubation. The main types of deproteinization involve extraction using phenol—chloroform, or phenol alone, or by centrifugation in either CsCl or guanidine thiocyanate solutions (46,54). In the authors' experience the most consistent results are obtained by extracting the RNA with phenol−chloroform as described in *Table 14*. Variations of this procedure, aimed at improving the yield of RNA, have included carrying out the phenol—chloroform extraction at 65°C and digestion with 100 μg/ml of protease K for 30 min at 37°C prior to phenol−chloroform extraction.

The RNA obtained by this procedure is contaminated by DNA which may interfere with subsequent analytical procedures. The DNA can be removed by ultracentrifugation at 100 000 g for 24 h in 50% (w/w) CsCl gradients with an initial density of 1.55 g/ml, this allows the RNA to pellet while the DNA bands in the solution; this

Table 14. Isolation of mitochondrial RNA.

1.	Add 0.1 vol of 20% (w/v) SDS to the mitochondrial suspension. Note that potassium ions precipitate SDS and should be omitted if possible, otherwise use 2% laurylsarcosine instead of SDS.
2.	Add an equal volume of redistilled phenol:chloroform:isoamyl alcohol (50:50:2 by vol) equilibrated with 10 mM Tris−HCl, pH 7.4, 1 mM EDTA. Extract nucleic acids for 10 min with mixing of the phenol and aqueous phases every minute for 10 sec.
3.	Separate the phenol−chloroform and aqueous phases by centrifugation at 10 000 *g* for 5 min at 8°C in a swing-out rotor.
4.	Immediately remove the aqueous phase (using a Pasteur pipette taking care not to disturb the interface) and mix it with a half volume of phenol:chloroform:isoamyl alcohol. Repeat the extraction as in step 2.
5.	Repeat the separation in step 3 and remove the aqueous phase. Add 0.1 vol of 25% (w/v) sodium acetate. pH 5.0.
6.	Add 2.5 vol of absolute ethanol chilled to −20°C. Leave the nucleic acids to precipitate overnight at −20°C.
7.	Pellet the precipitated RNA by centrifugation at 20 000 *g* for 10 min at 0°C. For very small amounts of RNA it will be necessary to centrifuge at 60 000 *g* for 30 min to ensure a good recovery of RNA.

method is only recommended for high molecular weight RNA. An alternative approach is to digest the total nucleic acids with 200 μg/ml of RNase-free DNase I at 37°C for 5 min in 10 mM $MgCl_2$, 1 mM EDTA, 10 mM vanadyl ribonucleoside, 10 mM Tris−HCl, pH 7.5, followed by re-extraction with phenol−chloroform (*Table 14*); RNase-free DNase I is commercially available. Another problem is that often the RNA obtained is contaminated with large amounts of unincorporated isotope. This acid-soluble material is not readily removed by simple washing of the precipitated RNA with 70% ethanol, especially if the RNA is contaminated with [^{32}P]phosphate. The most effective way to remove this material is to pass the RNA dissolved in 0.1% SDS, 10 mM NaCl, 10 mM Tris−HCl, pH 7.4, over a column of Sephadex G-50(F) as described in *Table 15*. The concentration of the final RNA solution can be determined spectrophotometrically; a 1 mg/ml solution of RNA has an optical density of 25 at 260 nm.

6.3.2 *Systems used to study transcription*

A number of systems can be used to study transcription in mitochondria. A significant amount of work has been carried out using isolated mitochondria purified free of nuclear and other cellular contaminants by differential centrifugation and, usually, isopycnic gradient centrifugation (see Section 4). Isolated, intact mitochondria incubated under optimal conditions (*Table 16*) will incorporate radioactive nucleoside triphosphates (NTPs) into RNA. The extent of transcription can be measured either in terms of the rate of synthesis or, after isolation of the RNA, by assessing the nature of the transcripts.

The advantage of *in organello* studies is that, as there is little disruption of the interior organization of the mitochondrion, including the nucleoid structure of the mtDNA, one can expect that the pattern of transcription will accurately reflect that occurring *in vivo*. The difficulty with using this system is that the mitochondrial membranes may be selectively permeable to components needed for transcription. Similarly, the presence of intact mitochondrial membranes may lead to the formation of endogenous pools of ions and other compounds within the mitochondrion which introduce difficulties, not

Table 15. Spin column chromatography with Sephadex G-50.

1.	Swell Sephadex G-50(F) in 10 mm Tris−HCl, pH 7.4, 1 mM EDTA, 0.1% SDS and autoclave at 110°C for 20 min.
2.	Take a 1 ml sterile syringe and remove the plunger. Place a small quantity of baked glass wool in the bottom of the syringe.
3.	Pipette a quantity of the swollen Sephadex G-50(F) into the syringe and place it inside a 15 ml Corex glass tube. Centrifuge at 1000 *g* for 5 min at 4°C in a swing-out rotor.
4.	Sephadex G-50(F) will pack down in the syringe. Add more Sephadex G-50(F) and repeat the centrifugation in step 3 until the column reaches the 1 ml calibration mark.
5.	Wash the column twice with 100 μl of 10 mM Tris−HCl, pH 7.4, 1 mM EDTA, 0.1% SDS.
6.	After the final wash remove the syringe and fix a sterile MCC tube to the end of the syringe to collect the nucleic acids.
7.	Replace the syringe and add 100 μl[a] of the nucleic acids sample in 10 mM Tris−HCl, pH 7.4, 1 mM EDTA, 0.1% SDS to the column.
8.	Repeat the centrifugation as in step 3. Either precipitate RNA with 3 vol of absolute ethanol and store at −20°C or cap the microcentrifuge tube and store in distilled water at −70°C.

[a]Sample volume should not exceed 0.1 vol of the Sephadex G-50(F) bed volume, i.e. 100 μl of sample RNA per 1 ml column.

only in defining exactly what are the precise requirements of the mtRNA polymerase, but also in interpreting other experimental results.

One possible solution to this problem is to 'permeabilize' the mitochondrial membranes; however the method used must be chosen with care. Triton X-100 and other non-ionic detergents can be used to lyse mitochondria at a concentration of 1−2% but, although this method works well in terms of solubilizing membranes, detergent lysis often inhibits transcription in mitochondria; a similar inhibition has been found if one uses macrolide antibiotics to permeabilize yeast mitochondrial membranes. Ultrasonication does distrupt mitochondria but overall it is not advisable to use this approach in that it may disrupt the integrity of mtDNA which is loosely packaged inside the mitochondrion. The method used by the authors is to permeabilize the mitochondria by osmotic shock (47). This method involves gently resuspending the mitochondria in a very hypotonic buffer (e.g. 1 mM EDTA, 20 mM Tris−HCl, pH 7.9) and the subsequent swelling of the mitochondria permeabilizes the membranes.

In an attempt to define the transcriptional mechanisms and its controls better, other systems have been devised. The obvious system, that of analysing the transcriptional pattern of the nucleoid, has been relatively neglected because of the difficulties in obtaining purified nucleoids and the likelihood of losing transcriptional factors during the isolation procedure. Instead studies have tended to concentrate on the isolation of mitochondrial protein fractions that exhibit RNA polymerase activity using either artificial templates or cloned fragments of mtDNA containing known transcription promoters (48−51). The cloned DNA can be transcribed *in situ* in the plasmid or after excision using restriction nucleases.

6.3.4 *Methods to study the rates of synthesis and initiation in mitochondria*

(i) *Measurement of the rate of transcription.* The 'rate' of transcription measures the rate of initiation and elongation of transcripts as well as, in some cases, nucleolytic

Table 16. RNA synthesis in isolated yeast mitochondria.

1.	In a sterile conical glass test tube prepare the following assay solutions (total volume 200 μl).
	0.6 M sorbitol
	20 mM Tris−HCl, pH 7.9
	10 mM MgCl$_2$
	50 mM NaCl
	0.5 mM ATP
	0.2 mM GTP and CTP
	To each assay solution add 25 μl of mitochondrial suspension ($\sim$2 mg/ml).
2.	Pre-incubate the assay at 28°C for 2 min to pre-equilibrate the mixtures.
3.	Initiate transcription by adding 25 μl of either 2 mM [α-^{32}P]UTP or [^{3}H]UTP (50 Ci/mmol, 1.85 TBq/mmol) and continue the incubation for another 10 min[a].
4.	Add 1 ml of 0.8 M perchloric acid, 100 mM sodium pyrophosphate and leave on ice for a minimum of 10 min.
5.	Pipette the solution onto a GF/C filter (25 mm diameter) on a vacuum filtration apparatus.
6.	Wash filters twice with 5 ml of 10% trichloracetic acid containing 10 mM sodium pyrophosphate and once with 5 ml of 5% TCA, 10 mM sodium pyrophosphate. Finally wash each filter with 2 ml of an ethanol:ether mixture (1:1, v/v).
7.	Place each filter in a scintillation vial and dry for at least 2 h at 45°C. Cool the vials and add 4.5 ml of toluene-based scintillator.

[a]Background radioactivity is estimated by preparing identical non-incubated assays kept in ice. The isotope is added followed directly by 0.8 M perchloric acid, 0.1 M sodium pyrophosphate. Acid-precipitable material is collected as in steps 5 and 6.

degradation of the RNA. Measurements of the rate of RNA synthesis are readily made simply by measuring the amount of radioactively labelled NTP incorporated into acid-precipitable material. For this type of assay mitochondria are incubated in the required incubation medium; *Table 16* gives the conditions used by the authors for yeast mitochondria. After an appropriate time interval the incubation is terminated by the addition of 0.8 M perchloric acid containing 100 mM sodium pyrophosphate (*Table 16*); this precipitates all polynucleotides of five nucleotides or longer.

(ii) *Measurement of initiation by capping of primary transcripts using [α-^{32}P]GTP and guanyltransferase.* RNA initiation and processing produce two different types of 5' termini. Initiation sites are unique in that they retain their 5' triphosphate terminus. Those 5' termini that are the product of processing lack this distinctive feature. Mitochondrial RNA is not capped *in vivo* and so capping is particularly suitable for measuring the level of initiation using this method first developed by Levens *et al.* (52). Guanyltransferase from *vaccinia virions* can be used to catalyse the capping of 5' triphosphate termini with [α-^{32}P]GTP thereby identifying unique sites of RNA initiation. The guanyltransferase required is now commercially available; the procedure originally described is a slight modification of the method developed by Monroy *et al.*(53).

The protocol for capping the mitochondrial RNA is as follows.

(i)   Dissolve 3−7.5 μg of mtRNA in 2.5 μl of double-distilled water. Add methyl mercuric hydroxide to a final concentration of 5 mM and incubate at room temperature for 10 min to denature the RNA.

(ii)   Adjust to a final volume of $20-40$ μl with 50 mM Tris−HCl, pH 7.5, 1 mM MgCl$_2$, $20-40$ μM [α-^{32}P]GTP ($\sim$2500 Ci/mmol, 100 TBq/mmol), 1 mM dithiothreitol (DTT).

(iii)   Add $10-12$ units of guanyltransferase and incubate for 15 min at 37°C.

(iv)   Terminate the incubation by addition of an equal volume of 2% SDS, 20 mM EDTA, 20 mM Tris−HCl, pH 7.5, containing $40-100$ μg of *Escherichia coli* tRNA.

(v)   Add an equal volume of re-distilled phenol equilibrated with 1% SDS, 10 mM EDTA, 10 mM Tris−HCl, pH 7.5.

(vi)   Separate the phenol and aqueous phases by centrifugation at 13 000 g in a microcentrifuge for 10 min at 4°C.

(vii)   Remove the aqueous phase and apply the capped RNA to a 5 ml (bed vol) Sephadex G-50 column equilibrated with 0.2% SDS, 2 mM EDTA, 10 mM Tris−HCl, pH 7.5.

(viii) Wash the column with equilibration buffer and locate the eluted RNA either with a Geiger counter or by acid precipitating aliquots followed by filtration and liquid scintillation counting.

(ix)   Pool the peak fractions and adjust to 1 M ammonium acetate. Precipitate the RNA by the addition of 2.5 vol of absolute ethanol chilled to -20°C and leave overnight at -20°C.

(x)   Pellet the precipitated RNA by centrifugation in a microcentrifuge for 10 min at 13 000 g at 4°C.

(xi)   Re-dissolve the RNA in a small volume of sterile double-distilled water and store at -80°C until required.

One possible problem in the use of this technique is that the triphosphate terminus may be degraded by phosphatase activity and hence these transcripts will not be detected. It is also necessary to denature the sample RNA completely before carrying out the capping reaction because RNA secondary structure has to be kept to a minimum to ensure that the 5′ termini are freely available to the guanyltransferase.

This technique does have considerable advantages if transcripts of mitochondrial origin are to be identified because the cytoplasmic mRNA is already capped and the cytoplasmic rRNA does not have a triphosphate terminus, thus neither of these likely contaminants interferes with this method. The capped mitochondrial RNAs can be analysed by a number of methods. In the absence of pre-existing transcripts determination of the incorporation of [α-^{32}P]GTP, that is the number of transcripts capped, gives a direct measure of the rate of initiation; whereas capping of RNA in *in organello* assays measures the steady-state levels of transcripts. However, more sophisticated analyses can also be carried out. The isolated RNA can be simply electrophoresed on a denaturing formaldehyde−agarose gel (*Table 17*) and the capped species visualized by autoradiography. A useful method for the identification of the number of initiation sites is to digest the capped RNA with the RNase T1 which is specific for GMP residues. *Sacch. cerevisiae* mtDNA has an extremely low G+C content (18% G+C) and so the digestion of capped RNA with RNase T1 generates a number of short RNA species which can be readily sequenced. By comparing the RNA sequence with known DNA

Table 17. Denaturing formaldehyde−agarose gels.

1.	Denature the sample by mixing 4.5 μl of the sample solution containing up to 20 μg of DNA or RNA with 3.5 μl of formaldehyde, 10 μl of formamide and 2 μl of 5 × gel buffer containing 50 mM sodium acetate, 5 mM EDTA and 0.2 M Mops−NaOH, pH 7.0. Heat the sample at 65°C for 15 min.
2.	Prepare an agarose gel, usually the concentration should be between 0.8% and 1.5% agarose depending on the nature of the sample, in gel buffer (10 mM sodium acetate, 1 mM EDTA and 40 mM Mops−NaOH, pH 7.0) containing 6.6% formaldehyde.
3.	Mix the sample with 1/10 volume of 50% glycerol, 1 mM EDTA, 0.5% bromophenol blue and 0.5% xylene cyanol.
4.	Load the sample onto the gel and run overnight at 2 V/cm.
5.	Rinse the gel in several changes of distilled water for 2 h.
6.	If required stain the gel by soaking it in 2 μg/ml of ethidium bromide for 30 min, then wash the gel in distilled water for 60 min and view under UV light.
7.	Radioactive samples can be visualized by autoradiography or fluorography.

Both formamide and formaldehyde are oxidized in air; the former should be deionized by mixing it with a mixed-bed resin until it is neutral while the formaldehyde should be neutralized before use.

sequences it is then possible to identify those sequences at the site of the initiation of transcription.

It is very important to appreciate that the guanyltransferase will cap the 5′ termini of pre-existing as well as newly initiated RNA as long as the terminal triphosphate has been retained. Thus this method cannot be used for measuring *in vitro* initiation in whole mitochondria and other methods must be used.

(iii) *5′ End-labelling transcripts with [^{32}P]phosphate or [^{35}S]sulphate nucleoside triphosphates.* The retention of the triphosphate group at the 5′ terminus of RNA molecules enables one to end-label the 5′ terminus of transcripts initiated *in vitro* in the presence of β- or γ-phosphate-labelled ATP or GTP; unlike guanyltransferase capping, this assay for the initiation of RNA is not affected by the presence of pre-existing transcripts. However, β-labelled nucleoside triphosphates are not generally available and so γ-position labelled nucleoside triphosphates have been used; but there are a number of problems associated with their use. The γ-phosphate is incorporated from ATP into protein and lipid molecules by kinases. Measurement of the incorporation of this label into RNA therefore necessitates the deproteinization and purification of RNA. The use of [γ-^{32}P]ATP or GTP may give erroneous results because the 5′ triphosphate termini can be degraded by phosphatase activity. This problem can be reduced by using either [γ-^{35}S]adenosine-5′-O-(3-thiotriphosphate) (ATP-γ-S) or guanosine-5′-O-(3-thiotriphosphate) (GTP-γ-S). These nucleotides are more resistant to phosphatase cleavage. Work in the authors' laboratory has successfully used [γ-^{35}S]ATP-γ-S to study the rates of initiation *in vitro*. In this case the *in vitro* assay is terminated by addition of SDS to a final concentration of 2% and the RNA is extracted as described in *Table 14*. The incorporation of radioactivity into RNA is quantified by acid-precipitating an aliquot of the RNA solution (steps 4−7 of *Table 16*) followed by liquid scintillation counting. Alternatively, the extracted RNA can be analysed on a denaturing gel (*Table 17*). The ^{35}S-labelled transcripts are visualized by fluorography. To do this soak the gel in 10 vol of 1.0 M sodium salicylate solution, pH 7, for 30 min, transfer the gel onto wetted Whatman 3 MM paper and dry down under vacuum. Place the dried gel

Table 18. Denaturing 6% polyacrylamide gels.

1.	Mix together 10 ml of 30% acrylamide stock solution (29% acrylamide, 1% *N,N'*-methylenebis-acrylamide), 5 ml of 10 × TBE buffer (25 mM EDTA, 0.9 M Tris−borate buffer, pH 8.3) and 17 ml of distilled water.
2.	Dissolve 21 g of urea in the solution with stirring and adjust the volume to 50 ml.
3.	Filter and de-gas the solution.
4.	Add 0.25 ml of freshly made 10% ammonium persulphate solution and 50 μl of TEMED, mix and pour the gel immediately.
5.	Overlayer the gel with water and allow to set.
6.	Electrophorese the gel using TBE as buffer at ∼5 V/cm for 15 h at room temperature.

in contact with Kodak X-omat R film or equivalent at $-70°C$. The incorporation of radioactivity into RNA is an indication that the RNA polymerase is initiating transcription *in vitro* and that the incorporation of labelled NTPs is not just the result of 'run-off' synthesis.

6.3.5 *Determination of the accuracy of initiation of transcripts*

As described in Sections 6.3.6−6.3.8, hybridization analysis is a very powerful tool for studying the accuracy of initiation of transcripts using S1 nuclease mapping and primer extension methods. However, other relatively simple methods to show that the *in vitro* pattern of transcript reflects the *in vivo* pattern also exist. One simple approach is to isolate the *in vivo* labelled RNA and compare the relative sizes of the transcripts synthesized with those products of the *in vitro* assay. Molecular weight markers are run on a denaturing formaldehyde−agarose gel (*Table 17*) for accurate size determination (e.g. cytoplasmic rRNA, *E. coli* 23S and 16S rRNA or a restriction digest of lambda bacteriophage DNA if the gel is completely denaturing). This will only reveal those RNAs synthesized at a relatively high rate, usually the rRNAs and tRNAs, but it is still a good indicator of the accuracy of the initiation of transcription *in vitro* for all transcripts. However, transcripts are often subjected to extensive post-transcriptional processing and in this case the *in vitro* synthesized products may well be much larger as a result of the lack of accurate processing of the RNA *in vitro*.

The run-off assay is, in principle, also a very simple method for the determination of specific initiation in an *in vitro* assay. A cloned restriction fragment of mtDNA is chosen that is known to contain the site of initiation and to terminate within the gene. This restriction fragment is then used as a template to assay the specificity of protein extracts that exhibit mitochondrial RNA polymerase activity. *In vitro* transcription should generate RNA molecules of known length that can be accurately sized by electrophoresis on a denaturing gel by comparison with RNA molecules of known size and composition. Alternatively, the RNA molecules synthesized *in vitro* can be sequenced and compared with the DNA sequence of the restriction fragment. A method that has been used to generate RNA molecules of known size is to omit a specific NTP from the *in vitro* incubation mixture. In the absence of the particular NTP the mitochondrial RNA polymerase terminates transcription prematurely thus synthesizing a short RNA molecule of defined size. The size of the RNA can then be determined by gel electrophoresis (e.g. using a similar protocol to that in Table 18).

Table 19. Northern blotting of RNA.

1.	Dissolve 10 μg of mitochondrial RNA in 8 μl of 1 M deionized glyoxal[a], 50% (v/v) dimethyl sulphoxide, 10 mM sodium phosphate, pH 7.0. Incubate for 60 min at 50°C.
2.	Cool the mixture in ice and add 2 μl of 50% (v/v) glycerol, 10 mM sodium phosphate buffer, pH 7.0, 0.01% bromophenol blue.
3.	Separate the RNA on a 1.5% agarose gel in 10 mM sodium phosphate buffer, pH 7.0 at 90 V for 6 h. Circulate the buffer between electrodes to maintain the pH at neutrality.
4.	Place gel on two sheets of Whatman 3MM paper saturated with 20 × SSC[b].
5.	Soak nitrocellulose first in distilled water then in 20 × SSC. Place the nitrocellulose over the gel followed by two sheets of Whatman 3MM paper. Place several paper towels and a weight on top of the filter paper. Transfer is complete after 12−15 h.
6.	Air dry the RNA blots and bake in a vacuum oven at 80°C for 2 h.
7.	Soak the filter in 20 mM Tris−HCl, pH 8.0 at 100°C and allow it to cool to room temperature to remove any remaining glyoxal.
8.	Pre-hybridize RNA blots for 8−20 h at 42°C in 50% (v/v) formamide, 5 × SSC, 50 mM sodium phosphate, pH 6.5, 250 μg/ml sonicated salmon sperm DNA, 10 × Denhardt's solution[c]. Incubate with gentle agitation at 42°C for 8−20 h.
9.	Replace the pre-hybridization buffer with hybridization buffer containing four parts pre-hybridization buffer with one part 50% (w/v) dextran sulphate.
10.	Heat the labelled probe DNA ($>10^8$ c.p.m./μg) to 100°C for 5 min, then cool in ice before adding it to the hybridization buffer in the bag; re-seal the bag. Incubate the filters for 20 h at 42°C. The optimum hybridization temperature will vary depending on the base composition of the probe.
11.	Wash the filters with four changes of 2 × SSC containing 0.1% SDS for 5 min each at room temperature.
12.	Wash the filters with two changes of 0.1 × SSC containing 0.1% SDS for 15 min each at 50°C.
13.	Expose blots to pre-flashed Kodak X-Omat R X-ray film or equivalent at −70°C.
14.	Hybridized probe can be removed by washing the filters in 0.1−0.05 times wash buffer (1 times wash buffer is 50 mM Tris−HCl, pH 8.0, 2 mM EDTA, 0.5% sodium pyrophosphate, 10 × Denhardt's solution) for 1−2 h at 65°C.

[a]Deionize glyoxal by mixing 20 ml of 40% glyoxal with 20 g of AG501-X8 ion-exchange resin. Stir for 30 min and then decant the supernatant and mix with fresh ion-exchange resin for 30 min, continue until the final pH is between pH 5.5 and pH 6.0. Store at −20°C.
[b]SSC is 0.15 M NaCl, 15 mM sodium citrate, pH 7.0.
[c]100 × Denhardt's solution contains 2% BSA, 2% Ficoll, 2% polyvinylpyrrolidone and can be stored at −20°C.

6.3.6 *An introduction to hybridization analysis of transcription*

The technique of DNA−RNA hybridization has been used extensively to analyse the patterns of transcription of mtDNA. However, the type of analysis depends on the nature of the transcription assay. In the case of *in organello* assays most of the RNA represents pre-existing transcripts which can interfere with some types of hybridization analyses. In contrast, *in vitro* transcription assays of cloned mtDNA avoid this problem.

There are two strategies that can be used for hybridization analysis of *in vitro* transcribed RNA. The classical approach is solution hybridization in which the amount of hybrid formed is measured by acid precipitation or isolation of the hybrid followed by liquid scintillation counting of the amount of radioactive hybrid formed. However, the development of blotting techniques has led to the development of a range of semi-quantitative methods based on autoradiographic techniques.

For blot hybridization either the DNA or RNA can be immobilized by adsorption onto nitrocellulose or covalently attached to diazobenzyloxymethyl paper (DBM-paper) and then incubated with radioactively labelled DNA or RNA. The separation of DNA restriction fragments on agarose gels and the transfer of fragments to nitrocellulose or DBM-paper is termed Southern blotting. However, the converse approach is usually more appropriate for transcription studies and it involves separating the RNA by gel electrophoresis and then transferring the RNA to either DBM-paper or nitrocellulose. This is known as the Northern blot technique (*Table 19*). A variation of these techniques is to immobilize the RNA or DNA directly onto nitrocellulose in a series of dots. After hybridization the amount of radioactive hybrid associated with the dots can be determined by autoradiography or the dots cut out and the amount of radioactivity of each quantitated by liquid scintillation counting. Autoradiographic analysis of the Southern or Northern blots after hybridization can be used to analyse RNA transcribed *in vitro* both semi-quantitatively and qualitatively. By scanning the developed films with a densitometer it is possible to obtain a good estimate as to the relative rates of transcription of specific genes of the mtDNA by using the appropriate probes (55).

Within the very limited space available in this chapter it is impossible to describe in detail the whole range of hybridization techniques used to analyse RNA transcripts. For such an overview the reader is strongly recommended to consult the much more detailed descriptions of hybridization protocols given in *Nucleic Acid Hybridisation*, another volume in the Practical Approach series (56) which covers this topic in depth. This section of the chapter will confine itself instead to the techniques most frequently used for the characterization of mitochondrial RNA.

6.3.7 Use of hybridization analysis to study the initiation of transcription in vitro

(i) *S1 nuclease mapping.* The single strand-specific nuclease S1 has been used extensively by workers studying transcription. The principle of this technique involves solution hybridization of *in vitro* transcribed RNA to a DNA probe known to contain the transcriptional unit including the site of RNA initiation (57). After hybridization, the DNA−RNA hybrid is digested with nuclease S1 to remove DNA and RNA not in the hybrid. The size of the hybrid is determined on a non-denaturing gel and compared with the protected fragment generated between the probe DNA and the *in vitro* transcribed RNA. If they have identical sizes one can conclude that the mitochondrial RNA polymerase is initiating at the correct site *in vitro*. This method can also be used to

Table 20. Mapping with nuclease S1.

1. Hybridize 10−70 ng of S1 probe to 50 ng of mtRNA in 10 μl of 80% (v/v) formamide (deionized), 500 mM NaCl, 60 mM Hepes−NaOH, pH 7.2, 2.5 mM EDTA. Heat at 100°C for 90 sec, bring to 46°C and incubate for 2−10 h.
2. Dilute the mixture with 100 μl of 250 mM NaCl, 30 mM sodium acetate, pH 4.8, 1 mM ZnCl$_2$. 20 μg/ml denatured salmon sperm DNA and 50−100 Vogt units of nuclease S1.
3. Incubate the mixture at 30°C for 40 min.
4. To stop the reaction add 1/10 vol of 3 M sodium acetate, pH 5, 3 vol of ethanol and 10 μg of carrier tRNA.
5. Pellet nuclease S1-resistant hybrids and resuspend in deionized formamide with 0.05% bromophenol blue and xylene cyanol as marker dyes. Heat at 100°C for 90 sec.
6. Separate the protected fragments by electrophoresis through 4% (w/v) polyacrylamide, 7 M urea (see *Table 21*).

map the 3′ termini of RNA synthesized *in vitro* to establish the accuracy of termination. A protocol for nuclease S1 mapping is given in *Table 20*.

(ii) *Primer extension analysis.* This method involves the hybridization of the RNA transcript to a DNA probe that is known to be entirely contained within the gene of interest but *not* including the site of initiation. The DNA probe is end-labelled and hybridized in solution to the complementary RNA synthesized *in vitro* (57). The molar ratio of DNA probe to RNA should be in the region of 5- to 10-fold, for example 0.04 μg of a 50-nucleotide DNA primer hybridized to 0.2 μg of RNA 1 kb long; large excesses of primer should be avoided otherwise non-specific priming may occur. The DNA primer is then extended to the 5′ (i.e. initiating) nucleotide of the RNA using reverse transcriptase and unlabelled deoxyribonucleotide triphosphates. The products are then run on a denaturing gel (*Table 18*) which, after autoradiography, will show the primer extension product and the DNA probe. The product is accurately sized by comparison with molecular weight markers to identify the exact position of the 5′ terminus of the transcript. Alternatively, the end-labelled primer extension product can be sequenced and compared with the DNA sequence in the coding region of the gene. A method for primer extension is given in *Table 21*. Methods for sequencing DNA are described in detail in another volume of the Practical Approach series (59).

6.3.8 *Determination of the rate of transcription of specific genes*

(i) *Solution hybridization.* A very useful and rapid method for the hybridization of DNA and RNA in solution is described in *Table 22*. While typically, the hybridization of filter-bound DNA to labelled RNA can take from 8 to 48 h, the hybridization of a short DNA probe to RNA in solution is complete in 15−30 min (60). The hybrid molecule can then be digested with RNase A or nuclease S1 to remove non-hybridized, single-

Table 21. Primer extension analysis of transcripts.

1.	Hybridize the end-labelled DNA primer to 1 μg of mitochondrial RNA in 120 mM KCl, 10 mM Tris−HCl, pH 8.5 in a final volume of 10 μl. Incubate the mixture at 65°C for 20 min then slowly cool to 42°C over 60−90 min[a].
2.	Add 10 μl of 120 mM KCl, 100 mM Tris−HCl, pH 8.5, 20 mM MgCl$_2$, 20 mM DTT, 1.6 mM of each dNTP and 5−20 units of avian myeloblastosis virus reverse transcriptase.
3.	Incubate the extension mixture for 2 h at 42°C.
4.	At the end of the incubation add 3 vol of absolute ethanol chilled to −20°C. Leave the nucleic acids to precipitate for 2 h at −20°C.
5.	Collect nucleic acids by centrifugation in a microcentrifuge at 13 000 *g* for 10 min.
6.	Briefly dry the pellet under vacuum and resuspend it in 10−20 μl of 80% formamide, 20 mM EDTA, 0.04% bromophenol blue, 0.04% xylene cyanol.
7.	Heat the sample at 100°C for 2 min by cooling on ice.
8.	Separate the samples by electrophoresis in 6% polyacrylamide, 7 M urea, 90 mM Tris−borate, pH 8.3 (*Table 18*). Include suitable DNA size markers for accurate size determination of the primer extension product.
9.	Visualize the primer extension product by autoradiography.

[a]This suggested re-annealing temperature is for *Sacch. cerevisiae* RNA which has a very low content of G+C. For mitochondrial DNA with G+C content of 40−60% use a higher temperature, typically 55−60°C.

Table 22. Solution hybridization analysis.

1.	End-label the chosen DNA probe using polynucleotide kinase (Section 6.3.7) the probe should be small, typically 100−200 bp.
2.	Mix together the DNA probe with the purified mtDNA both in distilled water and lyophilize.
3.	Re-dissolve the sample in 20 μl of 0.3 M NaCl, 0.03 M sodium citrate, 8 mM Pipes−NaOH, pH 6.4 containing, if desired, 1% SDS.
4.	Incubate the solution at 65°C for 10 min in a sealed capillary.
5.	Cool the solution to the hybridization temperature, 45°C for *Sacch. cerevisiae*, and incubate for 30 min.
6.	*Either* Mix the hybridization solution with 4 μl of 35% Ficoll, 0.5% SDS, 0.1% bromophenol blue and 0.1% xylene cyanol and separate the mixture on a non-denaturing agarose or polyacrylamide gel. *or* Digest the non-hybridized RNA by digestion with S1 nuclease or RNase A, acid precipitate the hybrid with 10% trichloroacetic acid, collect on filters and measure the amount of radioactivity using a scintillation counter.

stranded RNA. The hybrid can then be separated on a non-denaturing gel followed by autoradiography or fluorography to identify the hybrid. Alternatively, the hybrid can be precipitated with cold trichloroacetic acid (TCA) and the amount of RNA hybridizing to the probe DNA quantified by liquid scintillation counting of the acid-precipitable material.

(ii) *Dot−blot methods.* These methods allow the measurement of the level of specific transcripts depending on the exact nature of the DNA probe used. Labelled RNA can be hybridized to DNA on filters or, alternatively, the RNA can be dotted onto the filters and incubated with an appropriately labelled DNA probe. In the former case the DNA is blotted onto the filter as described in Section 6.2.3 and then, after pre-hybridization, the filter is incubated with either labelled mRNA or, more usually, its cDNA copy. However, it is usually more convenient to bind RNA to the filter and to hybridize it to a radioactive DNA probe. Prior to dotting the RNA onto the filter the RNA (~ 5 μg/ml) is denatured by adding an equal volume of 34% deionized glyoxal (see *Table 19*), 20 mM sodium phosphate, pH 6.5 and incubating the mixture at 50°C for 60 min in a small sealed tube. Prepare a range of dilutions in 0.1% SDS and apply 4 μl aliquots to a dried nitrocellulose filter previously soaked in 20 × SSC. Bake the filter at 80°C for 2 h and then wash it in 20 mM Tris−HCl, pH 8.0 to 100°C to remove any residual glyoxal. Transfer the filter to a plastic bag for pre-hybridization prior to hybridizing it to the labelled DNA probe (final concentration ~ 5 μg/ml, 10^8 c.p.m./μg) using essentially the same procedure as described in *Table 19*.

6.4 Analysis of the synthesis of mitochondrial proteins

6.4.1 Introduction

The biogenesis of mitochondria is dependent on two quite separate protein synthesizing systems, namely that taking place on the cytoplasmic ribosomes which provides about 90% of the mitochondrial proteins and that taking place within the matrix of the mitochondrion which utilizes a distinct set of ribosomes and tRNAs. However, the mitochondrial translational system itself requires a large number of cytoplasmically synthesized proteins. The proteins synthesized by the two translational systems can be readily identified because mitochondrial protein synthesis is not affected by cyclohex-imide, an inhibitor of cytoplasmic protein synthesis, but it is sensitive to chloramphenicol which has no effect on cytoplasmic protein synthesis. Thus by studying protein synthesis in the presence of one of these inhibitors it is possible to study the synthesis of mitochondrial proteins by each of the systems. One area where a great deal of work has been done is in the import of proteins into mitochondria but unfortunately space limitations preclude a discussion of this and the reader should consult a more detailed text for a description of the methodology required for these studies (61).

6.4.2 *Systems used to study protein synthesis*

(i) *In vivo labelling of proteins.* Yeast will take up amino acids from the growth medium and incorporate them into proteins. However, if one uses ^{3}H-labelled or ^{14}C-labelled amino acids then metabolic randomization of the radio tracer can lead to its incorporation into a wide range of molecules. Thus the preferred amino acid is methionine but some care is necessary because this is a rare amino acid and some small proteins may not contain a methionine in their native form. In order to obtain good incorporation of methionine *in vivo* it is imperative to grow the yeast in a defined medium (see Appendix to this chapter) lacking methionine; it should be noted that as a general rule cells grow more slowly in defined medium than in a complete one such as YPglu and this may be reflected in changes in a wide variety of metabolic pathways. For pulse-labelling studies one can incubate the yeasts in a defined medium for $5 - 15$ min in the presence of 200 μCi/ml (7.5 Mbq/ml) of [^{35}S]methionine (sp. act. 800 Ci/mmol; 30TBq/mmol) at 28°C. For longer labelling periods it is easier and cheaper to label proteins by growing the yeast in the presence of [^{35}S]sulphate ($3-5$ μCi/ml; $0.1-0.2$ MBq/ml) which is taken up into the cell and incorporated primarily into methionine and cysteine. Yeast can be grown in a sulphate-free defined medium (see Appendix to this chapter) or, usually more conveniently, in a complete medium such as YPglu which has been depleted of sulphate (*Table 23*); in the latter medium cells grow at the normal rate while in defined media the doubling times tend to be much longer. In the authors' experience labelling with sulphate only works well for RC cells and so for methionine should be used for labelling *rho*⁻ cultures.

(ii) *Protein synthesis in vitro.* Isolated yeast mitochondria will synthesize proteins when incubated *in vitro* under the conditions given in *Table 24*. In this case it appears that the mitochondria are readily permeable to amino acids such as [^{3}H]leucine and there is no need to permeabilize the mitochondria by, for example, osmotic shock. It is extremely important to ensure that, as far as possible, all solutions used for *in vitro* protein synthesis are as free of bacteria as possible. All solutions should be autoclaved or where this is not possible, sterilized by Millipore filtration; failure to take these precautions may lead to contamination of the mitochondria with labelled bacterially-synthesized proteins.

The further dissection of the protein synthetic system is more problematical, due in the main to the lack of caps on the mitochondrial mRNA 5′ terminus and the variant genetic code of the mitochondrial genome. A great deal of work has been done on the

Table 23. Depletion of complete medium of inorganic sulphate (64).

1.	Prepare the required complete medium (e.g. YPglu) as described in the Appendix to the chapter but at a 2-fold concentration.
2.	Add 1 M BaCl$_2$ to give a final concentration of 40 mM and stir for 60 min.
3.	Remove the precipitate by centrifugation at 2000 g for 20 min at 5°C.
4.	Carefully decant the supernatant and make up to twice the original volume; note that media for *petites* should be buffered with 5 mM MES−NaOH, pH 6.5.
5.	Autoclave the medium at 110°C for 20 min,

Table 24. Incubation medium for incorporating amino acids into isolated mitochondria.

In a final volume of 1.0 ml	
Mannitol	600 mM
KCl	50 mM
KH$_2$PO$_4$	10 mM
MgCl$_2$	10 mM
ATP	5 mM
GTP	0.5 mM
Phosphoenolpyruvate	5 mM
Pyruvate kinase	6 units
α-Ketoglutarate	5 mM
Tris−HCl, pH 7.2	20 mM
L-leucine	20 mM
Amino acid mixture (minus leucine)	0.1 mM
BSA	1 mg
Cycloheximide	100 μg
L-4,5-[^{3}H]leucine (55.9 Ci/mmol)	80 μCi (3MBq)
(2 TBq/mmol)	

Derived from ref. 61.

Table 25. Preparation of the gel for the separation of mitochondrial proteins.

Resolving gel	*Stacking gel*
30 ml total volume	20 ml total volume
10 ml 36% acrylamide, 1% bisacrylamide	2.5 ml 1 M Tris−HCl (pH 6.8)
3.75 ml 3 M Tris−HCl (pH 8.8)	1.75 ml 40% acrylamide, 1% bisacrylamide
0.3 ml 10% SDS	5.15 ml water
	10 ml 8 M urea
10 μl TEMED	0.2 ml 10% SDS
15 ml 8 M urea	7.5 μl TEMED
0.5 ml, 5% ammonium persulphate	300 μl 5% ammonium persulphate

TEMED: *N,N,N',N*-tetramethylethylene diamine

in vitro synthesis and import of nuclearly encoded proteins and these methods are described in detail elsewhere (61).

6.4.3 *Analysis of synthesized proteins*

The usual method for the analysis of newly synthesized proteins irrespective of whether they are synthezised *in vivo* or *in vitro* is to analyse them by slab gel SDS−PAGE followed by either autoradiography or fluorography. Fluorography must be used for

³H-labelled proteins and it also greatly enhances the detection of ³⁵S- or ¹⁴C-labelled proteins. Usually one uses a discontinuous gel system with a 10 or 12% separating gel (see *Table 25*); protocols for such gels are widely distributed in the literature and another book in the Practical Approach Series (62) gives a number of suitable protocols. After electrophoresis the gels can be stained if necessary with 2.5% Coomassie Brilliant Blue in methanol:acetic acid:water (5:1:5 by vol), de-stained in the same solution minus the Coomassie Brilliant Blue and photographed. For fluorography the gels are soaked in 1 M sodium salicylate for 60 min prior to drying down onto Whatman 3MM filter paper. The dried gels are exposed to Amersham Hyperfilm or equivalent at −70°C. If a higher degree of resolution is necessary then this can be obtained by running a two-dimensional gel of the labelled proteins (63).

7. REFERENCES

1. Darley-Usmar,V.M., Rickwood,D. and Wilson,M.T. (eds) (1987) *Mitochondria − A Practical Approach.* IRL Press, Oxford.
2. Rickwood,D., Darley-Usmar,V.M. and Wilson,M.T. (1987) In *Mitochondria − A Practical Approach.* Darley-Usmar,V.M., Rickwood,D. and Wilson,M.T. (eds), IRL Press, Oxford, p. 1.
3. Lang,B.F., Ahne,F., Distler,S., Trinkl,H., Kandewitz,F. and Wolf,K. (1983) In *Mitochondria 1983, Nucleo-mitochondrial Interactions,* Schweyen,R.J., Wolf,K. and Kandewitz,F. (eds), De Gruyter, Berlin, p. 313.
4. Clark-Walker,G.D., McArthur,C.R. and Daley,D.L. (1981) *Curr. Genet., 4,* 7.
5. Dujon,D. (1983) In *Mitochondria 1983, Nucleo-mitochondrial Interactions,* Schweyen,R.J., Wolf,K. and Kandewitz,F. (eds). De Gruyter, Berlin, p.1.
6. Dujon,B. (1981) In *Molecular Biology of the Yeast Saccharomyces cerevisiae.* Strathern,J.N., Jones,E.W. and Broach,J.R. (eds), Cold Spring Harbor Laboratory Press, New York, Vol. 2, p.505.
7. De Zamaroczy,M. and Bernardi,G. (1985) *Gene, 37,* 1.
8. Stevens,B. (1981) In *Molecular Biology of the Yeast Saccharomyces cerevisiae.* Strathern,J.N., Jones,E.W. and Broach,J.R. (eds), Cold Spring Harbor Laboratory Press, New York, Vol. 2, p.471.
9. Michaelis,G., Douglass,S., Tsai,M.J. and Criddle,R. (1971) *Biochem. Genet., 5,* 487.
10. Deutsch,J., Dujon,B., Netter,P., Petrochilo,E., Slonimski,P.P., Bolotin-Fukuhara,M. and Coen,D. (1974) *Genetics, 76,* 195.
11. Putrament,A., Baranowska,H. and Prazmo,W. (1973) *Mol. Gen. Genet., 126,* 357.
12. Kocavova,V., Irmlerova,J. and Kovac,L. (1968) *Biochim. Biophys. Acta, 162,* 157.
13. Kotylak,Z. and Slonimski,P.P. (1977) In *Mitochondria 1977.* Bandlow,W., Schweyen,R.J., Wolf,K. and Kandewitz,F. (eds), De Gruyter, Berlin, p. 83.
14. Leaver,C.J. and Gray,M. (1982) *Annu. Rev. Plant Physiol., 133,* 373.
15. Wolf,K., Dujon,B. and Slonimski,P.P. (1973) *Mol. Gen. Genet., 125,* 53.
16. Blanc,H. and Dujon,B. (1986) *Proc. Natl. Acad. Sci. USA, 77,* 3942.
17. Tzagoloff,A., Akay,A. and Foury,F. (1976) *FEBS Lett., 65,* 391.
18. Kruszewska,A. and Szczesniak,B. (1980) *Genet. Res. Camb., 35,* 225.
19. Bolotin-Fukuhara,M. and Fukuhara,H. (1976) *Proc. Natl. Acad. Sci. USA, 73,* 4608.
20. Murphy,M., Choo,K.B., Mcreadie,I., Marzuki,S., Lukins,H.B., Nagley,P. and Linnane,A.W. (1980) *Arch. Biochem. Biophys., 203,* 260.
21. Alexander,N.J., Vincent,R.D., Perlman,P.S., Miller,D.S., Hanson,D.K. and Mahler,H.R. (1979) *J. Biol. Chem., 254,* 2471.
22. Coen,D., Deutsch,J., Netter,P., Petrochilo,E. and Slonimski,P.P. (1970) In *Symposium of the Society for Experimental Biology.* Miller,P.L. (ed.), Cambridge University Press, Vol. 24, p. 449.
23. Avner,P.R., Coen,D., Dujon,B. and Slonimski,P.P. (1973) *Mol. Gen. Genet., 125,* 9.
24. Dujon,B., Bolotin-Fukuhara,M., Coen,D., Deutsch,J., Netter,P., Slonimski,P.P. and Weill,L. (1976) *Mol. Gen. Genet., 143,* 131.
25. Conde,J. and Fink,G.R. (1976) *Proc. Natl. Acad. Sci. USA, 73,* 3651.
26. Slonimski,P.P. and Tzagoloff,A. (1976) *Eur. J. Biochem., 61,* 27.
27. Fangman,W.L. and Dujon,B. (1984) *Proc. Natl. Acad. Sci. USA, 81,* 7156.
28. Parry,E.M. and Parry,J.M. (1984) In *Mutagenicity Testing − A Practical Approach.* Venitt,S.M. and Parry,J.,M. (eds), IRL Press, Oxford, p. 119.
29. Rickwood,D. and Hayes,A. (1984) *Prep. Biochem., 14,* 163.

30. Lang,B., Burger,G., Doxiadis,I., Thomas,D.Y., Bandlow,W. and Kandewitz,F. (1977) *Anal. Biochem.*, **77**, 110.
31. Schatz,G. and Kovacs,L. (1974) In *Methods in Enzymology*. Fleischer,S. and Packer,L. (eds), Academic Press, New York, Vol. 31, p. 627.
32. Bandlow,W. and Bauer,P. (1975) In *Methods in Cell Biology*. Prescott,D.M. (ed.), Academic Press, New York, Vol. XII, p. 311.
33. Velours,J., Guerin,B. and Duvert,M. (1977) *Arch. Biochem. Biophys.*, **182**, 295.
34. Nicholls,D.G. (1982) *Bioenergetics.* Academic Press, New York.
35. Somlo,M. and Krupa,M. (1974) *Eur. J. Biochem.*, **42**, 429.
36. Guerin,B., Labbe,P. and Somlo,M. (1979) In *Methods in Enzymology*. Fleischer,S. and Packer,L. (eds), Academic Press, New York, Vol. 55, p. 149.
37. Nedergaard,J. and Cannon,B. (1979) In *Methods in Enzymology*. Fleischer,S. and Packer,L. (eds), Academic Press, New York, Vol. 55, p. 3.
38. Sen,K. and Beattie,D.S. (1985) *Arch. Biochem. Biophys.*, **242**, 393.
39. Capaldi,R.A. (1982) *Biochim. Biophys. Acta*, **694**, 291.
40. Hatefi,Y. (1985) *Annu. Rev. Biochem.*, **54**, 1015.
41. Wickner,R.B. (1975) In *Methods in Cell Biology*. Prescott,D.M. (ed.), Academic Press, New York, Vol. 11, p. 295.
42. Fukuhara,H. (1969) *Eur. J. Biochem.*, **11**, 135.
43. Wintersberger,U. and Wintersberger,E. (1970) *Eur. J. Biochem.*, **13**, 11.
44. Wintersberger,U. and Wintersberger,E. (1970) *Eur. J. Biochem.*, **13**, 20.
45. Wintersberger,U. and Blutsch,H. (1976) *Eur. J. Biochem.*, **68**, 199.
46. Birnie,G.D. (1978) In *Centrifugal Separations in Molecular and Cell Biology*. Birnie,G.D. and Rickwood,D. (eds), Butterworths, London, p. 169.
47. Docherty,R.C. and Rickwood,D. (1986) *Eur. J. Biochem.*, **156**, 185.
48. Winkley,C.S., Keller,M.J. and Jachning,J.A. (1985) *J. Biol. Chem.*, **260**, 14214.
49. Levens,D. and Howley,P.M. (1986) *Mol. Cell. Biol.*, **5**, 2307.
50. Kelly,J.L. and Lehman,R.I. (1986) *J. Biol. Chem.*, **261**, 10340.
51. Schinkel,A.H., Groot-Koerkamp,M.J.A., Van der Horst,G.T.J., Touw,E.P.W., Osinga,K.A., Van der Bliek,A.M., Veeneman,G.H., Van Boom,J.H. and Tabak,H.F. (1986) *EMBO J.*, **5**, 1041.
52. Levens,D., Ticho,B., Ackerman,E. and Rabinowitz,M. (1981) *J. Biol. Chem.*, **256**, 5226.
53. Monroy,G., Spencer,E. and Hurwitz,J. (1978) *J. Biol. Chem.*, **253**, 4481.
54. Bhat,K.S., Kantharaj,G.R. and Avadhani,A.G. (1984) *Biochemistry*, **23**, 1695.
55. Williams,J.G. and Mason,P.J. (1985) In *Nucleic Acid Hybridisation – A Practical Approach*. Hames,B.D. and Higgins,S.J. (eds), IRL Press, Oxford, p. 139.
56. Hames,B.D. and Higgins,S.J. (eds) (1985) *Nucleic Acid Hybridisation – A Practical Approach*, IRL Press, Oxford.
57. Yoza,B.K. and Bogenhagen,D.F.(1984) *J. Biol. Chem.*, **259**, 3909.
58. Chang,D.D. and Clayton,D.A. (1985) *Proc. Natl. Acad. Sci. USA*, **82**, 351.
59. Rickwood,D. and Hames,B.D., eds (1982) *Gel Electrophoresis of Nucleic Acids – A Practical Approach*, IRL Press, Oxford.
60. Nobrega,F.G., Dieckman,C.L. and Tzagoloff,A. (1983) *Anal. Biochem.*, **131**, 141.
61. Beattie,D.S. and Sen,K. (1987) In *Mitochondria – A Practical Approach*. Darley-Usmar,V., Rickwood,D. and Wilson,M. (eds), IRL Press, Oxford, p. 283.
62. Hames,B.D. and Rickwood,D. (eds) (1981) *Gel Electrophoresis of Proteins – A Practical Approach*. IRL Press, Oxford.
63. Sinclair,J.H. and Rickwood,D. (1981) In *Gel Electrophoresis of Proteins – A Practical Approach*. Hames,B.D. and Rickwood,D. (eds), IRL Press, Oxford, p. 89.

APPENDIX

Media for the Isolation and Characterization of Yeast Mitochondrial Mutants

1. COMPOSITION AND PREPARATION OF MEDIA

Only media directly relevant to mitochondrial genetics are described. Other yeast media are described in Appendix I. Quotations of manufacturers or commercial suppliers of chemicals do not imply that they are superior to others but only that they have been successfully used in the procedures described.

1.1 Complete media without antibiotics

All complete media listed in this section contain 10 g/l of yeast extract (Difco) and 10 g/l of bactopeptone (Difco). They only differ in the carbon source and the presence or absence of sodium potassium phosphate buffer.

YPglu	20 g/l of D-glucose
N0	Same as YPglu but buffered at pH 6.24 using 50 mM final concentration of sodium potassium phosphate buffer
YP10	100 g/l of D-glucose
YPdif	1 g/l of D-glucose plus 20 ml/l of glycerol
YPgal	20 g/l of D-galactose
YPgly	20 ml/l of glycerol
N3	Same as YPgly but buffered at pH 6.24 using 50 mM final concentration of sodium potassium phosphate buffer
N1	20 ml/l of ethanol (add after autoclaving). Buffer at pH 6.24 using 50 mM final concentration of sodium potassium phosphate buffer

All ingredients are added to their final concentrations prior to autoclaving (except for N1 medium). For solid media add 22 g/l of bacto agar (Difco) prior to autoclaving. Note that some agar brands may contain trace amounts of fermentable carbon sources and should not be used for YPgly, YPdif, N1 or N3 plates. Allow agar-containing media to cool down to about 60°C before pouring plates. When the agar is set, allow all the plates to dry for about 2−3 days before use. Alternatively, dry the open plates in a 50°C incubator for 10−15 min. Liquid or solid media can be stored for months in the refrigerator to prevent desiccation. Note that rich solid media containing glucose in Petri dishes are highly susceptible to bacterial contamination. Adding an anti-bacterial antibiotic such as sodium benzyl penicillin (100 000 units/l) after autoclaving and before pouring plates greatly diminishes the risk of contamination without affecting yeasts. If necessary, penicillin can be added to other media as well.

1.2 Complete media containing antibiotics or inhibitory drugs

The following antibiotics or drugs have been used for the isolation of mitochondrial mutants:

specific inhibitors of the mitochondrial ribosome: chloramphenicol, erythromycin, spiromycin, paromomycin;

Table 1. Antibiotics and drugs containing media for *Saccharomyces cerevisiae*.

This table gives the final concentrations in media (N3 or N1) of the antibiotics or drugs and of their carriers.

(i) Antibiotics added as dry powders

N3C	4 g/l of chloramphenicol (Roussel UCLAF, France) in N3 (ref. 1)
N3E	5 g/l of erythromycin base (Roussel UCLAF, France) in N3 (ref. 1)
N3S	5 g/l of spiramycin base (Rhône-Poulenc, France) in N3 (ref. 2)
N3P	2 g/l of paromomycin sulphate (Parke Davies, California) in N3 all buffered at pH 6.5 using 50 mM sodium potassium phosphate buffer (ref. 3)

(ii) Antibiotics or drugs added from stock solutions

N3O	3	mg/l of oligomycin + 0.5% (v/v) of methanol in N3 (ref. 4)
N3V	1	mg/l of venturicidin + 0.5% (v/v) of methanol in N3 (ref. 5)
N3oss	2	mg/l of ossamycin + 0.5% (v/v) of methanol in N3 (ref. 6)
N1A	0.1	mg/l of antimycin A + 0.1% (v/v) of ethanol in N1 (ref. 7)
N1F	5	mg/l of funiculosin + 1% (v/v) of ethanol in N1 (ref. 7)
N3D	35	mg/l of diuron + 2% (v/v) of acetone in N3 (ref. 8)
N3M	0.3	mg/l of mucidin + 0.1% (v/v) of ethanol in N3 (ref. 9)
N3myx	2	mg/l of myxothiazol + 2% (v/v) of ethanol in N3 (ref. 10)

These media correspond to the standard concentrations of antibiotics or drugs used for growth of resistant strains and for distinguishing between resistant and sensitive colonies by replica plating. If other concentrations are used for particular purposes keep the final concentration of carrier constant.

(iii) Preparation of stock solutions

(a) *Oligomycin*: dissolve oligomycin (mix of oligomycins A, B and C; Sigma, Missouri) in methanol at a final concentration of 600 µg/ml. Add 5 ml of stock solution/l of N3 medium.

(b) *Venturicidin*: dissolve venturicidin A (BDH Biochemicals, UK) in methanol at a final concentration of 0.2 mg/ml. Add 5 ml of solution/l of N3 medium.

(c) *Ossamycin*: dissolve ossamycin (Bristol-Myers Co.) in methanol to a final concentration of 0.4 mg/ml. Add 5 ml of solution/litre of N3 medium.

(d) *Antimycin*: dissolve antimycin A (Serva, FRG or Boehringer, FRG) in ethanol at a final concentration of 100 µg/ml. Add 1 ml of stock solution/l of N1 medium.

(e) *Funiculosin*: dissolve funiculosin (Sandoz A.G., Switzerland) in ethanol at a final concentration of 0.5 mg/ml. Add 10 ml/l of N1 medium.

(f) *Diuron*: dissolve 3-(3,4-dichlorophenyl)-1,1-dimethylurea (E.I.duPont de Nemours and Co.) in acetone at a final concentration of 7 mg/ml. Take 5 ml of stock solution/l of N3 medium. Add 15 ml of acetone, mix and pour in medium.

(g) *Mucidin*: dissolve the 'Mucidermin' spray solution (Spofa, Prague, Czechoslovakia) in ethanol to a final concentration of 0.3 mg/ml of mucidin. Add 1 ml of solution for 1 litre of N3 medium.

(h) *Myxothiazol*: dissolve myxothiazol (Gesellschaft für Biotechnologische Forschung, Braunschweig, FRG) in ethanol to a final concentration of 0.1 mg/ml. Add 20 ml of solution/l of N3 medium.

specific inhibitors of the ATP synthase complex: oligomycin (or rutamycin), venturicidin, ossamycin;
specific inhibitors of the cytochrome bc_1 complex: antimycin A, funiculosin, diuron, mucidin, myxothiazol.

Always use buffered media (N3 or N1) since many antibiotics or drugs are only effective on yeast mitochondria within a narrow pH range. Final concentrations are given in *Table 1*.

Table 2. Supplements to minimal media corresponding to the most commonly used auxotrophic markers in yeast mitochondrial genetics.

1. Prepare stock solutions as below.
2. Autoclave.
3. Store in the refrigerator or freeze.
4. Add 10 ml of stock solution/l of minimal medium after autoclaving when temperature is below 60°C.

	Stock solutions	*Auxotrophic mutations commonly used in mitochondrial genetics*
adenine:	2 mg/ml in 50 mM HCl	*ade1, ade2*[a]
uracil:	2 mg/ml in 0.5% (w/v) NaHCO$_3$	*ura1, ura3*
L-arginine	1 mg/ml in water	*arg4*
L-histidine:	1 mg/ml in water	*his1, his4C*
L-isoleucine:	6 mg/ml in water	*ilv5*
L-leucine:	6 mg/ml in water	*leu1, leu2*
L-lysine − HCl:	1 mg/ml in water	*lys2*
L-methionine:	1 mg/ml in water	*met13*
L-phenylalanine:	2 mg/ml in water	*phe*
L-tryptophan:	2 mg/ml in water	*trp1, trp2*
L-tyrosine:	1 mg/ml in 50 mM HCl	*tyr6*
L-valine:	2 mg/ml in water	*ilv5*

[a]*ade1* and *ade2* can be particularly useful for mitochondrial genetics as *rho*[+] cells turn red while *rho*[−] cells remain white (*mit*[−] mutants are generally pink).

(i) *Media in which antibiotics are added as dry powders.* This is the case for N3C, N3E, N3S and N3P media. Prepare N3 medium, autoclave and let the medium cool down to about 60°C prior to adding antibiotics. Weigh the dry antibiotic powder and add to the medium directly to its final concentration. Note that these antibiotics are used at final concentrations close to their limit of solubility. Stir the medium with a magnetic stirrer until the powder is completely dissolved (it is convenient to place a magnetic stirring bar in the medium prior to autoclaving). Dissolution is quick and clumping is prevented if the powder is absolutely dry. It is therefore advisable to store the antibiotic powders in a dry container at room temperature; they are stable for years under such conditions.

(ii) *Media in which antibiotics or drugs are added from stock solutions.* This is the case for all other media listed in *Table 1* which contain antibiotics or drugs not soluble in media without any carrier. Prepare stock solutions as indicated (stock solutions can be stored at −20°C if properly sealed). Prepare N3 or N1 medium, autoclave and allow medium to cool down to below 60°C prior to adding antibiotics or drugs. Add the stock solution as indicated in *Table 1* and stir immediately.

1.3 Minimal media

In principle, minimal media can be prepared with a variety of carbon sources as for the complete media. In practice, however, yeast strains grow very slowly on minimal media containing a non-fermentable carbon source as the sole carbon source. For this reason, only the glucose medium has been of general use for mitochondrial genetics.

For solid media, add 20 g/l of bacto agar (Difco). Note that some agar brands may contain trace amounts of metabolites and should not be used for minimal media.

(i) *Standard minimal media*

W0   6.7 g/l of Yeast−Nitrogen Base without amino acids (Difco) and 20 g/l of D-glucose

W10   Same as W0 but containing 100 g/l of D-glucose

Both ingredients are added to their final concentrations prior to autoclaving.

(ii) *Minimal media for specific applications (11)*

G0   See *Table 3* for composition

Ggal   Same as G0 but containing 20 g/l of D-galactose

All minimal media can be supplemented as described in *Table 2* for the most commonly used auxotrophic markers.

2. USE OF MEDIA AND PRECAUTIONS IN GROWING MITOCHONDRIAL MUTANTS

2.1. **YPglu**

This is the most common medium providing maximum growth rate and highest final yield of cells. Wild-type cells as well as all types of mitochondrial mutants can grow in it. However, growth rates always favour RC cells over RD ones, hence leading to an under-representation of the latter type; this may become a problem if quantitative measurements are needed. The relative proportion of RD cells is even more significantly reduced when culture is extended until stationary phase since growth on this medium shows a biphasic pattern.

During the first phase of growth, when metabolizing glucose, yeasts produce ethanol but do not use it as long as the glucose concentration is high enough. During this first phase of growth both RC and RD cells can grow. When glucose comes near to exhaustion, a second phase of growth (usually 2−3 additional cell doublings) is possible only for the RC cells, using the previously produced ethanol as the carbon source.

The major concerns in growing yeast cells in this medium are therefore as follows. Firstly, for *rho*+ strains: spontaneous *rho*− mutants (which are frequent, see below) are not eliminated and tend to accumulate in the culture. Secondly, for *mit*− mutants: spontaneous *mit*+ revertants (when they can arise) will be favoured (because they are RC) and will rapidly outgrow the *mit*− population which grows more slowly. This medium is useful both as a liquid medium and for plates.

2.2 **N0**

This is similar to YPglu. It is useful only when pH becomes important, such as in ethidium bromide mutagenesis or inhibition of mitochondria by antibiotics or drugs.

2.3 **YP10**

Since the glucose concentration of this medium is extremely high, the stationary phase is reached before glucose exhaustion. It follows that a biphasic growth pattern is not

Table 3. Preparation of G0 minimal medium with or without sulphate.

A. Prepare separately the following stock solutions

10 × Salts: weigh the following, dissolve in 1 litre final volume with water and autoclave.

	With sulphate	*Without sulphate*
$(NH_4)H_2PO_4$ (mol. wt 115.03)	60 g	60 g
$MgSO_4 . 7H_2O$ (mol. wt 246.48)	5 g	—
$MgCl_2 . 6H_2O$ (mol. wt 203.31)	—	4.1 g
$(NH_4)_2SO_4$ (mol. wt 132.14)	20 g	—
NH_4Cl (mol. wt 53.50)	—	8.1 g
KH_2PO_4 (mol. wt 136.09)	10 g	10 g
NaCl (mol. wt 58.45)	1 g	1 g
$CaCl_2$ (mol. wt 110.99)	1 g	1 g

Trace elements: weigh the following, dissolve in 1 litre final volume with water and autoclave.

	With sulphate	*Without sulphate*
H_3BO_3 (mol. wt 61.83)	500 mg	500 mg
$CuSO_4 . 5H_2O$ (mol. wt 249.68)	60 mg	—
$CuCl_2 . 2H_2O$ (mol. wt 170.48)	—	40 mg
KI (mol. wt 166.01)	100 mg	100 mg
$MnSO_4 . H_2O$ (mol. wt 169.00)	400 mg	—
$MnCl_2 . 4H_2O$ (mol. wt 197.91)	—	470 mg
$Na_2MoO_4 . 2H_2O$ (mol. wt 241.95)	200 mg	200 mg
$ZnSO_4 . 7H_2O$ (mol. wt 287.54)	400 mg	—
$ZnCl_2$ (mol. wt 136.29)	—	190 mg

Ferric chloride: prepare a stock solution at 0.2 mg/ml of $FeCl_3$ (mol. wt 162.21) in water and filter sterilize

Vitamins and co-factors:

1. Weigh the following and dissolve in 99 ml final volume with water:
   calcium pantothenate (mol. wt 476.53)   40 mg
   thiamine hydrochloride (vit.B1) (mol. wt 337.28)   40 mg
   pyridoxine hydrochloride (vit.B6) (mol. wt 205.64)   40 mg
   nicotinic acid (niacin) (mol. wt 123.11)   10 mg
2. Add 1 ml of a 0.4 mg/ml stock solution of biotin (mol. wt 244.31) in 50 mM $NaHCO_3$.
3. Filter sterilize.

Inositol: dissolve 400 mg of inositol (mol. wt 180.16) into 100 ml of water and filter sterilize

B. Preparation of final medium

1. Measure 100 ml of '10 × Salts' stock solution with or without sulphate as appropriate.
2. Add 20 g of D-glucose.
3. Dissolve and complete to 950 ml with water.
4. Autoclave.
5. Allow the medium to cool down to 60°C and then add:
   5 ml of vitamins and co-factors stock solution
   5 ml of inositol stock solution
   1 ml of trace elements stock solution with or without sulphate as appropriate.
   1 ml of ferric chloride stock solution
   10 ml of each stock solution of required supplements (*Table 3*) and/or sterile water to complete to 1 litre final volume.
6. Mix.

observed and that RC cells are not so strongly favoured as in YPglu medium. For this reason YP10 is to be preferred to YPglu when revertible *mit⁻* mutants are grown or when a quantitative estimate of the frequency of *rho⁻* mutants is needed. It is useful both as a liquid medium and for plates. It is recommended that liquid medium cultures are not shaken to reduce further the selective advantage of RC cells.

2.4 YPgly, N3 and N1

These media allow the growth of RC cells only. Growth is slower than on YPglu or YP10 media but all spontaneous *rho⁻* mutants are immediately stopped. These media are therefore recommended if pure *rho⁺* cultures are needed. They are useful both as liquid media and for plates. Ensure good aeration of liquid cultures by vigorous agitation.

2.5 Antibiotic-containing media

These media allow the growth of RC antibiotic-resistant mutants only. These media are therefore recommended when a pure culture of such mutants is needed (e.g. as a pre-culture to start an experiment). They are useful both as liquid media and for plates.

2.6 YPdif

This is a differential medium which increases differences of growth between RC and RD cells. The limited amount of glucose is rapidly exhausted while glycerol permits the growth of RC cells only. It is useful only for plates to distinguish between *rho⁺* and *rho⁻* colonies. Incubation should be increased to about 5−6 days at 28°C to maximize differential growth. *Rho⁺* colonies are large, thick and yellowish, while *rho⁻* colonies are small, flat and white (hence the frequent designation of *rho⁻* mutants as *petite* mutants). Extended incubation of such plates for about 10 days may result, under specific genetic conditions, in the formation of a cluster of RC cells in an RD colony. These are easily visible as abscessed colonies.

2.7 YPgal

This medium is used to eliminate the effect of glucose repression. Both RC and RD cells can grow in it. This is useful as a liquid medium and for plates. Some strains may not grow on this medium if they are from glucose-grown cultures. In this case grow the cells first in YPgal containing 0.1% glucose prior to transferring them to YPgal.

2.8 W0 and W10

These are the minimal media used to select prototrophic diploids formed by crosses of auxotrophic parents. They are useful as liquid media and for plates. Both RC and RD cells grow in these media. However, the growth of RD cells is significantly slower than that of RC cells, resulting in a very strong counter-selection of RD cells. If quantitative estimates of the relative frequencies of RC and RD cells are needed (e.g. in crosses involving *rho⁻* or *mit⁻* mutants), use only W10 liquid medium without shaking and avoid prolonged incubation.

2.9 G0

This has the same properties as W0 but is much more laborious to prepare! Use this medium only for specific requirements such as minimal medium without sulphate (see *Table 3*).

3. REFERENCES

1. Coen,D., Deutsch,J., Netter,E., Petrochilo,E. and Slonimski,P.P. (1970) In *Symposium for the Society for Experimental Biology,* Miller,P.L. (ed.), Cambridge University Press, Vol. 24, p. 449.
2. Netter,P., Petrochilo,E., Slonimski,P.P., Bolotin-Fukuhara,M., Coen,D., Deutsch,J. and Dujon,B. (1974) *Genetics,* **78**, 1063.
3. Wolf,K., Dujon,B. and Slonimski,P.P. (1973) *Mol. Gen. Genet.,* **125**, 53.
4. Avner,P.R., Coen,D., Coen,B. and Slonimski,P.P. (1973) *Mol. Gen. Genet.,* **125**, 9.
5. Lancashire,W.E. and Griffiths,D.E. (1975) *Eur. J. Biochem.,* **51**, 403.
6. Lancashire,W.E. and Mattoon,J.R. (1979) *Mol. Gen. Genet.,* **176**, 255.
7. Pratje,E. and Michaelis,G. (1977) *Mol. Gen. Genet.,* **152**, 167.
8. Colson,A.M., Luu,T.V., Convent,B., Briquet,M. and Goffeau,A. (1977) *Eur. J. Biochem.,* **74**, 521.
9. Subik,J., Kovacova,V. and Takacsova,G. (1977) *Eur. J. Biochem.,* **73**, 275.
10. Thierbach,G. and Michaelis,G. (1982) *Mol. Gen. Genet.,* **186**, 501.
11. Galzy,P. and Slonimski,P.P. (1957) *C.R. Acad. Sci. Paris,* **245**, 2423.

CHAPTER 10

Membranes and lipids of yeasts

ANTHONY H.ROSE AND FELICITY J.VEAZEY

1. INTRODUCTION

The popularity of yeasts in cell-biology studies has grown phenomenally in the past decade and they are now often deemed to be model eukaryotes (1). Much of this interest has stemmed from the invention of recombinant DNA techniques and the facility with which they can be applied to yeasts. At the same time, the tempo of research into vectorial metabolism in eukaryotes continues unabated, and has focused increasing interest on yeast membranes, their isolation and composition. The present chapter describes methods and techniques which are currently favoured in the isolation of, and analysis of lipids in, yeast membranes. Although yeasts, as eukaryotes, possess nuclear, vacuolar, plasma and, when grown aerobically, mitochondrial membranes, the present account is confined to plasma and vacuolar membranes and lipid vesicles. Despite the popularity of yeasts in studies in cell biology, the vast majority of reports on these microorganisms are confined to a handful of species—in particular *Saccharomyces cerevisiae*—a tiny representation of some 600 yeast species currently recognized (2).

2. ISOLATION AND CHARACTERIZATION OF YEAST MEMBRANES

Three classes of method have been used to isolate plasma membranes from yeasts. Firstly, there is mechanical disruption followed by differential and density-gradient (3) or zonal centrifugation to isolate membranes from the homogenate. A second method also employs mechanical disruption of cells, but one less severe and which leads to formation of pieces of plasma membrane with wall fragments attached; these structures are then enzymically treated to remove the wall fragments (4). A third method involves preparing spheroplasts from a population of yeast cells, lysing the spheroplasts and isolating plasma membranes, vacuoles and lipid vesicles by appropriate centrifugation procedures (5). Although not exclusively used, this last method is without doubt currently the method of choice and it is the only one which is described in this chapter. It is worth remembering, however, that the second method could prove very useful in any attempt to study whether and to what extent overlying attached wall material regulates the composition of the underlying plasma membrane. Vacuoles are isolated from spheroplasts after gentle lysis. Their membranes are, in turn, obtained by lysis of the organelles.

2.1 Preparation of spheroplasts

This subject has been reviewed in two articles which have been published in the past decade or so (6,7). The term 'spheroplast' is preferred to 'protoplast' since the structures obtained from yeasts are rarely, if ever, totally free from wall material.

2.1.1 *Wall lytic enzymes*

The structural component of the walls in probably all yeasts is β-linked glucan molecules, so that β-glucanases are required to convert yeast cells into spheroplasts. A large number of such enzymes have been used for this purpose with varying success. Currently, there are some five commercially available enzymes which have consistently proved successful in preparing yeast spheroplasts.

The longest established of these preparations is a preparation from the alimentary canal of the Roman snail, *Helix pomatia*. Some 30 different enzyme activities have been detected in snail-gut juice (8). Snail-gut juice is marketed under two brand names: Helicase is retailed by Reactifs Biologique Française, Villeneuve-la-Garenne, France and Glusulase by Endo Laboratories Inc., Garden City, NY, USA. Both preparations are listed in the catalogue of the Sigma Chemical Co. It has been claimed that Glusulase is about four times more active than Helicase, which was shown to contain phospholipase activity (9). Both Helicase and Glusulase have been used in recent publications reporting isolation of yeast plasma membranes (10,11). A third source of β-glucanase has steadily gained acceptance in recent years, not least because it contains many fewer contaminating enzymes than Helicase or Glusulase. This is Zymolyase, a preparation obtained from culture filtrates of *Arthrobacter luteus* (12), and manufactured by the Kirin Brewery Co. Ltd, Takasaki, Gumma Prefecture, Japan. It can be obtained directly from Kirin, or from agencies in various countries (e.g. ICN Biochemicals Ltd, Castle Street, High Wycombe, Bucks, UK). Zymolyase is now the most widely used β-glucanase preparation for preparing yeast spheroplasts. Since it was first placed on the market in 1974, several new and increasingly active preparations have been developed by Kirin. Currently, the best preparation available is Zymolyase 100T (100 000 units/g), one which is partially purified by affinity chromatography; one unit of activity is defined by Kirin as that amount of enzyme which lyses 3 mg dry weight of stationary-phase *Sacch. cerevisiae* in 10 ml. Kirin also retail the proportionately less active (one-fifth) Zymolyase 20T. Zymolyase contains, in addition to β(1−3)-glucanase activity, some protease activity and it has been claimed that both are necessary for cell-wall lysis (13). Other commercially available β-glucanase preparations which still have their adherents for converting yeast cells into spheroplasts are Lytic Enzyme L1, obtained from a *Cytophaga* sp. and available from BDH Chemicals Ltd, Poole, Dorset, UK and Lyticase, a preparation produced from cultures of *Oerskovia xanthineolytica* (14).

2.1.2 *Methods for spheroplast formation*

The procedures described in this section of the chapter are those used to obtain spheroplasts from the small number of yeasts that have been studied in this context. Researchers turning to less commonly studied yeast species should experiment by altering reagent concentrations and times of incubation. Spheroplasts are most easily obtained from early- to mid-exponential phase cultures. Where the need arises to obtain membranes from cells from stationary-phase cultures, imaginative extension of the described protocols is required.

Some strains of yeast species are more easily converted into spheroplasts than others. Experience from our laboratory is that *Sacch. cerevisiae* NCYC 366 is among the easiest of strains of this species to convert into spheroplasts. With more intractable strains, pre-treating the cell populations with various reagents can pre-dispose the cells to action of the lytic enzymes. Two such types of reagent have been used. The first are thiol compounds, principally cysteine, β-mercaptoethanol, β-mercaptoethylamine, thioglycollate and dithiothreitol (6). After harvesting, wash the cells twice with water and suspend them in buffered osmoticum (see below) containing the thiol reagent at 1 mM and incubate with shaking at $25-30\,°C$ for $10-30$ min. The thiol reagents probably act by breaking disulphide bonds in wall proteins, thereby giving the β-glucanase easier access to the wall β-glucans. Sodium metabisulphite, at 0.1 mM, has also been used for this purpose. The second group of pre-treatment reagents that have been employed are proteolytic enzymes, usually pronase at 1.0 mg/ml. Again, they probably act by allowing the β-glucanase easier access to the wall β-glucans.

Spheroplasts are prepared by incubating yeast cells in buffered osmoticum containing the β-glucanase preparation and possibly other supplements. Non-metabolizable sugars and salts are used to stabilize the spheroplasts osmotically. While mannitol and raffinose were widely used some years ago, their place has been very largely taken by sorbitol. At the concentration used, around 1.0 M, mannitol tends to crystallize out of solution and for that reason now hardly ever finds favour. Salts, especially KCl and $MgSO_4$, were also once used, but are less so nowadays. Sorbitol is used usually in the range $1.0-1.2$ M. While this concentration range may be suitable for some yeasts, experience in our laboratory has been that other concentrations, usually approaching 1.5 M, should be examined when there is any evidence of spheroplast instability. A variety of buffers have been used. Most commonly advised in recent publications, when using snail-gut juice or Zymolyase, is a Tris$-$HCl buffer, in the $20-25$ mM concentration range for Tris. The pH value to which the Tris buffer is adjusted depends on the pH optimum for the β-glucanase preparation selected for use. The concentration of β-glucanase preparation to be included in the suspension is often recommended by the supplier. The Kirin Brewery's recommended dosage for Zymolyase 100T has already been indicated; with the less potent Zymolyase 20T, the dosage is proportionately increased. Helicase has been used at a concentration of approximately 1 mg/mg dry weight of cells. β-Glucanase preparations used to prepare yeast spheroplasts are often expensive. Experience from our laboratory is that the recommended concentration can often be excessive. We suggest that, for reasons of economy, examination of a range of concentrations below that recommended is advisable.

Various other supplements are often used in the buffered osmoticum in addition to yeast cells and β-glucanase preparation. Many workers prefer to include thiol compounds in the osmoticum rather than in a pre-treatment suspension. They are used in the osmoticum at the concentrations already referred to. Our colleague Charles Cartwright informed us that inclusion of sodium metabisulphite in the osmoticum, although very advantageous when using stationary-phase cells, should be viewed with caution. He found that this compound can inhibit the action of Zymolyase 100T when used with *Sacch. cerevisiae* NCYC 431. Another supplement which some workers include in the osmoticum is EDTA. During enzymic removal of the yeast cell wall, phosphomannan, a wall polymer which in the past has been referred to as yeast gum, is released by the action of β-glucanases. In the hands of some researchers, this apparently causes clumping of spheroplasts, which can be avoided by inclusion of EDTA in the osmoticum at a concentration of around 10 mM. However, EDTA should be included only when there is clear evidence of clump formation by spheroplasts in the suspension.

To obtain spheroplasts, incubate the buffered cell suspension, containing the β-glucanase preparation and possibly other supplements, at or around 30°C with shaking. The time-course of spheroplast formation is followed in either or both of two ways. The proportion of cells converted into spheroplasts, in a sample taken from the incubation suspension, can be assessed, albeit approximately, using the light microscope. With oval yeasts, conversion to spherical spheroplasts is easily assessed. With cells that are near-spherical, e.g. *Sacch. cerevisiae* NCYC 366, formation of spherical spheroplasts is harder to judge. Fortunately, spheroplasts can be identified easily using the phase-contrast microscope by their birefringence which can be quickly established by gentle focusing up and down of the microscope. Cells do not exhibit such birefringence. Also, adding a small amount of water to the slide preparation causes osmotically sensitive spheroplasts to burst. Having decided that spheroplasts exist in the suspension using this method, it does not necessarily follow that wall removal is complete. An alternative method of monitoring spheroplast formation is to dilute heavily portions of the suspension into buffered sorbitol and water, shaking the diluted suspension occasionally for 5 min and then measuring the optical density at 600 nm. When diluted into water, spheroplasts lyse and the optical density of the suspension falls. However, spheroplasts diluted into buffered sorbitol do not lyse, so that the optical density of these suspensions shows little change. When the optical density of the water suspension ceases to decrease, the suspension is incubated for a further 45 min to ensure that spheroplast formation is complete. *Figure 1* gives details of this method for following spheroplast formation in a strain of *Sacch. cerevisiae*.

Spheroplasts are usually isolated from the suspension by centrifugation and then washed two or three times in buffered sorbitol. Spheroplasts need to be handled with care, especially during resuspension in buffered sorbitol. Aldermann and Höfer (10) recommended an alternative method for separating spheroplasts from the suspension. This involves centrifugation on a discontinuous sucrose density gradient at 27 500 g for 30 min. The gradient consists of 3 ml 65%, 7 ml 60%, 15 ml 40% and 10 ml 20% (w/v) sucrose solutions.

2.2 Treatments to spheroplasts

Before they are lysed and plasma membranes isolated from the lysate, it is frequently desirable to treat spheroplasts in various ways, either in order to stabilize the plasma membrane during its separation, or in order that the plasma membrane can be unequivocally characterized after its isolation.

2.2.1 Concanavalin A

This plant lectin combines with specific carbohydrate residues in microbial wall poly-saccharides and, when reacted with spheroplasts, serves to stabilize the plasma membrane in these organelles when the spheroplasts are lysed. It was first used by Scarborough (15) for isolation of plasma membranes from *Neurospora crassa* and was subsequently applied in the isolation of plasma membranes from *Sacch. cerevisiae* (16). To use a typical experimental procedure (11), supplement a suspension of yeast spheroplasts containing 2×10^8 cell equivalents/ml with 0.25 mg concanavalin

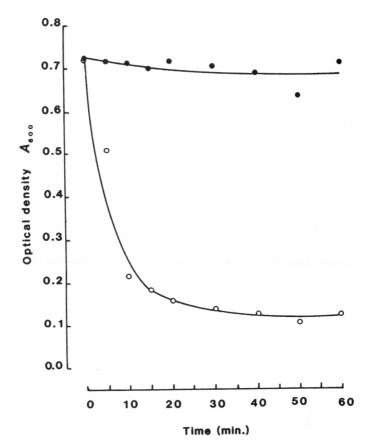

Figure 1. Time-course of spheroplast formation by *Sacch. cerevisiae* NCYC 366. In this experiment 100 mg dry weight organisms were suspended in 8 ml buffered sorbitol (20 mM Tris containing 1.2 M sorbitol and 10 mM $MgCl_2$, adjusted to pH 7.2 with HCl). Zymolyase 100T (0.5 g) was dissolved in 2 ml buffered sorbitol. After holding at 30°C for 5 min, the suspension and Zymolyase solution were mixed and incubated at 30°C with reciprocal shaking (120 r.p.m.). To follow spheroplast formation, portions (0.1 ml) of suspension were removed at the times indicated and diluted into 2.9 ml water (○) or buffered 1.2 M sorbitol(●). The diluted suspensions were incubated for 5 min at room temperature with occasional shaking and their optical density then measured at 600 nm (1 cm light path). Spheroplast formation was taken to be complete 45 min after the optical density measurements of water-diluted portions ceased to decline.

A (Con A)/ml and incubate the suspension for 10 min at room temperature. This causes the spheroplasts to aggregate. They are then washed and lysed. When lysates are fractionated on gradients, plasma membranes that have been treated with Con A band more tightly than those that have not been so treated.

2.2.2 Cationic microbeads

A recently introduced treatment of spheroplasts which greatly aids in separation of plasma membranes from spheroplast lysates involves treatment with cationic silica microbeads. Silica microbeads, 50 nm in diameter, are coated with a cationic polymer which, when mixed with spheroplasts, adheres to the negatively charged spheroplast surface. The high density (2.5 g/cm) of the microbeads enables rapid separation of coated plasma-membrane sheets from lysates, thereby speeding up the separation process and minimizing enzymic degradation of plasma-membrane components during the separation process. The cationic bead technique was invented by Jacobson and his colleagues. It has been applied in isolation of plasma membranes from several different types of cell including the yeasts *Sacch. cerevisiae* (17,18). and *Metschnikowia reukaufii* (19). To date, cationic microbeads are not available commercially, and all users appear to have been generously supplied by Dr Jacobson, who informed us (personal communication) that, while he will continue to supply beads on a personal basis, should demand for the beads increase, Sigma Chemical Co. have undertaken to retail them on a commercial basis.

To prepare cationic beads (19) carry out the following procedure:

(i)   Treat colloidal silica (Nalco 1060, supplied by Nalco Chemical Co., Chicago, IL, USA; other sources of colloidal silica can also be used) with a nominal diameter of 50 nm with an aluminium chlorohydroxide complex (Chlorhydrol; supplied by the Reheis Chemical Co., Berkeley Heights, NJ, USA) to give a surface positive charge.

(ii)   Add the Chlorhydrol complex (35 g: 50% w/w) to 100 ml of water in a Waring Blendor and supplement with 450 g Nalco 1060 colloidal silica in 90 ml of water.

(iii)   Blend the mixture for 2 min at high speed, incubate it at 90°C for 30−45 min and stir manually.

(iv)   After cooling, leave the suspension at room temperature for 16 h.

(v)   Add 35 ml of 0.2 M NaOH and again blend.

(vi)   Finally adjust the pH value to 4.4−4.8 with indicator paper (colloidal silica plays havoc with pH electrodes).

The cationic beads are routinely washed before use (17) and remain stable for many months.

A typical procedure for challenging a spheroplast population with cationic silica microbeads is as follows (18):

(i)   Take a buffered suspension containing 1.5×10^8 spheroplasts/ml, mix with a suspension of microbeads (3% w/v, in buffer) in the ratio 2:1 and incubate the suspension at 4°C for 3 min.

(ii)   After centrifugation wash once in coating buffer (1.2 M sorbitol containing 25 mM sodium acetate and 0.1 M KCl; pH 6.0) and suspend the coated spheroplasts in the same volume of coating buffer.

(iii)   Mix with an equal volume of coating buffer containing 2 mg of polyacrylic acid (M_r 90 000). Polyacrylic acid serves to block free cationic groups on the microbeads.

(iv)   Wash the spheroplasts in coating buffer.

2.2.3 *Iodination*

Identification of plasma-membrane preparations after fractionation of lysates can be carried out either by assaying the fraction for marker enzymes (see Section 2.5.1) or by labelling the outer surface of spheroplasts and identifying the label in the fractionated lysate. Two main labelling techniques are used, namely iodination and dansylation.

Iodination involves labelling tyrosine residues exposed on proteins on the outer surface of spheroplasts with ^{125}I, in a reaction catalysed by lactoperoxidase. The following procedure has been used with spheroplasts from *Sacch. cerevisiae* (5).

(i)   Take washed spheroplasts from 500 mg dry weight of cells and suspend them in 48 ml of buffered sorbitol (1.2 M containing 20 mM Tris and 10 mM MgCl$_2$; adjust the pH value to 7.2 with HCl).

(ii)   Supplement the suspension with 0.7 mg lactoperoxidase and 25 μCi Na^{125}I (618 MBq ^{125}I/μg iodine) in buffered sorbitol and make it up to 50 ml with the same buffer.

(iii)   Iodination is carried out at room temperature by adding H$_2$O$_2$ to a final concentration of 8 μM. Add equal amounts of 0.5 mM H$_2$O$_2$ in buffered sorbitol to the suspension at 1 min intervals up to 5 min.

(iv)   Wash the labelled spheroplasts four times with buffered sorbitol.

A closely similar protocol was recently described for iodination of spheroplasts derived from *M. reukaufii* (10).

2.2.4 Dansylation

The reagent 1-dimethylaminonaphthalene 5-sulphonyl chloride, known more often by the trivial name dansyl chloride, reacts with amino groups exposed on proteins located on the outer surface of spheroplasts. As with iodination, a radioactively labelled reagent is used. A typical procedure is that recently recommended for use with spheroplasts derived from *M. reukaufii* (10).

(i)   Supplement a 20 ml suspension of spheroplasts (20%, v/v) in 1 M sorbitol containing 0.3 M potassium phosphate buffer (pH 9.2) with 20 μCi [^{3}H]dansyl chloride and incubate for 30 min at 20°C.

(ii)   Wash the spheroplasts six times in buffered sorbitol before lysing them.

2.3 Lysis of spheroplasts

The obvious way in which to lyse spheroplasts is to submit them to hypo-osmotic conditions. This indeed is one way in which lysis can be effected and with successful results, often aided by pre-treatment with Con A, which leads to isolation of large areas of plasma membrane. However, osmotic lysis needs to be rapid, extensive in terms of the decline in concentration of the osmoticum and usually carried out with the assistance of mechanical agitation. Thus, when spheroplasts of *Candida utilis* were rapidly diluted 100-fold in 0.5−1.0 M KCl, without mechanical agitation, significant lysis was not observed. Moreover, when spheroplasts of this yeast were subjected to dialysis, they were able to remain intact in the presence of low concentrations of protective compounds (19). Lysis of spheroplasts of *Sacch. cerevisiae* sufficiently effective to permit subsequent isolation of plasma membranes and low-density vesicles, is achieved as follows (5):

(i)   Wash the spheroplasts from 240−260 mg dry weight of cells gently in buffered sorbitol three times.

(ii)   Suspend in 9 ml of buffered mannitol (5 mM Tris buffer containing 0.3 M mannitol, adjusted to pH 7.2 with HCl).

(iii) Subject the suspension to a combination of osmotic lysis and gentle mechanical disruption by incubating for at least 20 min in an ice−water mixture (to make the membrane more rigid) and disrupt with six strokes of a Teflon-glass hand homogenizer (0.1 mm clearance).

Preference is often given to the use of less drastic treatments employing virtually iso-osmotic conditions. This technique is mandatory if the objective is to isolate yeast vacuoles (20). The technique can be effected by metabolic lysis, which involves adding a metabolizable sugar (e.g. 30 mM glucose) to a suspension of spheroplasts in sorbitol or mannitol (22), the lytic effect being accentuated by also including chelating agents in the suspension (21). An even gentler method for lysing yeast spheroplasts was pioneered by Wiemken, Schwencke and their colleagues. This involves including in the spheroplast suspension polycationic polymers, such as poly-lysine or DEAE dextran (23). Another such polymer which has been exploited is protamine (at 25 μg/mg cells). Lysis using polycationic polymers, which interact with negatively charged phospholipids in the plasma membrane, can be used in combination with metabolic lysis (24).

2.4 Separation of organelles from lysates

The number of different protocols which have been recommended for isolating organelles from spheroplast lysates of yeasts is legion. Basically, they can be divided into two classes. There are those which are used when the objective is to isolate just one type of yeast organelle; the other sets of protocols are adopted when more than one type of organelle is sought.

When the aim is to isolate just one type of organelle, the first procedure to be adopted with the spheroplast lysate is differential centrifugation, which involves submitting the lysate to regimes of increasing gravitational force and isolating the organelles which are sedimented after each regime. Suitable choice of centrifugation regimes can permit isolation of fractions rich in each of the major membrane-bound intracellular organelles in yeasts. Nevertheless, it is still essential to subject the sedimented material to further purification procedures invariably involving density gradients. These latter procedures are described in detail later in this section, albeit when used with unfractionated lysates.

Reference has already been made to the use of cationic silica beads in isolation of plasma membranes from yeast spheroplasts. This undoubtedly now represents the easiest way of isolating these organelles from yeasts. A procedure for challenging spheroplasts with the beads has already been described.

(i) After mixing with coating buffer containing polyacrylic acid, wash the spheroplasts once with unsupplemented coating buffer (18).

(ii) Suspend the spheroplasts in lysis buffer (5 mM Tris−HCl containing 1 mM EGTA; pH 8.0) to 10^8 spheroplasts/ml and vortex the suspension for 5 min thereby bringing about 95% lysis of spheroplasts.

(iii) Centrifuge the lysate for 5 min at 1000 g, and wash the membrane-bead preparation three times in lysis buffer before examination or analysis.

The remainder of this section describes two protocols for separating various types of organelle from lysates of spheroplasts from *Sacch. cerevisiae* that have not been treated with cationic beads. The first deals primarily with isolation of plasma membranes and low-density vesicles and the second with obtaining vacuole preparations. These are representative protocols and readers should experiment with their component procedures when wishing to isolate organelles from yeasts other than *Sacch. cerevisiae*, or when researching on other organelle preparations.

2.4.1 *Isolation of low-density vesicles and plasma membranes*

The following procedure is adapted from that described by Henschke *et al.* (15). It can be scaled up or down appropriately when dealing with different quantities of lysate from spheroplasts.

(i)   Take a lysate from 240−260 mg dry weight of yeast, make up to 10 ml with buffered mannitol (50 mM Tris buffer containing 0.3 M mannitol, adjusted to pH 7.2 with HCl) and add 3 ml of 62% (w/v) sucrose in buffered mannitol.

(ii) Place the suspension in a 40 ml centrifuge tube and onto it layer a discontinuous gradient of sucrose made up successively with 2-ml portions of 7, 6, 5, 4, 3, 2 and 1% (w/v) sucrose in buffered mannitol. Overlay the gradient with 2 ml buffered mannitol and centrifuge the tube in a swing-out rotor at 11 000 g for 1 h at 4°C.

(iii) A milky layer forms on top of the gradient, this being made up of low-density vesicles. Remove this layer with a hypodermic syringe. This preparation is usually uncontaminated with other subcellular material and for that reason is not normally washed before examination.

(iv) Having removed the low-density vesicle fraction, discard the remainder of the gradient to leave a pellet containing membranes that include plasma membranes.

(v)   Wash the surface of the pellet gently with buffered mannitol and suspend it in 10 ml of buffered mannitol.

(vi) Homogenize the suspension by submitting it to 4−6 strokes in a Teflon-glass hand homogenizer (0.1 mm clearance).

(vii) Make the suspension up to 30 ml with buffered mannitol and incubate it for at least 20 min in an ice−water mixture.

(viii) Supplement the suspension with 9 ml of 62% (w/v) sucrose in buffered mannitol and centrifuge it for 30 min at 11 000 g in a swing-out rotor.

(ix) Remove the small low-density vesicle population which appears on top of the gradient and homogenize the membrane fraction in 2 ml of buffered mannitol.

Do not attempt to store this suspension, but proceed immediately to obtain purified plasma membranes. This can conveniently be done using a modification of the sucrose density-gradient method of Hossack (25).

(i)   Dilute the 2-ml suspension of membranes with a further 2 ml of buffered mannitol and layer on the suspension a discontinuous gradient of sucrose solutions; make up the gradient by adding on top of the suspension 5 ml each of 62% and then 50% (w/v) sucrose in buffered mannitol, followed by 3 ml each of 45, 40, 35, 30, 20 and 10% (w/v) sucrose in buffered mannitol.

(ii) Centrifuge the tube for 90 min at 4°C and 24 000 g in a swing-out rotor.

(iii) Several bands of membraneous material can be seen in the gradient after centrifugation; remove these with a peristaltic pump.

(iv) After removing the remaining supernatant liquid, resuspend the pellet in 35 ml

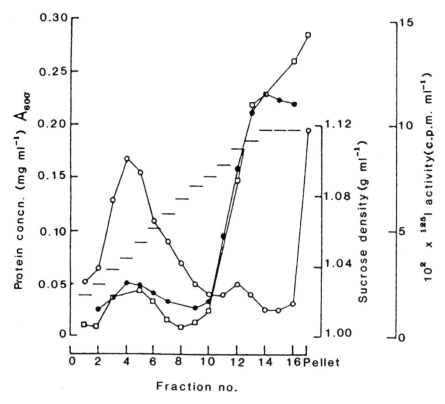

Figure 2. Sucrose density-gradient separation of plasma-membranes from spheroplast lysates of *Sacch. cerevisiae* NCYC 366 (5). Techniques for preparing spheroplasts and iodinating and lysing them are described in the text. Fractions (2 ml) were collected from the gradient with a peristaltic pump. Protein content (□) and ^{125}I activity (●) were measured in material retained by a 0.2 μm membrane filter (Nucleopore). ○ indicates optical density values of fractions. Horizontal lines indicate sucrose density.

of buffered mannitol, centrifuge and resuspend the pellet in 2 ml of buffered mannitol.

(v)   Supplement the suspension with 62% (w/v) sucrose in buffered mannitol, to a refractometer value of 30% (w/v) sucrose equivalent.

(vi)   Layer onto this suspension portions (2 ml) of 21% (w/v) through to 1% (w/v) sucrose in 50 mM Tris−HCl buffer (pH 7.2) in 2% step decreases in concentration.

(vii)   Centrifuge the gradient for 4.5 h at 4°C and 11 000 *g* in a swing-out rotor. The pellet is considered to be purified plasma membrane.

If the above procedures are carried out using spheroplasts in which the membranes have been dansylated or iodinated then the position of the membrane on the gradient may be shown by removing 2-ml fractions through the gradient, filtering through a 0.2 μm membrane filter, transferring the filter to scintillation vials containing 7.5 ml optiphase safe scintillation fluid (LKB) and counting in a scintillation counter. *Figure 2* shows a typical separation of a spheroplast lysate with most of the membrane in or just above the pellet but with a small peak higher in the gradient where the membrane is associated with vesicles. The position of the bands may also be demonstrated by optical density (A_{600}) and protein content of the fractions.

2.4.2 *Isolation of vacuolar membranes*

(i)   To obtain vacuoles, gently lyse spheroplasts by suspending them in sodium citrate buffer (10 mM; pH 6.8) containing 0.1 M mannitol.

(ii)   Place the lysate down a hypotonic density gradient made up of equal proportions of 7.0, 7.5 and 8.0% Ficoll.

(iii)   Centrifuge the gradient at 3000 *g* for 30 min.

(iv)   The soluble fraction in the lysate, together with some vacuoles and low-density vesicles, collect at the bottom of the tube, while at the top there appears a creamy layer made up of a mixture of vacuoles and vesicles. Remove this top layer with a hypodermic syringe and mix it with 4 vol of 10 mM sodium citrate buffer (pH 6.8) containing 0.6 M sorbitol.

(v)   Centrifuge the suspension at 3000 *g* for 20 min.

(vi)   Low-density vesicles float on top of the suspension after centrifugation and are removed. The remainder of the suspension contains vacuoles free from other organelles.

The vacuolar membrane, often known as the tonoplast, is a single unit membrane. To obtain preparations of this membrane, submit vacuolar suspensions to liquid-shear stress as described for spheroplasts. Centrifuge the suspension of disrupted vacuoles for 30 min at 5000 *g* to sediment membranes.

2.5 Examination and characterization of organelles

Establishing the purity of a subcellular organelle preparation is a problem which perpetually faces all cell biologists, not least those interested in yeasts. Principal among these problems is the task of establishing clearly defined categories of organelle, an attempt to delimit one from another. Unfortunately, cells do not entertain such clear delimitations, thereby contributing to the cell biologist's problems. Accepted wisdom dictates that two types of experimental evidence be used to establish whether or not yeast plasma membranes, low-density vesicles and vacuole membranes, isolated by the procedures described already in this chapter, are indeed what they are purported to be and to what extent they are contaminated with other subcellular material. Characterization involves examination in the electron microscope and assaying for the presence of marker activities or components, the former usually being enzymes.

2.5.1 *Electron microscopy*

Examination of plasma-membrane, vesicle and vacuole-membrane preparations in the transmission electron microscope can provide almost definitive evidence for the absence from these preparations of other subcellular structures, especially mitochondria. This evidence must, however, be evaluated alongside that of marker components and enzymes.

Plasma-membrane and vacuole-membrane preparations are prepared for transmission electron microscopy as follows:

(i) Mix a suspension of plasma membrane with an equal volume of 2% (w/v) osmium tetroxide in 50 mM sodium cacodylate and leave for 60 min at 4°C with occcasional shaking.

(ii) Centrifuge the preparation at 2500 *g* for 5 min, discard the supernatant and resuspend the pellet in 50 mM sodium cacodylate.

(iii) Wash the pellet three times with 50 mM sodium cacodylate and resuspend in 2% (w/v) agar.

(iv) Take this suspension into a Pasteur pipette and extrude it, as it sets, onto a glass plate.

(v) Cut the sample into small sections and dehydrate the sections in the following sequence of acetone (v/v) concentrations: 10 min 15%, 10 min 50%, 10 min 90% and three times for 15 min in 100%.

(vi) Transfer the sections into 50% (v/v) Taab EM resin (Taab Laboratories Equipment, Reading, Berks, UK): acetone mixture and rotate for 24 h.

(vii) Transfer the sections to 90% resin:acetone mixture and rotate for a further 24 h and then transfer to 100% resin before being rotated for a further 24 h.

(viii) Transfer sections to Taab capsules, add fresh resin and polymerize at 60°C for 3 days.

(ix) Cut gold sections on a suitable ultramicrotome and stain the sections for 10 min in 70% (v/v) ethanol saturated with uranyl acetate, followed by 10 min in Reynolds (26) lead citrate solution. Samples are then ready for examination using an electron microscope.

In the microscope, plasma membranes usually appear to be associated with the low-density vesicles (*Figure 3a*) which serve to supply precursors for growth of the plasma membrane and cell wall (5).

To fix vesicles carry out the following protocol:

(i) Mix the preparation with an equal volume of 2% (w/v) osmium tetroxide in water and leave for 60 min at 4°C with occasional agitation.

(ii) Wash three times with 50 mM sodium cacodylate, resuspend in 2% (w/v) agar (clarity as high as possible) and extrude from the Pasteur pipette as before.

(iii) Stain the vesicle preparation with 1% (w/v) uranyl acetate for 1 h at 20°C in the dark.

Vesicle preparations are prepared for examination in the electron microscope starting with the dehydration sequence already described for membranes. Their spherical shape in the transmission electron microscope and the absence of contaminating material from these preparations are illustrated in *Figure 3b*.

2.5.2 *Assays for enzyme activities and chemical composition*

There has as yet not been a report of the presence of a unique chemical component in plasma membranes of yeasts. However, it is now generally agreed that the best manner in which to characterize these organelles, after an examination of their appearance in the transmission electron microscope, is to assay their activity of an ATPase that is inhibited by orthovanadate, diethylstilboestrol and *N,N'*-dicyclohexylcarbodiimide

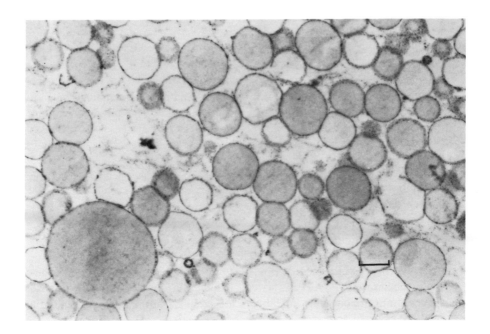

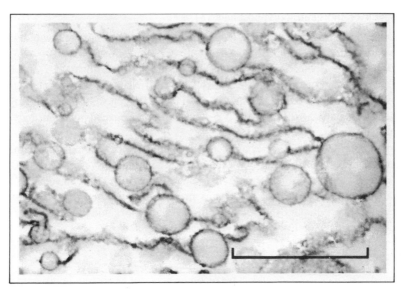

Figure 3. Transmission electron micrographs of a section through a population of low-density vesicles from *Sacch. cerevisiae* NCYC 366 (**a**; bar indicates 0.5 μm) and through a preparation of plasma membrane showing vesicular associations (**b**; M.J.White, unpublished data; bar indicates 0.5 μm).

(DCCD) but not by oligomycin. Take precautions when using these toxic chemicals, especially diethylstilboestrol. Yeast mitochondria and vacuoles also contain an ATPase. However, orthovanadate and diethylstilboesterol do not inhibit the mitochondrial enzyme (27). All three enzymes are inhibited by DCCD, although the mitochondrial and vacuole-membrane enzymes are approximately ten times more sensitive to this inhibitor than the plasma-membrane enzyme (28).

Plasma-membrane ATPases, like other ATPases, are assayed by following release of inorganic phosphate from ATP.

(i)   Prepare a reaction mixture (1 ml) consisting of 100 mM MES−Tris buffer (pH 6.0) containing 80 mM KCl, 6.0 mM $MgCl_2$ and plasma membrane preparation containing 25−50 μg protein. The protein content of plasma-membrane preparations and indeed, of other yeast membrane preparations, is best assayed using the Bio-Rad protein assay based on the dye-binding technique of Bradford (29).

(ii)   Start the reaction by adding 6.0 mM ATP (sodium salt) and incubate for 30 min at 30°C.

(iii)   Stop the reaction by adding 2.0 ml of 2.0% (v/v) H_2SO_4 containing 0.5% (w/v) ammonium molybdate and 0.5% (w/v) sodium dodecyl sulphate (SDS).

(iv)   Add to the reaction mixture 0.02 ml of 10% (w/v) ascorbic acid and allow the colour to develop over 5 min at 30°C.

(v)   The absorbance of the mixture is related to concentration of inorganic phosphate by a standard curve. ATPase activity is quoted as μmol inorganic phosphate released/mg protein/min. Inhibitors are included in the reaction mixture at 100 μM.

Morphological characterization of low-density vesicles can be corroborated by assaying their content of inorganic polyphosphate (5).

(i)   Dilute suspensions of vesicles (in buffered mannitol) appropriately with buffered mannitol and make 0.5 M with H_2SO_4.

(ii)   Divide the suspension into two and boil one of the portions for 7 min (30).

(ii)   Determine the orthophosphate content of each suspension after centrifugation, as already described for the assay for ATPase. Polyphosphate contents of low-density vesicles are of the order 1.5 μmol orthophosphate equivalent/mg vesicle protein (5).

Vacuolar membranes can be characterized chemically by demonstrating that the ATPase activity in the membranes is of the order of ten times more sensitive to DCCD compared with the plasma-membrane ATPase. Activities of ATPase in vacuolar membranes are assayed as described for plasma membranes. Further characterization of vacuoles can be carried out by assaying their content of inorganic polyphosphate. This is done as described for low-density vesicles.

3. EXTRACTION AND ANALYSIS OF LIPIDS

This section describes methods which are recommended for extracting lipids from whole cells of yeasts and from yeast organelles, as well as analytical methods which are best suited to the portfolio of lipid classes found in yeasts.

3.1 Extraction of lipids

A necessary prerequisite to an analysis of the lipids in yeast cells and organelles is extraction of these compounds from the material under examination. Basic though it is to yeast lipid analysis, curiously little critical attention has been paid to the extraction processes used. Several largely unaddressed problems arise. The first concerns the efficiency with which lipids are extracted. The general approach in extracting lipids from yeast cells and organelles is to expose the material to mixtures of methanol and chloroform. Very rarely indeed does the worker attempt to assess whether all lipids have been extracted from the material. This can be done by saponifying the residual material after extraction and analysing for the presence of long-chain fatty acids by GLC (see Section 3.1.1) The variable values reported for lipid contents of yeast cells could well be attributed, in part at least, to inefficient extraction. Especially when measures are taken to promote efficient extraction (see Section 3.1.1), there is the likelihood that largely hydrophobic proteins are extracted along with lipids. This possibility has been virtually ignored by workers concerned with extraction of lipids from yeast cells and organelles. Another problem is that of selectivity, the possibility that the various lipid classes (with different proportions of hydrophilic and hydrophobic character) are extracted to different extents by an extraction regime. When determinations of the total lipid content of yeasts are required, whole-cell preparations must clearly be the starting material. It is strongly recommended that freshly harvested cells be used for extraction. Although supporting data are not to hand, it is generally believed that the efficiency of extraction is lowered and phospholipases activated when preparations of freeze-dried cells are used, convenient though this may be. However, when the aim is to acquire an overall lipid analysis of yeast cells, it is strongly recommended, although by no means universal practice, that extraction be carried out on preparations of disrupted cells to maximize the efficiency. Using whole cells or subsequently disrupted cells, a preliminary extraction procedure which has much to recommend it, but which is far from universally used, involves treating cells with hot ethanol in order to inactivate lipolytic enzymes (31).

(i)   To a portion of cells (400−500 mg dry weight) add 40 ml of ethanol (80%, v/v) at 80°C.

(ii) Maintain the suspension at 80°C for 15 min and filter it through Whatman no. 44 filter paper.

(iii) Combine the ethanol extract with the chloroform−methanol extracts obtained later in the extraction process.

To extract lipids from 500 mg dry weight equivalent of yeast cells carry out the following procedures (32).

(i)   Mix freshly harvested and washed cells with 10 ml of methanol containing an antioxidant such as butylated hydroxytoluene (15 mg), any necessary internal standards such as heptadecanoic acid (2.5 mg) or dihydrocholesterol (2.5 mg) and an inhibitor of lipase activity such as *p*-chloromercuribenzoate (1 mM).

(ii) Transfer this mixture to a Braun bottle (B.Braun, Melsungen, FRG) together with 35 g glass beads (Sigma; 0.45−0.55 mm diameter) and disrupt in a Braun homogenizer for six periods each of 15 sec at speed 2 (4000 r.p.m.). The temperature is kept low during homogenization by cooling with liquid CO_2.

(iii) Add 20 ml of chloroform to the mixture and transfer the entire contents of the bottle to a 100-ml stoppered conical flask (Quickfit).

(iv) Leave the mixture to stir on a flat-bed magnetic stirrer for 2 h at room temperature

(20−24°C) and filter under vacuum through a Whatman no. 44 filter paper.
(v)   Collect the filtrate, store it at −20°C and repeat the extraction procedure on the residue, including the paper, twice.
(vi)   Finally combine the three filtrates, rotor evaporate to dryness and take up in petroleum ether (60−80°C).

Irrespective of the manner in which they extract lipids from yeast cells, the majority of workers clean their extracts using the procedure of Folch *et al.* (33). To do this, mix the extract with 0.25 vol 0.88% (w/v) KCl, shake and separate either by leaving the suspension at −20°C overnight or by centrifugation. Carefully remove the upper lipid layer.

Lipids are extracted from isolated organelles by the same procedures, except that disruption in a homogenizer is not required. Because it is not always possible to obtain the equivalent of 500 mg dry weight of an organelle, the procedure needs to be scaled down appropriately.

3.2 Analysis of lipids

Although the lipid composition of yeasts is not unusual as far as eukaryotic organisms are concerned, nevertheless, in terms of numbers of individual types of lipid molecule, there is decided complexity. There are some half dozen major phospholipid classes, each of which may contain a variety of fatty-acyl chains. The same applies to sterols and sterol esters. The fifth major class of lipid, namely triacylglycerols, also presents a vast variety of molecules, again depending on the nature of the fatty-acyl chains (34). In practical terms, however, most workers are interested in either the overall fatty-acyl composition of yeast lipids, or the proportions of the types of molecule in the main lipid classes. The remainder of this chapter is devoted to practical details for these analyses.

3.2.1 *Fatty-acyl composition of total cellular lipids*

A frequently required analysis is a determination of the fatty-acyl composition of the total cellular lipids in a yeast. This analysis has little physiological significance, although some workers require it to assess, for example, the overall fluidity of yeast membranes. In addition, it has been used as an aid in yeast classification.

To avoid contaminating a GLC column with underivatized sterols the extracted lipids should be saponified before derivatization.

(i)   Lipids from approximately 100 mg dry weight of cells are evaporated under nitrogen gas to dryness in a suitable container.
(ii)   Add 8 ml of 2 M KOH in 95% (v/v) methanol and 2 ml of benzene, fill the headspace with nitrogen gas and incubate the mixture at 80°C for 3 h.
(iii)   If this procedure is scaled up, the mixture may be refluxed.
(iv)   After cooling, add an equal volume of methanol and extract the non-saponified fraction by shaking with three 5-ml washes of petroleum ether (60−80°C).
(v)   The lower methanol layer may then be acidified with 6 M HCl so releasing the saponified lipids which may be extracted with petroleum ether as already described.

The extracted fatty acids are normally derivatized before analysing by GLC. This may be done by the preparation of methyl esters using 14% BF_3 in methanol.

(i) Evaporate to dryness in a small reaction vial the saponifiable lipid fraction from 100 mg dry weight of cells.

(ii) Add 1 ml of BF_3 in methanol, fill the headspace with nitrogen gas, seal and heat at 80°C for 1 h.

(iii) Tip the contents into 5 ml of water in a Quickfit test tube and extract three times by shaking with 5 ml of petroleum ether (60−80°C).

(iv) Concentrate the pooled extracts and examine by GLC.

Methyl esters of fatty acids may be separated easily on any polar column and detected with a flame ionization detector (FID). We recommend use of a 25 m BP21 capillary column, internal diameter 0.53 mm, in a Pye Unicam GCD chromatograph. Samples (0.1−0.2 μl) are applied via an on-column injection system (SGE Ltd). Hydrogen is used as a carrier gas. The column temperature is 135°C for 5 min and increases to 180°C at a rate of 8 ml/min. The detector temperature is 280°C and the injection temperature 240°C. Should a capillary system not be available, a conventional column using, for example, 15% EGSS-Y supported on 100−200 mesh Gas Chrom P is quite adequate but may not separate isomers of the longer chain acids quite so well. Do not attempt to use a temperature programme with packed columns unless a dual-column system is available as the base line will rise as the column material is heated and breaks down slightly. It is common practice to express the fatty-acyl composition of yeast lipids as a percentage of the total content of residues or by reference to a known weight of an internal standard such as heptadecanoic acid. As all fatty acid methyl esters tested affect the FID to the same extent, each peak area may be directly related to the amount of residue present. The area under a peak may be calculated manually but an integrator will calculate this more consistently and save a great deal of time.

3.2.2 *Analysis of lipid classes*

Yeast physiologists are usually interested in the size and composition of the major classes of lipids found in yeasts. There are two groups of these compounds, namely polar phospholipids and sterols and non-polar neutral triacylglycerols and sterol esters. A certain amount of free fatty acid is always found. These can conveniently be separated from one another prior to determining the amount of each class in a lipid extract or, except with free sterols, the fatty-acyl composition of the lipid class or of its components.

Preparative separation of the major classes in yeast lipid extracts is carried out by TLC. Plates pre-coated with silica gel are usually used, such as silica gel G (0.25 mm thick) supplied by Whatman.

(i) Apply the lipid extract to the plate as a band.

(ii) Develop the plate with the solvent mixture; petroleum ether (40−60°C):diethyl ether:acetic acid (70:30:1, by vol).

(iii) After development, visualize the bands of lipid classes either by exposing a sample strip to iodine vapour in a fume cupboard or by spraying the plate with 2′,7′-dichlorofluorescein in ethanol and viewing under UV radiation (254 nm).

(iv) Whichever method is used to visualize the bands, mark them on the plate with a pencil and scrape off each band into separate bijou bottles.

3.2.3 *Phospholipids*

Fatty-acyl residues in phospholipid fractions may be derivatized with 14% BF_3 in methanol as described for total lipids. Saponification is not necessary.

(i)   Put 1 ml of BF_3 in methanol directly onto the scraped-off silica-gel band and allow to react under nitrogen gas as before. The silica gel will be left behind when the mixture is extracted with petroleum ether.

(ii)   Quantify residues by addition of a known quantity of heptadecanoic acid to the bijou bottle before methylation.

Total phospholipids in extracts are assayed by measuring the phosphate content of the extract (35).

(i)   Elute the phospholipid band with petroleum ether ($60-80°C$) and evaporate to dryness in a Kjeldahl flask.

(ii)   Add 0.1 ml of ashing solution (3 parts concentrated H_2SO_4 mixed with 2 parts 70% perchloric acid) and digest in a microKjeldahl system for 15 min.

(iii)   Allow to cool and adjust the volume to 4 ml with water.

(iv)   Add 4 ml of freshly made reagent [1 part 12 M H_2SO_4, 1 part 2.5% (w/v) ammonium molybdate, 1 part 10% (w/v) ascorbic acid and 2 parts water].

(v)   Cover and incubate at 37°C for 2 h.

(vi)   Measure the absorbance at 820 nm against a reagent blank and relate the value to a standard curve.

Some workers prefer to add the ascorbic acid solution separately at the start of the reaction. The assay is accurate within the range of $1-10$ μg of phosphate. Values for phosphorus content are multiplied by 25 to give the total phospholipid content.

3.2.4 *Sterols*

Total sterols are assayed as follows (36,37).

(i)   Place up to 0.3 mg of sterol in a reaction vial and evaporate to dryness under nitrogen gas.

(ii)   Add 6.0 ml of glacial acetic acid, mix and add 4.0 ml of ferric chloride solution [4 ml 2.5% $FeCl_3.6H_2O$ (w/v) with 85% (v/v) orthophosphoric acid taken up to 50 ml with concentrated H_2SO_4].

(iii)   Vortex, cool and allow to stand at room temperature for 10 min.

(iv)   Measure the absorbance of the solution at 550 nm against a reagent blank and relate to a standard curve prepared from dilutions of a pure sterol up to 0.5 mg/ml.

To assay individual sterols, they are first silylated and then separated by GLC on a non-polar column.

(i)   Elute the sterols from the silica gel with three 5-ml portions of petroleum ether.

(ii)   Collect and pool the eluate, concentrate to one 1 ml, transfer to a small reaction vial and evaporate to dryness.

(iii)   Add 0.2 ml of bis-trimethylsilyltrifluoracetamide in pyridine, flush the headspace with nitrogen gas and heat for 15 min at 80°C.

(iv)   The silylated sterols should be analysed immediately by GLC or stored at $-20°C$ preferably for not longer than a week.

As with fatty-acid analysis, GLC is open to adaptation to suit the facilities available. Our experience is that silylated sterols are best analysed on a 25 m SE30 wall-coated quartz capillary column (SGE). The instrument used is a Pye Unicam PU 4500 with a split ratio of 30:1 using helium as gas carrier with a flow rate through the column of 1 ml/min. The temperature programme is from 250°C to 300°C rising at 16°C/min. Do not attempt to programme a single conventional column. The detector temperature is 350°C and the injection temperature 345°C.

An internal standard sterol may be used, for example dihydrocholesterol. However, comparing known quantitative standards of cholesterol and ergosterol with dihydro-cholesterol has shown us that these sterols do not all evoke the same response with a FID. Although modern integrators are well able to deal with this variation, suitable standards of yeast sterols are unfortunately not available commercially to indicate what these response factors are. This means that, while the ergosterol peak may be quantified by comparison with known concentrations of pure ergosterol, or by calculation from the size of the internal standard with a known response factor, it cannot be assumed that other peaks obtained may be quantified with any certainty. For the same reason identification of some of the peaks obtained may not be possible without reference to GLC/MS.

3.2.5 *Free fatty acids*

These are analysed easily by the techniques already described for total cellular lipids and phospholipids. Quantification is quite straightforward as any fatty acid internal standard applied to the original extraction will appear in this fraction, preference being given to heptadecanoic acid. There are usually only low levels of free fatty acids present in extracts and some of these may have arisen from the breakdown of other lipids during extraction.

3.2.6 *Triacylglycerols*

Triacylglycerols may be methylated directly as already described for phospholipids to yield a mixture of fatty-acid methyl esters. Again quantification is probably best carried out by adding a known amount of heptadecanoic acid to the vial containing the scraped-off band. Whole triglycerides may be separated by their degree of saturation using t.l.c. on silica gel G plates impregnated with $10-30\%$ AgNO$_3$ in a solvent system of chloroform:methanol (99.2:0.8%, v/v) (38,39). The separated triglycerides are visualized with dichlorofluorescein and may be eluted with chloroform:methanol (9:1, v/v).

3.2.7 *Sterol esters*

The sterol-ester band must be saponified as described for total lipid analysis so that sterols and fatty acids may be analysed separately by the techniques already described. Great care must be exercised throughout as very low levels of sterol esters are found in yeast lipid extracts. This band may be contaminated by butylated hydroxytoluene if used as an anti-oxidant. The anti-oxidant will appear on the fatty acid chromatogram in the same area as the C$_{14}$ acids and must be eliminated from any calculation which expresses fatty-acid content as a percentage of the total. Again the use of an internal standard added to the band eliminates this possible error as each peak area will be compared directly with that of the standard.

3.2.8 *Separation and assay of individual phospholipid classes*

Individual phospholipid clases may be separated by one- or two-dimensional TLC techniques. We have found a one-dimensional technique to be adequate and consider this is preferable as larger quantities of individual phospholipids may be collected for fatty-acid analysis.

(i)   Apply a continuous band of eluted phospholipid to the bottom of a silica gel G TLC plate and develop in a solvent system of chloroform:methanol:acetic acid (65:25:8, by vol); (40).

(ii)   Dry the plates and visualize the bands with molybdenum blue reagent (0.65%, v/v, molybdenum oxide in 4.2 M H_2SO_4) (41).

(iii) The bands may be identified by comparison with standards such as phosphatidylcholine, phosphatidylethanolamine, phosphatidylinositol and phosphatidylserine. They may be quantified to some extent using a scanning densitometer (wavelength 626 nm, aperture width 0.1−0.3 mm; Chromoscan 3, Joyce Loebl, Gateshead, UK). Should the fatty-acyl content of the individual phospholipids be required, visualize with 2',7'-dichlorofluorescein (0.2%, w/v, in ethanol) and methylate as already described.

Should suitable equipment be available, individual phospholipids may be separated using HPLC techniques (as can be many other larger lipid fractions). To separate whole phospholipids on a PU 4100 binary gradient system with a column oven, elute the band in chloroform:methanol, evaporate the chloroform and take up in methanol. All solvents must be of HPLC grade. The mobile phase requires an acetonitrile:phosphate buffer gradient (pH 5.0) at 2.0 ml/min on a diol column. Well-separated analysis takes about 20 min.

3. REFERENCES

1. Rose,A.H. (1981) In *Current Developments in Yeast Research*. Stewart,G.G. and Russell,I. (eds), Pergamon Press, Toronto, p. 645.
2. Kreger-van Rij,N.J.W. (1984) *The Yeasts, a Taxonomic Study*. Elsevier, Amsterdam.
3. Nurminen,T., Taskinen,T. and Suomalainen,H. (1976) *Biochem. J.*, **154**, 751.
4. Nurminen,T., Oura,E. and Suomalainen,H. (1970) *Biochem. J.*, **116**, 61.
5. Henschke,P.A., Thomas,D.S., Rose,A.H. and Veazey,F.J. (1983) *J. Gen. Microbiol.*, **129**, 2927.
6. Kuo,S.C. and Yamamoto,S. (1975) In *Methods in Cell Biology*. Prescott,D.M. (ed.) Academic Press, New York, Vol. 11, p. 169.
7. Arnold,W.N. (1981) In *Yeast Cell Envelopes: Biochemistry, Biophysics and Ultrastructure*. Arnold,W.N. (ed.) Vol. II, p. 93.
8. Myers,F.L. and Northcote,D.H. (1958) *J. Exp. Biol.*, **35**, 639.
9. Santos,E., Villanueva,J.R. and Sentandreu,R. (1978) *Biochim. Biophys. Acta*, **508**, 39.
10. Aldermann,B. and Höfer,M. (1984) *J. Gen. Microbiol.*, **130**, 711.
11. Bussey,H., Saville,D., Chevallier,M.R. and Rank,G.H. (1979) *Biochem. Biophys. Acta*, **553**, 185.
12. Kitamura,K. and Yamamoto,Y. (1972) *Arch. Biochem. Biophys.*, **153**, 403.
13. Kitamura,K., Kaneko,T. and Yamamoto,Y. (1974) *J. Gen. Appl. Microbiol.*, **20**, 323.
14. Scott,J.H. and Schekman,R. (1980) *J. Bacteriol.*, **142**, 414.
15. Scarborough,G.A. (1975) *J. Biol. Chem.*, **250**, 1106.
16. Duran,A., Bowers,B. and Cabib,E. (1975) *Proc. Natl. Acad. Sci. USA*, **72**, 3952.

17. Schmidt,R., Ackerman,R., Kratky,Z., Wasserman,B. and Jacobson,B. (1983) *Biochem. Biophys. Acta,* **732**, 421.
18. Cartwright,C.P., Veazey,F.J. and Rose,A.H. (1987) *J. Gen. Microbiol.,* **133**, 857.
19. Chaney,L.K. and Jacobson,B.S. (1983) *J. Biol. Chem.,* **258**, 10062.
20. Indge,K.J. (1968) *J. Gen. Microbiol.,* **51**, 433.
21. Schwencke,J. (1988) In *The Yeasts.* Rose,A.H. and Harrison,J.S. (eds) Academic Press, London, 2nd edition, Vol. 3, in press.
22. Indge,K.J. (1968) *J. Gen. Microbiol.,* **51**, 425.
23. Dürr,M., Boller,T. and Wiemken,A. (1975) *Arch. Microbiol.,* **105**, 319.
24. Schlenk,F., Dainko,J.L. and Svilha,G. (1970) *Arch. Biochem. Biophys.,* **140**, 228.
25. Hossack,J.A. (1975) Ph.D. Thesis, University of Bath.
26. Reynolds,E.S. (1963) *J. Cell Biol.,* **17**, 208.
27. Serrano,R. (1977) *Eur. J. Biochem.,* **22**, 51.
28. Okorokov,L.A., Kulakovskaya,T.V. and Kulaev,I.S. (1982) *FEBS Lett.,* **145**, 160.
29. Bradford,M.M. (1976) *Anal. Biochem.,* **72**, 248.
30. Dürr,M., Urech,K., Boller,T., Wiemken,A., Schwencke,J. and Nagy,M. (1979) *Arch. Microbiol.,* **121**, 169.
31. Letters,R. (1968) In *Aspects of Yeast Metabolism,* Mills,A.K. (ed.) Blackwell, Oxford, p. 306.
32. Calderbank,J., Keenan,M.H.J., Rose,A.H. and Holman,G.D. (1984) *J. Gen. Microbiol.,* **130**, 2817.
33. Folch,J., Lees,M. and Sloane Stanley,G.H. (1957) *J. Biol. Chem.,* **226**, 497.
34. Rattray,J.B.M., Schibeci,A. and Kidby,D.K. (1975) *Bacteriol. Rev.,* **39**, 197.
35. Chen,P.S., Toribara,T.Y. and Warner,H. (1956) *Anal. Chem.,* **28**, 1756.
36. Courchaine,A.J., Miller,W.H. and Stein,D.B. (1959) *Clin. Chem.,* **5**, 609.
37. Zlatkis,A., Zak,B. and Boyle,A.J. (1963) *J. Lab. Clin. Med.,* **41**, 486.
38. Kuksis,A. and Marai,L. (1967) *Lipids,* **2**, 217.
39. Privett,O.S. and Nutter,L.J. (1967) *Lipids,* **2**, 149.
40. Kramer,R., Kopp,F., Niedermeyer,W. and Fuhrmann,G.F. (1978) *Biochim. Biophys. Acta,* **507**, 369.
41. Dittmer,J.C. and Lester,R.L. (1964) *J. Lipid Res.,* **5**, 126.

Standard media for cultivation of yeasts

I.CAMPBELL

1. SUPPLIERS OF MEDIA

The following media are most easily prepared by reconstituting commercially prepared dehydrated media. Principal suppliers in the UK are:

Difco Laboratories Ltd, PO Box 14B, Central Avenue, East Molesey, Surrey KT8 0SE, UK

Oxoid Ltd, Wade Road, Basingstoke, Hampshire RG24 0PW, UK

2. ROUTINE NUTRIENT MEDIA

2.1 Malt extract broth (or agar)

Dehydrated malt extract	3 g
Mycological peptone	0.5 g
Agar (if required)	1.5 g in 100 ml; pH ~5.5

2.2 MYGP broth (or agar) (1)

Dehydrated malt extract	0.3 g
Dehydrated yeast extract	0.3 g
Glucose	1 g
Mycological peptone	0.5 g
Agar (if required)	1.5 g in 100 ml; pH ~5.5

For YEP−glucose medium, omit the malt extract and increase the yeast extract content to 0.5 g/100 ml.

2.3 Sabouraud's glucose broth (or agar)

Glucose	4 g
Mycological peptone	1 g
Agar (if required)	1.5 g in 100 ml; pH ~5.5

2.4 Wickerham's chemically defined media (1)

Although commercially available (Difco), if large volumes are required it may be more economical to prepare your own stock solutions for preparation of these media.

Yeast Carbon Base: defined medium containing glucose as carbon source but lacking nitrogen source, to which the chosen nitrogen source is added, usually at 0.5 g/100 ml.

Yeast Nitrogen Base: defined medium containing ammonium sulphate + asparagine as nitrogen source but lacking carbon source, to which the chosen carbon source is added, usually at 1.0 g/100 ml.

Yeast vitamin-free medium: defined medium lacking growth-factor mixture.

The quantities listed below are for preparation of 1 litre, probably the minimum practicable quantity. In practice, only the carbon and the main nitrogen source have to be weighed out each time: the supplementary nitrogen source and growth-factor mixtures (both kept at 0–2 °C as sterile-filtered concentrated solutions) and salts + trace elements concentrated mixture (which can be kept at room temperature) can be kept as stock solution. The pH of the media, mainly due to the salts mixture, is approximately 5.5. For solid medium, use purified agar, e.g. Oxoid agar no. 1.

Carbon source:	Glucose	10 g
Nitrogen source:	Ammonium sulphate	3.5 g
	L-Asparagine	1.5 g

Supplementary nitrogen source (added to all media)

L-Histidine	10 mg
DL-Methionine	20 mg
DL-Tryptophan	20 mg

These three amino acids are added to ensure the growth of various yeast strains unable to synthesize L-his, L-met or L-trp. The amounts (10 mg/l of the L-amino acid in each case) are too small to function alone as a satisfactory N source. Note the use of synthetic (DL-) met and trp: the L-amino acids from natural sources could contain sufficient biotin to invalidate results with 'vitamin-free' media.

Salts mixture:	KH_2PO_4	1 g
	$MgSO_4.7H_2O$	0.5 g
	NaCl	0.5 g
	$CaCl_2.6H_2O$	0.5 g

Trace elements mixture:		
	H_3BO_3	500 μg
	$MgSO_4.4H_2O$	400 μg
	$ZnSO_4.7H_2O$	400 μg
	$FeCl_3.6H_2O$	200 μg
	$Na_2MoO_4.2H_2O$	200 μg
	KI	100 μg
	$CuSO_4.5H_2O$	40 μg

We find it convenient to prepare a single mixture of salts + trace elements concentrate for addition by pipette as required.

Growth factor mixture:		
	myo-Inositol	10 mg
	Calcium pantothenate	2 mg
	Niacin or nicotinic acid	400 μg
	Pyridoxine–HCl	400 μg
	Thiamine–HCl	400 μg
	p-Amino benzoic acid	200 μg

Riboflavin	200 μg
Biotin	20 μg
Folic acid	2 μg

A 10× concentrated solution, filter-sterilized and kept at 0−2 °C in sterile bottles, is a convenient vitamin source for defined media.

It is common practice to sterilize all ingredients (other than the vitamin supplement) as a single medium for 10 min at 114 °C (10 p.s.i.). The sterile vitamin solution is added after sterilization. The slight caramelization is not normally important, but can be prevented by sterilizing the sugar solution separately (by autoclave or filter).

3. SPORULATION MEDIA

3.1 Acetate agar

Two versions are in common use: according to Fowell (2) and McClary (see Section 2.3 of Chapter 4). Fowell's medium is simpler, and in my experience gives satisfactory results. Alternative sporulation media based on plant extracts are used by some workers; see (1).

Dissolve 0.5 g of sodium acetate trihydrate in 100 ml distilled water and adjust the pH to 6.5−7.0, preferably 6.8. This is outside the buffering range of acetate solutions, so add HCl dropwise to avoid excessive change of pH. Then dissolve 1.5 g of agar, dispense in approximately 7 ml amounts in McCartney bottles, sterilize for 15 min at 120 °C and allow to solidify as slants. Fowell's acetate agar is not a nutrient medium, so a heavy inoculum of culture from malt extract agar should be spread on the slant. Incubate with the cap slightly loose, since good aeration is required for sporulation. Best sporulation is usually observed on the upper half of the slant. Spores may appear within 2 days at 25 °C, but with the more 'difficult' strains incubation for at least 2 weeks may be necessary.

4. REFERENCES

1. Van der Walt,J.P. and Yarrow,D. (1984) In *The Yeasts, a Taxonomic Study*, 3rd Edition. Kreger-van Rij,N.J.W (ed.), Elsevier-North Holland, Amsterdam, p. 45.
2. Fowell,R.R. (1952) *Nature*, **170**, 578.

Preservation and transport of yeast cultures

J.F.T.SPENCER and DOROTHY M.SPENCER

1. INTRODUCTION

Requirements for storage vary, since cultures may merit short-term, medium-term or long-term storage, and the methods which are feasible are very different for each. Similar methods are suitable for storage of bacteria. These are described because *Escherichia coli*, in particular, is widely used for cloning of yeast genes.

2. SHORT-TERM STORAGE (1 – 2 weeks)

This does not have very stringent requirements. The cultures can be stored in Petri dishes of normal media, preferably in plastic bags, in the refrigerator. The bags should be unsealed, and handled with care to avoid accumulation of moisture in the bags, leading to contamination of the cultures. Some workers add 2% soluble starch to media for preservation of cultures in this way, but we have not observed as good preservation as these investigators claim.

3. MEDIUM-TERM STORAGE (up to 1 year)

Cultures, of *Saccharomyces* species in particular, can usually be stored in the refrigerator for up to 1 year on yeast extract-peptone-glucose agar slants, with or without soluble starch, or on malt extract agar, in screw-capped tubes or McCartney or universal bottles. The method is akin to the older method of preservation of fungal and bacterial cultures on normal media, under sterilized mineral oil, but is not so messy. Cultures have been known to survive for 5 – 6 years in screw-capped containers as described, but subculturing at 6 monthly or yearly intervals is advisable. Bacterial cultures may be stored as stab cultures in nutrient agar, inoculated by a needle dipped in the culture.

Cultures are incubated with the cap loosely closed. After incubation, the cap is screwed on tightly for storage.

4. LONG-TERM STORAGE (over 1 year)

Cultures may be preserved for periods of at least 10 – 20 years by the following methods.

4.1 Lyophilization

This method has been used for at least 50 years, and involves, in general, suspending the cultures in ampoules, generaly plugged with cottonwool, in an appropriate medium (skim milk is often used), freezing to a low temperature in a bath of acetone + dry ice (solid CO_2), and then drying under high vacuum. After drying, the tubes are sealed under vacuum, and stored either at room temperature or in the refrigerator. There is some possibility of mutations arising in this method also. Survival percentages are low, but this is not important, as viable cells can be obtained for a great many years in properly lyophilized cultures. The tubes are opened by scratching with a file across the centre of the plug, breaking under aseptic conditions, and transferring the dry pellet to the appropriate medium, either broth or agar. Broth may be introduced into the ampoule with a Pasteur pipette and the culture incubated in the original tube if desired. The culture should be restreaked when growth occurs. Several types of equipment are available commercially for lyophilization of cultures.

4.2 Storage under liquid nitrogen

The cells are suspended in an appropriate medium, cooled gradually in the vapour above the liquid nitrogen itself, and finally transferred to the liquid. Storage life is indefinite. The major disadvantage is that an assured supply of liquid nitrogen is essential.

4.3 Freezing in glycerol solutions

This method has the advantage of simplicity and requires only a freezer, preferably at $-70°C$. The cultures are grown for $28-48$ h in normal medium, a suitable aliquot is taken, and made to 50% glycerol concentration (or 15% for bacterial cultures), slanted and frozen. Storage time is indefinite, and if the freezer fails at any time, the culture can be transferred and refrozen later from a fresh culture as before. Routine culturing is done by scraping a little of the culture from the frozen surface, and inoculating fresh medium with it. The rest of the culture is returned to the freezer immediately. Cultures stored at $-70°C$ survive considerably longer than those stored at $-20°C$, but the latter will generally survive for several years. It is important to keep several tubes of each culture.

5. TRANSPORT OF YEAST CULTURES

If the cultures are in lyophil tubes, they can simply be sent by mail in a protective box or envelope. Cultures can also be shipped in culture tubes or screw-capped vials of any desired size, if the transit time is not excessive, and if they are not likely to be exposed to abnormally high temperatures. Low temperatures are less likely to affect them.

The method of shipping on filter paper has come into use fairly recently, and is probably the least expensive. Filter paper is cut into small squares, usually not more than 1 cm a side, and sterilized. Squares of metal foil, $2-3$ cm a side are also prepared, pre-folded and sterilized. The cultures are grown for $24-48$ h on a slope of YEP−glucose medium. The metal foil squares are unfolded on a sterile surface, a square of filter paper is moistened with sterile water or any other desired suspending medium, and placed on the open foil. A loopful (alternatively, use a sterile toothpick) of yeast cells is spread on the filter paper, which is immediately wrapped in the foil, using sterile (alcohol-flamed) forceps and microspatula, and labelled. The package can be stuck to a card, or directly to the covering letter, and mailed. The cultures will survive for several weeks at ambient temperature, and can be revived by placing the paper in liquid medium or on the surface of a plate of YEP−glucose agar.

Principal sources of yeast cultures

I.CAMPBELL

UK	National Collection of Yeast Cultures, Food Research Institute, Colney Lane, Norwich NR4 7UA
Netherlands	Centraalbureau voor Schimmelcultures, Yeast Division, Julianalaan 67a, 2628 BC Delft
Czechoslovakia	Czechoslovak Collection of Yeasts, Institute of Chemistry, Slovak Academy of Sciences, Dubravska cesta, 809 33 Bratislava
USA	American Type Culture Collection, 12301 Parklawn Drive, Rockville, MD 20852
	Yeast Genetics Stock Culture Center, Donner Laboratory, University of California, Berkeley, CA 94720

Media and buffers for yeast genetics

J.R.JOHNSTON

CIP buffer ($\times$10)

    0.5 M Tris−HCl buffer, pH 9.0

    10 mM MgCl$_2$

    1 mM ZnCl$_2$

    10 mM spermidine

L broth

    10 g bactotryptone

    5 g bacto yeast extract

    10 g NaCl

    1 litre distilled water

LA solution

    0.25 M lithium acetate in TE buffer, pH 8.0

MS (medium-salt) buffer

    50 mM NaCl

    10 mM Tris−HCl buffer, pH 7.5

    10 mM MgCl$_2$

    1 mM DTT

Regeneration agar

    1 mM sorbitol, 100 ml

    0.67 g Difco Yeast Nitrogen Base without amino acids

    2.0 g glucose

    3.0 g agar no. 1

    and addition of amino acids or bases, as appropriate for auxotrophy

SCE buffer

    1.0 M sorbitol

    0.1 M sodium citrate buffer, pH 5.8

    60 mM EDTA

SEC buffer
    1.0 M sorbitol
    100 mM EDTA
    10 mM citrate−phosphate buffer, pH 5.8

SET buffer
    20 % (w/v) sucrose
    50 mM EDTA
    50 mM Tris−HCl buffer, pH 7.6

SP (spheroplasting) buffer
    1 M sorbitol
    50 mM potassium phosphate buffer, pH 7.5
    15 mM β-mercaptoethanol

SSC buffer
    0.15 M NaCl
    15 mM sodium citrate buffer, pH 5.8

ST buffer
    25% (w/v) sucrose
    50 mM Tris−HCl buffer, pH 8.0

STC buffer
    1 M sorbitol
    10 mM Tris−HCl buffer, pH 7.5
    10 mM $CaCl_2$

TE buffer
    50 mM mM Tris−HCl buffer, pH 8.0
    20 mM EDTA

TEN buffer
    10 mM Tris−HCl buffer, pH 7.6
    1 mM EDTA
    10 mM NaCl

TES buffer
    30 mM Tris−HCl buffer, pH 8.0
    50 mM EDTA
    5 mM NaCl

INDEX